Elastomeric Proteins

Structures, Biomechanical Properties, and Biological Roles

Elastomeric proteins occur in a wide range of biological systems where they have evolved to fulfil precise biological roles. The best-known include proteins in vertebrate muscles and connective tissues, such as titin, elastin and fibrillin, and spider silks. However, other examples include byssus and abductin from bivalve molluscs, resilin from arthropods, and gluten from wheat. Interest in elastomeric proteins is currently high for several reasons – first, because of their biological and medical significance, particularly in human diseases. Second, the unusual properties of proteins, such as spider silks, provide opportunities to develop novel materials. Third, the development of scanning probe microscopy makes it possible to study structures and biomechanical properties of these proteins at the single molecule level. This book will be of interest to graduate students and researchers from a broad range of disciplines, working on any aspect of elastomeric proteins.

Peter R. Shewry is the Associate Director and Head of Crop Performance and Improvement Division at Rothamsted Research, Harpenden, UK. He received the Rank Prize for Nutrition in 2002. Professor Shewry has edited ten books, including *Plant Protein Engineering* (Cambridge University Press, 1996).

Arthur S. Tatham is a Senior Research Fellow at the Department of Agricultural Sciences, University of Bristol, and Principal Research Scientist at the Institute of Arable Crops Research, Long Ashton Research Station. For twenty years, he has studied the elastomeric proteins of wheat, comparing these proteins with other known elastomeric proteins to understand their elastic mechanisms.

Allen J. Bailey is currently Head of Collagen Research Group, University of Bristol, which studies the basic structure of the collagen molecule and the chemistry of stabilising cross-links, including glycation cross-links. His research applies to aging, osteoarthritis, and diabetes. He has won several awards, including the Royal Society of Chemistry Senior Medal in food chemistry.

T0215112

Elastomeric Proteins

Structures, Biomechanical Properties, and Biological Roles

Edited by

PETER R. SHEWRY
Rothamsted Research

ARTHUR S. TATHAM
Long Ashton Research Station

ALLEN J. BAILEY
University of Bristol

CAMBRIDGE UNIVERSITY PRESS
Cambridge, New York, Melbourne, Madrid, Cape Town, Singapore,
São Paulo, Delhi, Dubai, Tokyo

Cambridge University Press
The Edinburgh Building, Cambridge CB2 8RU, UK

Published in the United States of America by Cambridge University Press, New York

www.cambridge.org
Information on this title: www.cambridge.org/9780521128483

First published in *Philosophical Transactions of the Royal Society of London* (2002) B 357, 117-234
This revised edition published by Cambridge University Press 2003
This digitally printed version 2009

A catalogue record for this publication is available from the British Library

Library of Congress Cataloguing in Publication data

Elastomeric proteins : structures, biomechanical properties, and biological
  roles / edited by Peter R. Shewry, Arthur S. Tatham, Allen J. Bailey.

   p.   cm.

   Includes bibliographical references and index.

   ISBN 0-521-81594-0

   1. Elastin.   2. Gluten.   3. Muscle proteins.   4. Elastomers.   5. Proteins.
   I. Shewry, P. R.   (Peter R.)   II. Tatham, Arthur S.   III. Bailey, Allen J.

   QP552.E4 E437   2003
   572′.6 – dc21   2002035018

ISBN 978-0-521-81594-9 Hardback
ISBN 978-0-521-12848-3 Paperback

Additional resources for this publication at www.cambridge.org/9780521128483

Contents

Preface

Elastic proteins occur in a wide range of biological systems, from plants and invertebrates to humans, where they have evolved to fulfil precise functional roles. The majority of these proteins possess rubber-like elasticity, undergoing high deformation without rupture and then returning to their original state on removal of the stress, with virtually all the energy stored on deformation being returned. The second stage of this process is passive and does not require an input of energy. Such an entropic mechanism is ideal for elastomers that are required to last a long time (e.g., aortic elastin undergoes millions of stress-strain cycles in a human life span). However, not all biological protein elastomers have purely entropic mechanisms.

In simple terms, chains of a protein elastomer must be flexible and independently free to respond rapidly to an applied force and must also form a network of monomers stabilised by cross-links between non-elastic domains. The elasticity will therefore vary with the length of the flexible domain and the extent of cross-linking.

The best-known and most widely distributed protein elastomer is elastin, which is responsible for the elasticity of the aorta and skin of mammals, and is also present in the ligamentum nuchae which is involved in raising the heads of grazing hoofed animals. More recently, fibrillin, which forms the scaffold for elastin in vertebrate tissues, has also been shown to have elastic properties. Resilin is also well known as the elastic ligament attaching the wings to the thoraxes of insects, where it is responsible for the rapid vibration of the wings, and, more recently, the properties of giant protein titin – which is responsible for the elasticity of vertebrate striated muscle – have been elucidated.

Elastic proteins also perform unusual and unique functions, for example: abductin ligaments in clams (bivalve molluscs) open the shell when the muscles relax, generating a primitive swimming action; byssus threads bind mussels to rocks and possess elastic and rigid domains to resist wave action; and spider webs comprise stiff dragline silks that form the spokes and extensible flagelliform silks that form the capturing spirals, absorbing the energy of the impacting insect without catapulting it out of the web.

Biologically active protein elastomers are unknown in plants, but the elastic properties of the gluten proteins of wheat grain have been well documented. Although there is no known biological role for gluten elasticity, it is responsible for the ability to make leavened bread. Thus, an entirely fortuitous property of a plant storage protein has a major impact on food processing. The collagen fibres of tendons are generally considered virtually inextensible but, like steel wire, they also have elastic properties. This elasticity is exploited by animals during rapid running and, in doing so, their tendons act as energy stores for rebounds, thus providing more efficient running.

The ability of proteins to exhibit rubber-like elasticity is provided by their molecular organisation. The relationships between mechanical properties and biochemical structure have intrigued scientists for many years, particularly since the seminal work of Miles Partridge on vertebrate elastin and Torkel Weis-Fogh on insect resilin in the 1960s.

The recent rapid explosion of our understanding of protein elasticity has been based on our ability to determine the molecular sequences and three-dimensional structures of the proteins and to relate them to the mechanical properties of macromolecular arrays and single molecules, the latter using new techniques, such as laser tweezers and atomic force microscopy. These techniques have led to rapid advances, but have inevitably also raised many more questions. For example, it is now considered that the elastic behaviour of elastin may be due to the presence of β-spiral structures rather than randomly free chains as in classical rubbers. In contrast, resilin – which is almost a perfect elastomer and exhibits an entropic mechanism like elastin – does not appear to contain β-spirals. Similarly, the stretching of titin appears to involve immunoglobulin-like domains. Furthermore, some elastomers may possess a two-stage mechanism operating at low and high extensions (e.g., the spider silks).

Although these various elastomers have been shown to have little or no amino acid sequence homology, it is possible to recognise some common properties. For example, most contain regions comprised of repeating blocks of particular sequences, which may adopt similar secondary structures capable of generating elasticity. Similarly, β-turns and β-spiral structures appear to occur in several unrelated protein elastomers and readily account for their elastic properties, but are by no means universal. It has, therefore, become clear that many different structural designs in proteins may be responsible for their elastic behaviour, with the requirements for variable levels of elastic behaviour being satisfied in different ways. This is hardly surprising in view of the diverse organisms in which elastomers occur.

The current interest in elastomeric proteins is enhanced by their importance in medicine (e.g., understanding the role of aortic elastin in atherosclerosis) and by opportunities to develop novel biomaterials with specific properties. Synthetic polypeptides based on the repeat sequence of elastin can be used to study platelet interactions and to prevent tissue adhesions in wounds, whereas synthetic structures based on spider silks are being studied to exploit their high strength and high energy-absorbing properties. Additional applications will arise from a greater understanding at the molecular level of the mechanisms that nature has evolved to confer elastic behaviour.

Most of the papers in this volume are based on those presented at a Discussion Meeting of the Royal Society held in London in May 2001 (and since published in *Philosophical Transactions of the Royal Society Series B* 2002, Volume 357, No. 1418, pp. 117–234). This meeting brought together for the first time leading researchers and covered the latest developments in the relationships between the molecular structures and mechanical properties of elastomeric proteins. However, these papers have been revised and expanded, whereas additional chapters (notably on spider silks, resilin, biomimetics, spectrin, and the comparative properties and functions of elastomeric proteins) have been commissioned to complete the coverage. It is hoped that these reviews will together stimulate additional researchers in protein chemistry, molecular biology, and biomechanics to enter this exciting field.

Contributors

R. McNeil Alexander. School of Biology, University of Leeds, Leeds LS2 9JT, UK.
E-mail: r.m.alexandra@leeds.ac.uk

S. O. Andersen. August Krogh Institute, Universitetsparken 13, DK-2100
Copenhagen, Denmark.
E-mail: soandersen@aki.ku.dk

A. J. Bailey. Collagen Research Group, Division of Molecular and Cellular
Biology, University of Bristol, Langford, Bristol BS40 5DS, UK.
E-mail: a.j.bailey@bristol.ac.uk

C. Baldock. School of Biological Sciences, University of Manchester,
Manchester M13 9PT, UK.

C. M. Bellingham. Cardiovascular Research Program, Research Institute,
The Hospital for Sick Children, Toronto, Ontario, M5G 1X8, Canada.

S. Belton. School of Chemical Sciences, University of East Anglia, Norwich
NR4 7TJ, UK.

E. Carrington. Department of Biological Sciences, University of Rhode Island,
Kingston, RI 02881, USA, and Department of Zoology, University of British
Columbia, 6270 University Boulevard, Vancouver, BC V6T 1Z4, Canada.

J. Dea. SPAWAR Systems Center, Code D711, San Diego, CA 92152-5001, USA.

D. E. Discher. Institute for Medicine and Engineering, School of Engineering
and Applied Science, and Structural Biology Program – The Wistar Institute,
112 Towne Building, University of Pennsylvania, Philadelphia, PA 16104, USA.
E-mail: discher@eniac.seas.upenn.edu

P. Fratzl. Erich Schmid Institute of Materials Science, Austrian Academy of
Sciences, and University of Leoben, Jahnstr. 12, 8700 Leoben, Austria.
E-mail: fratzl@unileoben.ac.at

H. Gaub. Ludwig-Maximilians-Universität, Lehrstuhl für Angewandte Physik,
Amalienstraße 54, D-80799 Munich, Germany.

J. Gosline. Department of Zoology, University of British Columbia, 6270
University Boulevard, Vancouver, BC V6T 1Z4, Canada.
E-mail: gosline@zoology.ubc.ca

P. Guerette. Department of Zoology, University of British Columbia, 6270
University Boulevard, Vancouver, BC V6T 1Z4, Canada.

N. Halford. Rothamsted Research, Harpenden, Hertfordshire AL5 2JQ, UK.
E-mail: nigel.halford@bbsrc.ac.uk

L. Hayes. Bioelastics Research Ltd., 2800 Milan Court, Suite 386, Birmingham,
AL 35211-6918, USA.

T. Hugel. Ludwig-Maximilians-Universität, Lehrstuhl für Angewandte Physik,
Amalienstraße 54, D-80799 Munich, Germany.

F. W. Keeley. Departments of Biochemistry and Laboratory Medicine and
Pathobiology, University of Toronto, Toronto, Ontario, Canada,
and Cardiovascular Research Program, Research Institute, The Hospital for
Sick Children, Toronto, Ontario, M5G 1X8, Canada.
E-mail: fwk@sickkids.on.ca

C. M. Kielty. School of Medicine, University of Manchester, Manchester M13
9PT, UK.
E-mail: cay.kielty@man.ac.uk

D. P. Knight. Zoology Department, University of Oxford, South Parks Road,
Oxford OX1 3PS, UK.
E-mail: knight@tegdown.u-net.com

D. Lee. School of Biological Sciences, University of Manchester, Manchester,
M13 9PT, UK.

R. Lewis. Department of Molecular Biology, University of Wyoming, Wyoming
NSF (National Science Foundation) EPSCoR (Experimental Program to
Stimulate Competitive Research), Box 3944, Laramie, WY 82071, USA.
E-mail: silk@uwyo.edu

M. Lille. Department of Zoology, University of British Columbia, 6270 University Boulevard, Vancouver, BC V6T 1Z4, Canada.

J. M. Lucas. Marine Science Institute and Department of Molecular, Cell and Developmental Biology, University of California, Santa Barbara, CA 93106, USA.

C. Ortlepp. Department of Zoology, University of British Columbia, 6270 University Boulevard, Vancouver, BC V6T 1Z4, Canada.

T. Parker. Bioelastics Research Ltd., 2800 Milan Court, Suite 386, Birmingham, AL 35211-6918, USA.

F. Prochazka. University of Minnesota, Twin Cities Campus, BioTechnology Institute, 1479 Gortner Avenue, St. Paul, MN 55108-6106, USA.

M. J. Rock. School of Biological Sciences, University of Manchester, Manchester, M13 9PT, UK.

K. Savage. Department of Zoology, University of British Columbia, 6270 University Boulevard, Vancouver, BC V6T 1Z4, Canada.

M. Seitz. Ludwig-Maximilians-Universität, Lehrstuhl für Angewandte Physik, Amalienstraße 54, D-80799 Munich, Germany.

L. Sheiba. SPAWAR SYSTEMS CENTER, Code D711, San Diego, CA 92152-5001, USA.

M. J. Sherratt. School of Biological Sciences, University of Manchester, Manchester, M13 9PT, UK.

P. R. Shewry. Rothamsted Research, Harpenden, Hertfordshire, AL5 2JQ, UK. E-mail: peter.shewry@bbsrc.ac.uk

C. A. Shuttleworth. School of Biological Sciences, University of Manchester, Manchester, M13 9PT, UK.

C. Sun. Marine Science Institute and MCDB Department, University of California, Santa Barbara, CA 93106, USA.

A. S. Tatham. Long Ashton Research Station, Department of Agricultural Science, University of Bristol, Long Ashton, Bristol BS41 9AF, UK. E-mail: arthur.tatham@bbsrc.ac.uk

G. H. Thomas. Departments of Biology, Biochemistry, and Molecular Biology, The Pennsylvania State University, 208 Erwin W. Mueller Laboratory, University Park, PA 16802, USA. E-mail: gxt5@psu.edu

J. Trinick. Astbury Centre for Structural Molecular Biology, School of Biomedical Sciences, University of Leeds, Leeds LS2 9JT, UK.
E-mail: j.trinick@leeds.ac.uk

L. Tskhovrebova. Astbury Centre for Structural Molecular Biology, School of Biomedical Sciences, University of Leeds, Leeds LS2 9JT, UK.

D. W. Urry. University of Minnesota, Twin Cities Campus, BioTechnology Institute, 1479 Gortner Avenue, St. Paul, MN 55108-6106, USA, and Bioelastics Research Ltd., 2800 Milan Court, Suite 386, Birmingham, AL 35211-6918, USA, and Ludwig-Maximilians-Universität, Lehrstuhl für Angewandte Physik, Amalienstraβe 54, D-80799 Munich, Germany.
E-mail: urryx001@tc.umn.edu

E. Vaccaro. Marine Science Institute and MCDB Department, University of California, Santa Barbara, CA 93106, USA.

J. F. V. Vincent. Department of Mechanical Engineering, University of Bath, Bath, BA2 7AY, UK.
E-mail: J.F.V.Vincent@bath.ac.uk

F. Vollrath. Zoology Department, Universitetsparken B135, 8000 Aarhus C, Denmark, and Zoology Department, University of Oxford, South Parks Road, Oxford OX1 3PS, UK.
E-mail: fritz.vollrath@zoo.ox.ac.uk

J. H. Waite. Marine Science Institute and MCDB Department, University of California, Santa Barbara, CA 93106, USA.
E-mail: waite@lifesci.ucsb.edu

J. Xu. Bioelastics Research Ltd., 2800 Milan Court, Suite 386, Birmingham, AL 35211-6918, USA.

Elastomeric Proteins

ONE

Functions of Elastomeric Proteins in Animals

R. McNeill Alexander

INTRODUCTION

Elastomeric proteins play many important roles in the lives of animals. They enable animals ranging from fleas to large mammals to jump further than would otherwise be possible. They save energy in locomotion for galloping horses, hopping kangaroos, flying flies, and swimming jellyfish. They make clam shells spring open when the muscles inside relax, they help to support the heavy heads of cattle, they smooth the flow of blood round our bodies, and they cushion the impact of our heels on the ground. This chapter attempts to explain briefly how they do all these things.

POWER AMPLIFIERS

Catapults are power amplifiers. The rubber is stretched in preparation for shooting, storing up strain energy. This can be done slowly; but when the catapult is released, the rubber recoils very rapidly, returning the stored energy as kinetic energy of the missile. The work done by the recoiling rubber is (almost) equal to the work previously done stretching the rubber; but, it is done in a much shorter time so the power (rate of doing work) has been amplified. Using a catapult, I can project a missile much faster than I can move my hand.

Catapults are useful because the power output that can be obtained from a muscle is limited. For example, Peplowski and Marsh (1997) made physiological measurements on a leg muscle of a tree frog and found that the highest power output obtainable from it was 240 W/kg. However, they calculated that the power required for the longest jumps, which it could make at the same temperature, was about 800 W/kg muscle. They concluded that a catapult mechanism must be involved.

The best-known examples of catapult mechanisms in animals are in jumping insects. The importance of the catapult principle was first demonstrated by Bennet-Clark and Lucey (1967) in a study of the flea *Spilopsyllus*. This insect is about 1.5 mm long. It jumps by rapidly extending its hind legs, accelerating itself

1

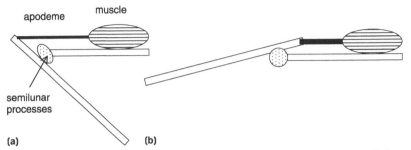

Figure 1.1: Diagrams of a locust knee (a) bent with the muscle taut and (b) extended. The extensor apodeme is stretched and the semilunar process is compressed in (a).

from rest to its take-off velocity of 1 m/s in a distance of 0.4 mm. This implies (assuming constant acceleration) that the legs extend in 0.8 m/s. No known muscle can complete an isolated contraction in as short a time as this (though the fibrillar flight muscles of some small insects are capable of oscillating the wings at 1000 Hz). Bennet-Clark and Lucey (1967) showed that the flea's springs are blocks of the rubbery protein resilin at the bases of the hind legs. They confirmed that these blocks were amply large enough to store the energy of the jump and demonstrated a locking mechanism that prevents the resilin from recoiling until the flea is ready to jump.

Subsequent research led to the discovery of catapult mechanisms in other insects, including locusts, click beetles, and flea beetles (see Gronenberg, 1996). The springs can be arranged in either or both of two principal ways. Both arrangements are found in locusts (Bennet-Clark, 1975) and are illustrated in Figure 1.1. The jump is powered by the extensor muscle of the 'knee' joint. One of the springs is the apodeme, through which the muscle attaches to the skeleton. The other is a structure in the joint known as the semilunar process. Both consist of chitin fibres in a protein matrix. Initially, the joint is locked by a catch, in the bent position [Figure 1.1(a)]. Contraction of the muscle stretches the apodeme and compresses the joint, storing up strain energy in both. When the catch is released, both springs recoil, allowing the animal to jump.

Springs can amplify the power available for jumping, even (as seems to be the case in frogs) when there is no mechanism to lock the joints. This point is illustrated by a simple mathematical model of jumping that is designed to be applicable to the whole range of jumping animals, from fleas to humans (Alexander, 1995). Knee extensor muscles with elastic tendons power this model's jumps. The forces that the muscles can exert fall as the rate of shortening rises, according to Hill's equation (see textbooks on muscle physiology). Figure 1.2 shows examples of computer simulations of three styles of jumping. In Figure 1.2(a),

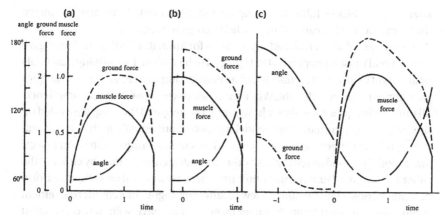

Figure 1.2: Computer simulations of standing jumps: (a) squat jump, (b) catapult jump, and (c) countermovement jump. The force exerted on the ground is expressed as a multiple of body weight, and the force exerted by the muscle that extends the leg is expressed as a multiple of its isometric force. Both these forces, and the angle of the knee joint, are plotted against a dimensionless time parameter. From Alexander (1995).

a squat jump, no catch is used. The animal starts with its knees bent. At time zero, it activates the knee extensor muscles. Initially, the muscles shorten quite rapidly, stretching their tendons as they develop force. As the tendons stretch, the force exerted by the muscles rises and their rate of shortening falls accordingly. However, because the muscles are still shortening, the force can never reach the isometric level that can be exerted only when the rate of shortening is zero. As knee extension continues at an ever-increasing rate, the rate of shortening of the muscles has to increase again, the force they exert falls, and the tendon starts to recoil. Eventually, the muscles are no longer able to exert enough force to maintain the angular acceleration of the legs and keep the feet on the ground. The model calculates the speed at which the animal leaves the ground and hence the height of the jump.

Figure 1.2(b) shows a catapult jump. In this case, there is a catch that is initially set. It keeps the knee bent while the muscle develops tension. At time zero, with the muscle exerting its isometric force, the catch is released and the force exerted by the feet on the ground rises abruptly to a high value. The force on the muscle falls as the tendon recoils elastically, and the lower force allows the muscle as well as the tendon to shorten. Eventually, the feet leave the ground.

Figure 1.2(c) shows a countermovement jump. As in Figure 1.2(a), there is no catch. The action does not start with the knee bent, as in the other cases, but with it straight. The jumper allows his or her body to fall, but activates the knee muscles before the knees are fully bent. The time of activation is adjusted so that the angle of the knees, at the bottom of the fall, is the same as the initial knee angle in Figure 1.2(a,b). During the fall, the muscle force may rise above the isometric level because the muscle is being stretched.

Alexander (1995) performed simulations for animals of different sizes. I gave each animal muscle properties that I judged to be typical, for jumping mammals or insects of its size. If the tendons had no elastic compliance, all three techniques gave jumps of the same height. When the tendons were given elastic compliance, all three techniques gave higher jumps, and the height of the jump depended on the technique. A human-sized model jumped about equally high in a catapult jump and a countermovement jump, and less high in a squat jump. An insect-sized model jumped much higher by the catapult technique than by either of the other techniques. Humans have no catch mechanism, so they cannot perform catapult jumps. They can jump a few centimetres higher in a countermovement jump than in a squat jump. Jumping insects, as we have seen, rely on catapult mechanisms.

Power-amplifying functions have also been postulated for a few structures that are not involved in jumping. Aerts et al. (1987) argued that there must be an elastic mechanism that amplifies power in suction feeding by fish. Gronenberg (1996) has described spring-loaded jaws in ants.

ENERGY STORES FOR RUNNING

A bouncing ball loses kinetic energy when it hits the ground, then immediately regains it by elastic recoil. A perfect ball in a frictionless world would continue bouncing forever.

This principle is important in running for large mammals such as ourselves. At each footfall, a runner loses and regains kinetic energy and gravitational potential energy. Cavagna et al. (1964) measured these energy fluctuations for men running at different speeds. They compared the energy fluctuations with the metabolic energy cost of running, calculated from measurements of oxygen consumption. These comparisons seemed to show that the muscles were working with much higher efficiencies than had been measured for other activities, such as walking up and down slopes, or pedalling a bicycle ergometer. Cavagna and colleagues argued that some of the kinetic plus gravitational potential energy lost and regained in each footfall must be stored briefly as elastic strain energy in the muscles, and returned in an elastic recoil. Energy must be saved by the principle of the bouncing ball.

A paper by Cavagna et al. (1964) gave the impression that the elastic compliance was in the muscles themselves. However, Alexander (1974) filmed a dog and recorded the forces that its feet exerted on the ground as it took off for a jump. The joint angle changes shown in the film implied that the gastrocnemius and plantaris muscle–tendon complexes stretched and then shortened by more than 20 mm while their feet were on the ground. I calculated that the peak forces on the tendons were enough to stretch them by this amount. This suggested that tendons might be the important energy-saving springs in the legs. A similar study of hopping by kangaroos (Alexander and Vernon, 1975) led to the conclusion that the gastrocnemius and plantaris tendons were indeed very important energy-saving springs.

Recent technical advances have made it possible to investigate the role of tendon elasticity more directly and more precisely. Biewener et al. (1998) performed surgical operations on wallabies, fitting tendon buckles to measure the forces in tendons and implanting sonomicrometry crystals to record the length changes of muscle fascicles. They showed that the fascicles of the gastrocnemius and plantaris muscles remained almost constant in length, while their tendons stretched and recoiled during the part of the hopping stride in which the foot was on the ground. They calculated that at 6 m/s (the fastest speed investigated), 45% of the energy that would otherwise have been needed for hopping was saved by elastic storage in tendons. Roberts et al. (1997) had performed a similar experiment on turkeys, with similar results.

The importance of tendon elasticity in running has also been demonstrated in other moderate-to-large mammals [e.g., in camels and horses (Alexander et al., 1982; Dimery et al., 1986; Biewener, 1998)] and in ostriches (Alexander et al., 1979). Muscles whose tendons are important as springs tend to have very short muscle fascicles. If tendon stretching allows most of the movement required at a joint, short muscle fascicles are sufficient to provide the rest of the movement. The shorter the fascicles, the less metabolic energy is required to maintain tension in them. In the extreme cases of the camel and horse, the fascicles of some of the distal leg muscles have almost disappeared, leaving tendons that run virtually uninterrupted from one skeletal attachment to the other, and must function as passive springs. Ker et al. (1987) showed that, in human running, the ligaments of the arch of the foot are important energy-saving springs, as well as the Achilles tendon.

Although tendon elasticity seems to give substantial energy saving in running for most large mammals, it appears to have very little importance in kangaroo rats (mass about 100 g; Biewener and Blickhan, 1988). These small mammals hop like kangaroos, but the tendons in their legs stretch very little. The reason seems to be that the forces that act when these animals jump are much larger than the forces of hopping. The tendons have to be thick enough to withstand jumping forces, so the stresses that act in them in hopping are small.

Alexander et al. (1985) identified another energy-saving spring in mammals that gallop. In the 'gathered' stage of a galloping stride, all four legs have to reverse their directions of swing. The forelegs have been moving backwards and must be made to swing forward; and the hind legs have been moving forward and must be made to swing back. We showed that much of the kinetic energy, which is lost and regained as the legs reverse, could be stored in the aponeurosis of the longissimus muscle. (An aponeurosis is a sheet of tendon, and the longissimus is the principal extensor muscle of the back.) Some more of the energy would be stored in the vertebral column, which would function as a compression spring. The elastic compliance of the vertebral column in compression is located partly in the intervertebral discs and partly in the bony centra.

ENERGY STORES IN FLIGHT AND SWIMMING

Energy can be saved by springs in other oscillatory movements. Weis-Fogh (1960) discovered the rubber-like protein resilin in the thoraxes of some insects and suggested that it might serve as an energy-saving spring. Insects fly by beating their wings, often at very high frequencies. In each wing beat cycle, the wings and the added masses of air that move with them are accelerated in one direction, halted, accelerated in the opposite direction, and halted again. The work required to give them this kinetic energy is called inertial work to distinguish it from the aerodynamic work that is needed to keep the animal airborne.

The need for the muscles to do inertial work can, in principle, be eliminated by springs that give the system a resonant frequency matching the wing beat frequency. Such springs would take the kinetic energy from the wings at the end of one stroke and return it by elastic recoil for the next. This would save energy if the inertial work was large compared with the aerodynamic work. However, if the inertial work is small, the kinetic energy taken from the wings can be used to do aerodynamic work, and there may be no advantage in having springs in the system. Wilmott and Ellington (1997) estimated for a hovering moth (*Manduca*) that the aerodynamic power requirement was 18–26 W/kg, and the inertial requirement 23–38 W/kg. In this case, there is potential for substantial energy savings by an elastic mechanism.

Weis-Fogh (1973) calculated the aerodynamic and inertial work required for hovering by various insects and compared it with the metabolic rates obtained from measurements of oxygen consumption. He calculated the efficiencies with which the work must be done if there was no elastic storage and if there was perfect elastic storage that entirely eliminated the need for the muscles to do inertial work. He concluded that some species must have energy-saving elastic mechanisms, because otherwise the muscles would have to work with impossibly high efficiencies. However, the metabolic rates that he used seemed in some cases to have been much too low, because the measurements were made on tethered insects whose wing beats may not have been strong enough to generate the aerodynamic forces needed for flight.

More recent calculations of the efficiency of insect flight have not given unambiguous evidence for elastic mechanisms. Possibly the best evidence for energy saving by elastic mechanisms in insect flight comes from an ingenious experiment by Dickinson and Lighton (1995), who stimulated a tethered fruit fly to attempt to turn alternately to either side, by means of a moving pattern of lights. In trying to turn, it varied the frequency and amplitude of its wing beats, making the ratio of inertial to aerodynamic power vary. Dickinson and Lighton measured the fly's oxygen consumption and correlated the fluctuations of oxygen consumption with the fluctuations of wing beat frequency and amplitude. Their analysis led to the conclusion that the muscles worked with an efficiency of 0.11 and that modest energy savings were made by elastic storage.

In principle, elastic mechanisms could be important in the flight of birds, as well as insects. Wells (1993) calculated that the wing muscles of hovering hummingbirds must work with an improbably high (but not necessarily impossible) efficiency of 25%, unless energy was saved by elastic storage. However, no elastic mechanisms that could save energy in bird flight have been convincingly demonstrated. Pennycuick and Lock (1976) suggested that wing feathers might be important springs, but Alexander (1988) gave reasons for thinking this implausible. Jenkins et al. (1988) took cine X-ray pictures of flying starlings and showed that the furcula bends and recoils in each wing beat cycle. It might be supposed that it was a potentially useful elastic energy store, but it can be shown from their data that the elastic strain energy it stores is very much less than the kinetic energy of the flapping wings, so it cannot be important.

Elastic energy-saving mechanisms may be important in swimming for some animals. For example, dolphins swim by beating their tail flukes up and down. The fluke, and the added mass of water that moves with it, loses and regains kinetic energy twice in each tail beat cycle. The tail muscles insert through numerous long, slender tendons. Bennett et al. (1987) measured the elastic compliances of these tendons and the vertebral column, and discussed the possibility that they might have an energy-saving role. Our data (reinterpreted by Blickhan and Cheng, 1994, in light of an improved hydrodynamic analysis) indicate that the compliance of the tendons has approximately the value required to eliminate the need for muscles to do negative work. This implies that they are rather more compliant than would be optimal, according to the theory of Alexander (1997), which is outlined in the following section.

Scallops (*Pecten*, etc.) are exceptional among bivalve molluscs in being able to swim. They do this by rapidly and repeatedly opening and closing their shells, ejecting jets of water on either side of the hinge. The shell is closed by means of a muscle, but there is no muscle to open it; instead, there is an elastic hinge ligament just inside the hinge. This is a block of the rubber-like protein abductin. It is compressed when the shell closes and recoils elastically to open it. The resonant frequency of the animal, due to the interaction of the compliance of the hinge ligament with the masses of the valves of the shell and the added masses of water that move with them, matches the frequency of the swimming movements. It has been suggested that the hinge ligament functions as an energy-saving spring, like the others described in this section. However, Cheng et al. (1996)

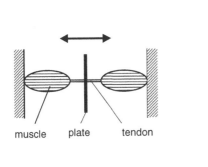
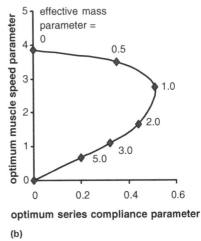

(a)   (b)

Figure 1.3: (a) Model of oscillatory motion powered by muscles. (b) Graph showing the combinations of muscle speed parameter and series compliance parameter that are predicted to minimise the metabolic power needed to drive the system for different effective mass parameters. Parameters are defined in the text.

showed that the inertial work of opening and closing the shell is tiny, compared with the hydrodynamic work required to power the jet; so, any energy-saving effect of the abductin is trivial.

Jellyfish also swim by jet propulsion, by repeatedly contracting and expanding their bells. Muscles contract the bell, but it expands by elastic recoil of the mesogloea. De Mont and Gosline (1988) showed that *Polyorchis* swims at its resonant frequency and that the inertial work requirements are large enough, compared with the hydrodynamic work, for the mesogloea to function as an energy-saving spring.

Alexander (1997) pointed out that all the activities that have been discussed in this section can be represented by the model shown in Figure 1.3(a). A pair of antagonistic muscles oscillate a plate in a fluid, doing inertial and hydrodynamic work as required. There are compliant tendons in series with the muscles, and there may also be a spring acting in parallel with the muscles. (The hinge ligament of scallops is an example of a parallel spring.) If the length changes of the muscles are calculated (taking account of tendon compliance), and also the forces that the muscles have to exert at each stage of a cycle, the metabolic rates of the muscles can be estimated. Unfortunately, my calculations of metabolic rate had to be based on physiological measurements on muscles making single contractions, because no suitable data on repetitive contractions were available. For this reason, rates calculated from the model may be to some extent misleading.

To use the model, it is necessary to assign values to three parameters:

- Effective mass parameter – If there is no parallel spring, this is the ratio of the peak inertial force to the peak hydrodynamic force. A parallel spring reduces the effective mass parameter by exerting some of the required inertial force. If this spring has the compliance required to match the resonant frequency of the system to the frequency of the motion, the effective mass parameter is zero.
- Series compliance parameter – This is the amount by which the peak hydrodynamic force would stretch the tendons, divided by the amplitude of the motion.
- Muscle speed parameter – This is the maximum speed at which the muscles are capable of shortening (generally referred to by physiologists as v_{max}), divided by the peak speed of the plate.

Calculations show that, for any given value of the effective mass parameter, there is an optimum combination of the other two parameters that minimise metabolic rate [Figure 1.3(b)]. When the effective mass parameter is zero, the optimum is a series compliance parameter of zero and a muscle speed parameter of 3.9; the muscles should not have compliant tendons and should be rather fast. As the effective mass parameter increases to about 1.2, the optimum compliance parameter rises to 0.5, and the optimum muscle speed parameter falls a little. As the effective mass parameter rises further, the optimum values for both the other parameters fall, approaching zero as the effective mass parameter approaches infinity.

Alexander (1997) compared the properties of muscles that drive oscillatory movements to the properties that the model predicted as optimal, in the few cases in which the data were available. Scallops swim at their resonant frequencies, so have an effective mass parameter of zero. The adductor muscle has no tendon, so the compliance parameter is zero; and measurements of its force/velocity properties gave muscle speed parameters of about four. Thus, this muscle seems to be very close to the predictions of the model. However, the effective mass parameter for a swimming dolphin is about 0.3, giving a predicted optimum compliance parameter of 0.22, whereas the actual value seems to be 0.67. (This discrepancy does not make much difference to the predicted energy cost of swimming, because the optimum is rather flat.) Because I know of no force/velocity experiments on dolphin muscle, I cannot calculate a muscle speed parameter. I do not have the data that would be needed to calculate either parameter for flying bumblebees (*Bombus*) or for hovering hummingbirds (*Amazilia*). The effective mass parameters for these two examples (assuming they do not have parallel springs) are about three and five, respectively.

The model illustrated in Figure 1.3 can be applied to running, as well as to swimming. As an animal runs, its legs oscillate backwards and forwards. Also, the deceleration and reacceleration of the body that occur in each step can be regarded as an isolated half-cycle of an oscillation. The aerodynamic forces are trivial, so the effective mass parameter is extremely high. The optimum design predicted by the model would incorporate very slow muscles, with tendons of the right compliance to do all the inertial work. As noted in the previous section, mammals such as kangaroos, horses, and humans have distal leg muscles with long tendons and very short muscle fascicles (implying a low maximum shortening speed). However, the model ignores the need for fast muscles for acceleration and jumping. Mammals have fast muscles with long fascicles principally in the proximal parts of their legs.

RETURN SPRINGS

Bivalve molluscs have adductor muscles to close their shells, but no muscles to open them. Instead, the shell springs open by elastic recoil of the hinge ligament. We have already seen how the hinge ligament of scallops gives these molluscs a resonant frequency matching the frequency of their swimming movements. In other bivalves, which do not swim, the hinge ligament functions simply as a return spring that enables the animal to open the closed shell. The hinge ligament of scallops has low hysteresis, suiting its function in swimming, but the hinge ligaments of other bivalves show much more marked hysteresis (Trueman, 1953).

There are many other examples of elastic materials that serve as return springs in animals. Sea anemones, such as *Metridium*, can contract the body to a small conical mass or expand it to form a remarkably large, thin-walled cylinder. They contract by means of muscles and expand by using cilia to inflate themselves with water. The jellylike mesogloea in the body wall has viscoelastic properties (Alexander, 1962). If the muscles and cilia are put out of action by anaesthesia, the animal slowly becomes a small cylinder, whatever its initial size and shape. This implies that, when it is unstrained, the mesogloea is a small cylinder. The high hysteresis of the mesogloea seems appropriate for an animal that lives on shores attached to rocks. It is exposed to waves, with periods of the order of 10 s. The mesogloea resists waves reasonably well, but is easily stretched over periods of the order of an hour by the small pressures generated by the ciliary pump (see also Koehl, 1977).

Cicadas sing by oscillating the tymbal, a sheet of thin cuticle reinforced by a series of sclerotised ribs. A muscle pulls the tymbal inward to generate each group of sound pulses, but the tymbal springs out again elastically. In this case, the behaviour of the elastic structures is much more complex and interesting than that of a simple return spring. As the muscle contracts, the tymbal ribs buckle inward in turn. Each buckling event excites vibrations of the tymbal at the resonant frequency of the system. Thus, a muscle contracting with a frequency of the order of 100 Hz (Josephson and Young, 1981) drives a system that emits sound at 4000 Hz. The tymbal functions as a frequency multiplier (Bennet-Clarke, 1997).

The ligamentum nuchae of ungulate mammals is another example of a return spring. It is a thick band of elastin connecting the back of the skull to the neural spines of the neck and thorax. It is stretched when the animal lowers its head to a feeding position, and recoils when the head is raised. The neck muscles are fully capable of supporting the weight of the head, but the ligament reduces the tension they have to exert and so saves metabolic energy. Dimery et al. (1985) found that the ligament was capable of supporting the whole weight of the head in a camel, but that some tension was required in neck muscles in deer and sheep. The strain and tension in the ligament are greater in the feeding position than in the alert position. As the moment of the weight of the head about the base of the neck is greater in the feeding position, this is appropriate.

SMOOTHING FLOW

The elastic compliance of the arteries of vertebrate animals smoothes flow in the circulation (see Caro et al., 1978). They owe their compliance to elastin in their walls. In each heartbeat, the heart ejects a small volume of blood at high pressure, distending the aorta. For example, in dogs, the diameter of the aorta is increased by about 10%. While the ventricle is refilling, the aorta returns to its initial diameter. Consequently, the velocity of the blood flowing through the rest of the circulation fluctuates much less than it would do if the walls of the aorta were rigid. This saves energy, because the work required to pump a fluid through a system of tubes is the product of the driving pressure and the volume of fluid. If the blood flowed in a series of discrete pulses, the pressure required while it was flowing would be greater. Shadwick et al. (1987) have shown that cephalopods as well as vertebrates benefit from distensible aortas.

CUSHIONING IMPACTS

The feet of running animals are brought suddenly to rest when they hit the ground. For example, the feet of human runners typically hit the ground at about 0.7 m/s. The mass of the foot of an adult man is only about 1 kg, but the lower leg and foot are sharply decelerated at impact. The effective mass that is brought to rest is about 3.6 kg (Ker et al., 1989). If the foot and the ground were both rigid, large, and potentially damaging forces would act.

The foot is cushioned by the compliance of the fatty heel pad in humans and by the paw pads of other mammals. Humans often run on artificial surfaces that are more rigid than most natural ones and use running shoes to supplement the intrinsic compliance of the foot. The natural heel pad consists of fat permeating a network of collagen fibres and has viscoelastic properties (Aerts et al., 1995).

The high compliance of spiders' webs enables them to absorb the kinetic energy of the impact of potential prey without breaking (Denny, 1976; Craig, 1987). The web has high hysteresis, reducing the danger that the prey will be thrown off by the elastic recoil, like a gymnast on a trampoline.

FORCE CONTROL

For some tasks, we need to be able to control the position of a limb precisely (e.g., when using a computer mouse). In others, we may be more concerned with controlling force (e.g., when picking up a fragile object such as an empty eggshell). Tendon compliance makes it more difficult to control position and easier to control force.

Most of the muscles that operate our fingers have their bellies in the forearm and are connected to the hand skeleton by long tendons. These tendons have much larger cross-sectional areas, relative to the physiological cross-sectional areas of their muscles, than the long tendons that serve as energy-saving springs (Cutts et al., 1991). Consequently, the stresses they experience are relatively low. Nevertheless, because they are so long, they may stretch by a millimetre or two in use. For example, the long flexor muscle of the thumb is capable of exerting forces large enough to stretch its tendon by 1.4 mm. As a 1.4 mm movement of the tendon's insertion on the distal phalanx of the thumb rotates the thumb through 11°, it has an appreciable effect on control. Rack and Ross (1984) showed that the compliance of the tendon makes it more difficult to hold the thumb in a constant position against a fluctuating force. However, it makes it easier to exert a constant force on a vibrating surface.

CONCLUSIONS

This chapter has shown how elastomeric proteins perform a wide variety of functions in living animals. Different functions require different elastic properties, and different molecular structures give different properties. These and many other topics are discussed in the chapters that follow.

REFERENCES

Aerts, P., Ker, R. F., de Clercq, D., Ilsley, D. W., and Alexander, R. McN. (1995) The mechanical properties of the human heel pad: A paradox resolved. *J. Biomech.* **28,** 1299–1308.

Aerts, P., Osse, J. W. M., and Verraes, W. (1987) Model of jaw depression during feeding in *Astrotilapia elegans* (Teleostei: Cichlidae): Mechanisms for energy storage. *J. Morphol.* **194,** 85–109.

Alexander, R. McN. (1962) Visco-elastic properties of the body-wall of sea anemones. *J. Exp. Biol.* **39,** 373–86.

Alexander, R. McN. (1974) The mechanics of jumping by a dog (*Canis familiaris*). *J. Zool.* **173,** 549–73.

Alexander, R. McN. (1988) *Elastic Mechanisms in Animal Movement.* Cambridge, UK: Cambridge University Press.

Alexander, R. McN. (1995) Leg design and jumping technique for humans, other vertebrates and insects. *Philos. Trans. R. Soc. Lond. B* **347,** 235–48.

Alexander, R. McN. (1997) Optimum muscle design for oscillatory movement. *J. Theoret. Biol.* **184,** 253–59.

Alexander, R. McN., Dimery, N. J., and Ker, R. F. (1985) Elastic structures in the back and their role in galloping in some mammals. *J. Zool. A* **207,** 467–82.

Alexander, R. McN., Maloiy, G. M. O., Ker, R. F., Jayes, A. S., and Warui, C. N. (1982) The role of tendon elasticity in the locomotion of the camel. *J. Zool.* **198,** 293–313.

Alexander, R. McN., Maloiy, G. M. O., Njau, R., and Jayes, A. S. (1979) Mechanics of running of the ostrich (*Struthio camelus*). *J. Zool.* **187,** 169–78.

Alexander, R. McN., and Vernon, A. (1975) Mechanics of hopping by kangaroos (Macropodidae). *J. Zool.* **177,** 265–303.

Bennet-Clark, H. C. (1975) The energetics of the jump of the locust *Schistocerca gregaria*. *J. Exp. Biol.* **63,** 53–83.

Bennet-Clark, H. C. (1997) Tymbal mechanics and the control of song frequency in the cicada *Cyclochila australasiae*. *J. Exp. Biol.* **200,** 1681–94.

Bennet-Clark, H. C., and Lucey, E. C. A. (1967) The jump of the flea: A study of the energetics and a model of the mechanism. *J. Exp. Biol.* **47,** 59–76.

Bennett, M. B., Ker, R. F., and Alexander, R. McN. (1987) Elastic properties of structures in the tails of cetaceans (*Phocaena* and *Lagenorhynchus*) and their effect on the energy cost of swimming. *J. Zool.* **211,** 177–92.

Biewener, A. A. (1998) Muscle-tendon stresses and elastic energy storage during locomotion in the horse. *Comp. Biochem. Physiol. B* **120,** 73–87.

Biewener, A. A., and Blickhan, R. (1988) Kangaroo rat locomotion: Design for elastic energy storage or acceleration? *J. Exp. Biol.* **140,** 243–55.

Biewener, A. A., Konieczynski, D. D., and Baudinette, R. V. (1998) *In vivo* muscle force-length behavior during steady-speed hopping in tammar wallabies. *J. Exp. Biol.* **201,** 1681–94.

Blickhan, R., and Cheng, J.-Y. (1994) Energy storage by elastic mechanisms in the tail of large swimmers – A re-evaluation. *J. Theoret. Biol.* **168,** 315–21.

Caro, C. G., Pedley, T. J., Schroter, R. C. and Seed, W. A. (1978) *The Mechanics of the Circulation.* Oxford, UK: Oxford University Press.

Cavagna, G. A., Saibene, F. P., and Margaria, R. (1964) Mechanical work in running. *J. Appl. Physiol.* **19,** 249–56.

Cheng, J.-Y., Davison, I. G., and DeMont, M. E. (1996) Dynamics and energetics of scallop locomotion. *J. Exp. Biol.* **199,** 1931–46.

Craig, C. L. (1987) The ecological and evolutionary interdependence between web architecture and web silk spun by orb web weaving spiders. *Biol. J. Linnean Soc.* **30,** 135–62.

Cutts, A., Alexander, R. McN., and Ker, R. F. (1991) Ratios of cross-sectional areas of muscles and their tendons in a healthy human forearm. *J. Anat.* **176,** 133–37.

De Mont, M. E., and Gosline, J. M. (1988) Mechanics of jet propulsion in the hydromedusan jellyfish, *Polyorchis penicillatus*, III. A natural resonating bell; the presence and importance of a resonant phenomenon in the locomotor structure. *J. Exp. Biol.* **134,** 347–61.

Denny, M. (1976) The physical properties of spiders' silk and their role in the design of orb-webs. *J. Exp. Biol.* **65,** 483–506.

Dickinson, M. H., and Lighton, J. R. B. (1995) Muscle efficiency and elastic storage in the flight motor of *Drosophila. Science* **268,** 87–9.

Dimery, N. J., Alexander, R. McN., and Deyst K. A. (1985) Mechanics of the ligamentum nuchae of some artiodactyls. *J. Zool. A* **206,** 341–51.

Dimery, N. J., Alexander, R. McN., and Ker, R. F. (1986) Elastic extensions of leg tendons in the locomotion of horses (*Equus caballus*). *J. Zool.* **210,** 415–25.

Gronenberg, W. (1996) Fast action in small animals: Springs and click mechanisms. *J. Comp. Physiol. A* **178,** 727–34.

Jenkins, F. A., Dial, K. P., and Goslow, G. E. (1988) A cineradiographic analysis of bird flight: The wishbone in starlings is a spring. *Science* **241,** 1495–98.

Josephson, R. K., and Young, D. (1981) Synchronous and asynchronous muscles in cicadas. *J. Exp. Biol.* **91,** 219–37.

Ker, R. F., Bennett, M. B., Alexander, R. McN., and Kester, R. C. (1989) Foot strike and the properties of the human heel pad. *J. Eng. Med.* **203,** 191–96.

Ker, R. F., Bennett, M. B., Bibby, S. R., Kester, R. C., and Alexander, R. McN. (1987) The spring in the arch of the human foot. *Nature* **325,** 147–49.

Koehl, M. A. R. (1977) Mechanical diversity of connective tissue of the body wall of sea anemones. *J. Exp. Biol.* **69,** 107–25.

Pennycuick, C. J., and Lock, A. (1976) Elastic energy storage in primary feather shafts. *J. Exp. Biol.* **64,** 677–89.

Peplowski, M. M., and Marsh, R. L. (1997) Work and power output in the hindlimb muscles of Cuban tree frogs *Osteopilus septentrionalis. J. Exp. Biol.* **200,** 2861–70.

Rack, P. M. H., and Ross, H. F. (1984) The tendon of flexor pollicis longus: Its effect on the muscular control of force and position at the human thumb. *J. Physiol.* **351,** 99–110.

Roberts, T. J., Marsh, R. L., Weyand, P. G., and Taylor, C. R. (1997) Muscular force in running turkeys: The economy of minimizing work. *Science* **275,** 1113–15.

Shadwick, R. E., Gosline, J. M., and Milsom, W. K. (1987) Arterial haemodynamics in the cephalopod mollusc *Octopus dofleini. J. Exp. Biol.* **130,** 87–106.

Trueman, E. R. (1953) Observations of certain mechanical properties of the ligament of *Pecten. J. Exp. Biol.* **30,** 453–67.

Weis-Fogh, T. (1960) A rubber-like protein in insect cuticle. *J. Exp. Biol.* **37,** 889–906.

Weis-Fogh, T. (1973) Quick estimates of flight fitness in hovering animals, including novel mechanisms for lift production. *J. Exp. Biol.* **59,** 169–230.

Wells, D. J. (1993) Muscle performance in hovering hummingbirds. *J. Exp. Biol.* **178,** 39–57.

Willmott, A. P., and Ellington, C. P. (1997) The mechanics of flight in the hawkmoth *Manduca sexta.* I. Kinematics of hovering and forward flight. *J. Exp. Biol.* **200,** 2705–22.

TWO

Elastic Proteins: Biological Roles and Mechanical Properties

John Gosline, Margo Lillie, Emily Carrington, Paul Guerette, Christine Ortlepp, and Ken Savage

INTRODUCTION

The objective of this chapter is to present an analysis of structural design in elastic proteins, to elucidate the functional role that these materials play in the lives of real organisms, and to discover whether molecular mechanisms in these materials could be exploited through biotechnology. One striking feature of the elastic proteins we will consider is that they exhibit an exceptionally broad range of material properties and functional roles. As a starting point, therefore, it may prove useful to explore some general features of mechanical design in elastic proteins to set the stage for the detailed analysis of the individual proteins in the chapters that follow.

It is frequently assumed that mechanical and biochemical devices in organisms represent perfect or near perfect solutions to the problems that organisms encounter in their lives. Although it is not clear if this optimistic view is strictly true for elastic proteins, or for any other systems in biology, it is likely that elastic proteins are relatively good designs because they have been tested and modified through eons of evolutionary history. For the purpose of our discussion, we will define 'design' as the relationship between the structure and the function of biological devices, as they exist in living organisms. There are two paths towards an understanding the design of elastic proteins. Most obvious, the direct analysis of microscopic and molecular structure will reveal details of molecular mechanisms in elastic proteins and will document structure–property relationships for these materials. This approach gets us only half the way to an understanding of design. The problem is that material properties alone do not specify the function of a mechanical device. Materials science offers a large number of properties that we can use to quantify the behaviour of a structural material; but to understand the function of a device, it is necessary to identify which of the properties are key to its design. Thus, it is essential that we can evaluate quality in light of the material properties that truly reflect the function of elastic proteins in living animals. It is this second path, the evaluation of material

15

properties in the context of the mechanical role, which forms the substance of this chapter.

MATERIAL PROPERTIES

Table 2.1 provides a convenient way to begin thinking about the problem at hand. It lists a number of functional attributes that can be assigned to structural materials, and it gives the associated material properties and units that can be used to quantify these attributes. Each of the attributes can be assessed by a mechanical test, usually some variation of a stress-strain test, and the key to evaluating design quality will be to decide which properties are dominant in the function of particular elastic proteins in life.

Figure 2.1 shows the results of mechanical tests conducted on a number of elastic proteins, proteins unfortunately restricted to those that can be obtained as macroscopic samples and tested by conventional methods. The mechanical tests were carried out at temperature and hydration states that correspond to in vivo conditions for the proteins. For example, spider silks were tested in air, byssus fibres were tested in seawater, and tendon collagen, elastin, and resilin were tested in water or dilute physiological saline. The stress-strain curves plotted are typical data that would be seen in a constant strain rate test to failure.

It is clear from Figure 2.1 that elastic proteins are remarkably diverse in their properties; but, unfortunately, the stress-strain curves in Figure 2.1 do not clearly reveal the full range of this diversity. This linear plot is dominated by the two strongest materials: dragline silk and viscid silk. The stress-strain curves for tendon collagen and mussel byssal fibres are squeezed together at the bottom of the graph at lower stresses, and the stress-strain curves for the rubber-like proteins, elastin and resilin, lie at such small stresses that they can barely be distinguished from the horizontal axis of the graph.

Table 2.1. Some Functional Attributes of Materials, and the Material Properties and Units Used to Quantify These Attributes

Functional Attribute	Material Property	Units
Stiffness	Modulus of elasticity, E_{init}	$N\ m^{-2}$
Strength	Stress at fracture, σ_{max}	$N\ m^{-2}$
Toughness	Energy-to-break	$J\ m^{-3}$
	Work of fracture	$J\ m^{-2}$
Extensibility	Strain at failure, ε_{max}	No units
Spring efficiency	Resilience	%
Durability	Fatigue lifetime	s to failure or cycles to failure
Spring capacity	Energy storage capacity, W_{out}	$J\ kg^{-1}$

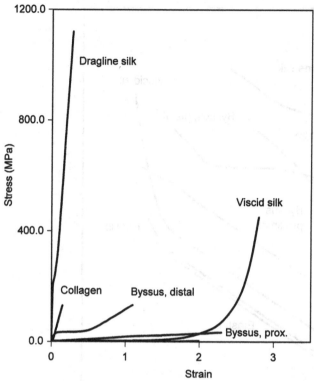

Figure 2.1: Stress-strain curves for seven elastic proteins. Plot is dominated by super-strong spider silks; and, as a consequence, the curves for the rubber-like proteins, elastin, and resilin are indistinguishable from the strain axis and have been left unlabelled. Sources for these stress-strain curves are listed in the legend to Table 2.2.

To remove the dominance of the silk curves, the data in Figure 2.1 have been replotted on logarithmic axes in Figure 2.2, and this makes it possible to see each individual stress-strain curve clearly. Stress-strain plots on linear axes, such as those in Figure 2.1, can be used directly to determine stiffness and other mechanical properties listed in Table 2.1, but the distortion of the transformed curves makes their interpretation a bit more difficult. However, the figure is useful because it clearly represents the enormous range of properties in this collection of proteins. The upper end of each curve indicates the failure point for each material, and this allows us to determine the tensile strength and extensibility for each material. Using these end points, it can be seen that strength varies by about 1,000-fold and extensibility by about 20-fold. Variation in stiffness, as determined by the initial modulus of elasticity (E_{init}), is more difficult to extract from the logarithmic plot. For all materials other than collagen, there is an initial linear region in the stress-strain curve, and this initial slope can

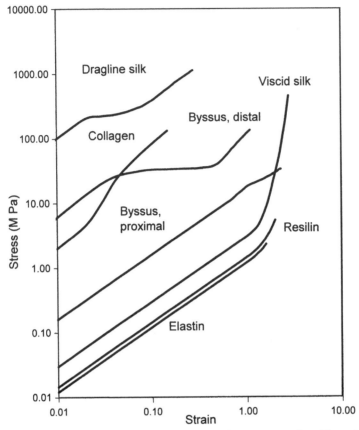

Figure 2.2: Stress-strain curves for the seven elastic proteins from Figure 2.1 are plotted on logarithmic axes to reveal the full diversity of their mechanical properties.

be used to estimate E_{init}. The stress achieved when the initial linear portion of each curve is extended to a strain of 1.0 gives the value of the E_{init}, and this analysis indicates that stiffness varies by a factor of about 10,000-fold. Finally, toughness, determined from the area under the stress-strain curve, can only be obtained from linear plots. Values for modulus, strength, extensibility, and toughness given in Table 2.2 are derived from the stress-strain curves in Figures 2.1 and 2.2.

Data in Table 2.2 give us little insight into functional significance or design for any of the materials listed. They simply list values for a range of properties that can be documented in mechanical tests. To understand design, we must think about the way that elastic devices work in living systems. Indeed, it is important to consider what is meant by the term 'elastic', as this may help us to understand

Table 2.2. Material Properties That Can Be Extracted from the Stress-Strain Curves in Figures 2.1 and 2.2

Material	Modulus, E_{init} (GPa)	Strength, σ_{max} (GPa)	Extensibility, ε_{max}	Toughness (MJ m^{-3})	Resilience (%)
Elastin (Bovine ligament)[a]	0.0011	0.002	1.5	1.6	90
Resilin (Dragonfly tendon)[b]	0.002	0.004	1.9	4	92
Collagen (Mammalian tendon)[c]	1.2	0.12	0.13	6	90
Mussel byssus, distal (Mytilus californianus)[d]	0.87	0.075	1.09	45	28[e]
Mussel byssus, proximal (Mytilus californianus)[d]	0.016	0.035	2.0	35	53[e]
Dragline silk (Araneus diadematus)[f]	10	1.1	0.3	160	35
Viscid silk (Araneus diadematus)[f]	.003	0.5	2.7	150	35
Kevlar[g]	130	3.6	0.027	50	
Carbon fibre[g]	300	4	.013	25	
High-tensile steel[g]	200	1.5	0.008	6	

References: [a]Aaron and Gosline (1981); [b]Weis-Fogh (1961) and Gosline (1981); [c]Pollock and Shadwick (1994); [d]Bell and Gosline (1996); [e]Waite et al. (2001); [f]Denny (1976) and Gosline et al. (1999); [g]Gordon (1988).

designs that use elastic proteins. Strictly speaking, elastic implies the physical phenomenon of *elasticity*, the subject of Hooke's law, which states that when a force is applied to an object, that object will deform in proportion to the magnitude of the applied force. Furthermore, when the force is removed, the object will return to its original state. Hence, the term elastic means reversible deformation. In addition, reversible deformation implies that the mechanical energy required to deform the object is stored as elastic strain energy and that all of this stored energy can be recovered in elastic recoil. That is, the elastic efficiency or resilience of a load-unload cycle should be 100%. An alternate meaning for the term elastic is less precise. Elastic is often taken to mean 'stretchy', like a rubberband, implying that elastic solids can be deformed to large strains with small forces. In the sections that follow, we will use these two criteria, reversible elasticity and stretchiness, to develop an understanding of the function, and hence the design, of elastic proteins.

THE FUNCTIONAL DESIGN OF RUBBER-LIKE PROTEINS

The rubberlike proteins, elastin and resilin, are elastic by both criteria described previously. Both proteins exhibit reversible deformation with high resilience. In addition, both proteins are stretchy, reaching maximal extensions in excess of 100%, with a very low modulus of elasticity. This suite of properties implies that a key function of resilin and elastin is to provide low stiffness, high strain, and efficient elastic energy storage components in animal devices. Elastin functions in association with collagen in vertebrate connective tissues where soft, reversible elasticity is required (e.g., in skin and elastic cartilage). In addition, elastin is a major component of arteries, where its stretchiness and ability to store elastic strain energy allow arteries to smooth the pulsatile flow of blood from the heart, lowering peak blood pressure and the mechanical work of the heart, and maintaining a relatively steady flow of blood through tissues. Thus, elastin is definitely an elastic protein, but interestingly it does not behave like one under all conditions.

Figure 2.3 provides data for the mechanical behaviour of elastin under a number of circumstances, including varying strain rate, hydration level, and temperature. Stretchy materials like elastin achieve their mechanical properties because they contain flexible molecules that can easily change their shape, or conformation, when stretched. The desirable properties of low stiffness, high extensibility, and high resilience that are key to elastin's function rely entirely on the ability of the molecules to change their shape faster than the macroscopic shape change imposed by an external force. Thus, these elastic properties are strongly affected by strain rate in a mechanical test. In addition, because conformational change in elastic proteins occurs only in hydrated proteins, elastic properties can also be strongly affected by hydration level. Finally, because

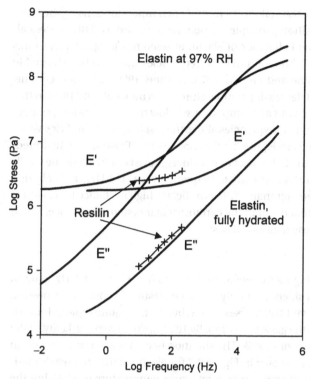

Figure 2.3: Dynamic mechanical properties of elastin and resilin. Two master curves for elastin were constructed at a reference temperature of 37°C. Redrawn from Gosline and French (1979) and Gosline (1980). Data for resilin (crosses) were obtained at room temperature and were redrawn from Gosline (1980). RH, relative humidity.

conformational changes are driven largely by thermal agitation, properties are also influenced by temperature. The effects of these environmental parameters are illustrated in Figure 2.3, which shows the results of dynamic mechanical tests on mammalian elastin.

Dynamic testing involves the application of small amplitude, sinusoidal deformation to a test sample at a range of test frequencies. Records of time-varying stress and strain are analyzed to provide a value for the Dynamic Modulus, E^*, which is the ratio of the amplitudes of dynamic stress and strain waveforms. In addition, measurement of the phase shift between the stress and strain waveforms allows the dynamic stiffness to be separated into two components. These are, Storage Modulus, E', which is the stiffness associated with the storage of elastic energy, and Loss Modulus, E'', which is the stiffness associated with molecular friction and energy dissipation. E' and E'' for elastin are plotted in Figure 2.3 as a function of test frequency. The plots are master curves that were constructed by combining dynamic tests taken at a number of temperatures, using the time-temperature superposition principle for polymeric materials. This process allows one to predict the behaviour of elastin at reference temperature, in this case 37°C, over a much broader range of frequencies than can be achieved in laboratory tests (Gosline and French, 1979; Gosline, 1980; Lillie and Gosline, 1990). The logic is that decreasing temperature slows molecular motion, so that mechanical tests carried out at a temperature below the reference temperature will reveal the behaviour at frequencies above the test frequency, and vice versa.

Look first at the lines in Figure 2.3 that are labeled 'Elastin, fully hydrated'. These curves predict the behaviour of elastin as it exists in living tissues, fully hydrated at 37°C. Note that at cardiac frequencies (ca. 1–3 Hz), E' is about 2 orders of magnitude greater than E''. This indicates high resilience for elastin at rates of deformation that occur in the cardiovascular system. Resilience, R, can be calculated from these data as follows:

$$R = e^{-2\pi\delta},$$

where δ is the damping factor, equal to the ratio $\delta = E''/E'$. At 1 Hz, data in Figure 2.2 indicate a resilience for fully hydrated elastin of about 90%. However, as frequency rises above 1 Hz, E' rises slowly, but E'' rises quite rapidly, indicating elastin's entry into the glass transition. By 100 Hz, δ has risen markedly, to the point where resilience is only 50%. The situation becomes even worse if elastin hydration is reduced, as shown in Figure 2.3 for elastin at 97% relative humidity. This is another master curve with a reference temperature of 37°C, but the water content is reduced by about 50% (from about 0.55 g water per g elastin to 0.29 g per g). The resulting loss of molecular mobility causes a dramatic change in properties, effectively shifting the master curve for fully hydrated elastin to the left by about three decades on the frequency scale. Now, efficient elastic behaviour only occurs at frequencies below 10^{-2} Hz. At cardiac frequencies, the resilience is well below 50%. Thus, increasing frequency or reducing water content dramatically reduces resilience, and this will limit the utility of using elastin in strain energy storage devices. This may explain the absence of elastin in the flight system of hummingbirds.

Hovering flight is an energetically expensive process, and elastic energy storage systems can be used to minimize the cost of flight (Weis-Fogh, 1972). There is strong circumstantial evidence that hummingbirds function as harmonic oscillators, and that they flap their wings at their resonant frequency to reduce the inertial costs of accelerating and decelerating the mass of the rapidly oscillating wings (Wells, 1993a, 1993b; Chai et al., 1996). However, even though hummingbirds do have elastin, it is not used in their flight system, probably because its resilience is too low at their wing beat frequencies (40–70 Hz). Wing elasticity may be provided by the flight muscles or by the tendons connecting these muscles to the wings.

A similar situation is found in the flight systems of insects. There is compelling evidence that insects whose flight is powered by asynchronous flight muscles function as resonant harmonic oscillators (Josephson et al., 2000). Insects do not have elastin, but there is another rubber-like protein, resilin, that is found in the wing hinges of some insects (Weis-Fogh, 1960). As shown in Figure 2.3, the dynamic properties of resilin are quite similar to those of elastin (Jensen and Weis-Fogh, 1962; Andersen and Weis-Fogh, 1964; Gosline, 1980). The curves for E' and E'' run essentially parallel to those for fully hydrated elastin, indicating that resilin, like elastin, is not capable of serving as an efficient elastic energy storage material for the flight system of insects at high frequencies. Interestingly, resilin's prevalence in insect wing hinges is limited to insects that fly at modest wing beat frequencies using synchronous flight muscles. Locust and dragonflies, for example, have considerable resilin in their wing systems, but their wing frequencies are at or below 25 Hz, where resilin's resilience is more than 70%. Insects with asynchronous flight muscles have wing beat frequencies in the 100- to 700-Hz range; at these frequencies, resilin is not a significant component of the wing oscillator. As a consequence, elastic energy storage must be provided by other sources: the flight muscle and the rigid thoracic cuticle.

It appears that elastin's function as a strain energy store is restricted to low-frequency load cycles and conditions where elastin can maintain full hydration. It remains for us to quantify its elastic energy storage capacity under these conditions. To do this, we need information on the durability of elastin in long-term loading to establish the maximum stress level that can be used safely in elastin-based energy storage devices. We have been investigating the fatigue lifetime of arterial elastin to assess the possibility that structural fatigue might contribute to the mechanical degradation of arterial elastin with age and disease. Elastin is an unusual protein in that it apparently does not turn over during the lifetime of an animal (Shapiro et al., 1991; Davis, 1993). Elastin synthesised during development remains in place through the full life span of the organism, which for humans is 60–80 years. Thus, elastin must be an extremely durable material.

Figure 2.4 shows results from our experiments to establish the fatigue lifetime of elastin, based on the behaviour of purified pig arterial elastin. The procedure involves failure tests performed at a wide variety of constant strain rates, which allow us to establish a correlation between failure stress and failure time. One then predicts the fatigue lifetime at in vivo stress by finding the failure time associated with that stress level. The problem for elastin is that the fatigue life-time must be of the order of the lifetime of the animal, which in the case of pig elastin is of the order of 10 years, and it is virtually impossible to conduct fatigue

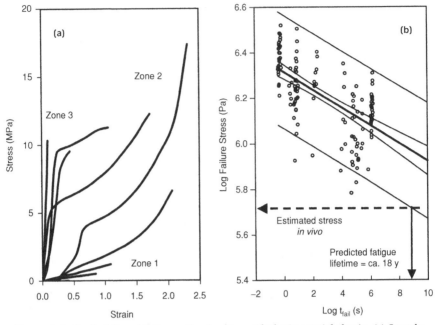

Figure 2.4: Static fatigue lifetime estimates for purified, pig arterial elastin. (a) Sample stress-strain curves for constant strain-rate tests. Samples tested under full hydration at low strain rates show typical rubber-like behaviour (zone 1); but, as strain rate is increased, elastin passes through its glass transition, first becoming tough (zone 2) and then becoming a brittle polymeric glass (zone 3). (b) Correlation of failure stress and failure time for fully hydrated elastin at a broad range of strain rates. Extrapolation of this failure envelope to long times provides an estimate of static fatigue lifetime at in vivo stress, as described in the text.

tests of this duration. We have used the time-temperature shifting procedures described previously for the creation of the master curves shown in Figure 2.3 to extend the time scale of our experiments, and then have extrapolated the trend to times that include the full lifetime of animals.

Figure 2.4(a) shows typical stress-strain tests to failure for arterial elastin ring samples, tested over an enormous range of effective strain rates. The mechanical properties, including failure stress and failure strain, are strongly influenced by test conditions. At low strain rates, high temperatures and full hydration elastin behaves as a typical rubber, as shown by the low stiffness and high extensibility of the curves in zone 1. Under these test conditions, failure stresses are low, but the extensions and times to break are high. High strain rates, low temperatures, and low hydration give elastin the properties of a rigid polymeric glass, with very high stiffness and low extensibility (zone 3). Failure stresses here are high, but the extensions and times to break are extremely low. At intermediate test conditions, elastin is in the middle of its glass transition where it is a tough, semirigid polymer that fails at high stress and high strain (zone 2).

To assess the fatigue lifetime of elastin in its natural form in Figure 2.4b, we have plotted data from 104 samples that showed zone 1 type behaviour. This plot shows the correlation between failure stress and failure time that allows us to estimate durability. The correlation between failure stress and failure time is robust, but the large variation in failure stress limits the accuracy of our predicted fatigue lifetime. In this figure, the heavy line is the predicted regression, the inner set of thinner lines is the 95% confidence interval of the regression, and the outer set of thinner lines is the 95% confidence interval for the prediction. It is the intersection of the lower 95% confidence interval for the prediction that probably establishes the minimum estimate of the fatigue lifetime. We estimate that, in vivo, arterial elastin fibres are stretched by 50% and hence experience a stress of about 0.55 MPa. This level is indicated by the dashed, horizontal arrow in Figure 2.4(b). The vertical arrow indicates the predicted fatigue lifetime as about 18 years, essentially equal to the lifetime of the animal from which this elastin was isolated. This result provides two important conclusions. First, elastin fibres in life appear to function very close to the upper stress limit allowed by its fatigue lifetime. Second, this stress level provides the information needed to estimate the elastic energy storage capacity of elastin.

Elastic energy storage capacity is calculated as follows. The energy per unit volume required to deform elastin to its stress and strain in use, W_{in}, is calculated as

$$W_{in} = 2(\sigma_{in\ use} \times \varepsilon_{in\ use}),$$

where $\varepsilon_{in\ use}$ is estimated from the ratio of the stress-in-use divided by the modulus of elasticity. The elastic energy recovered in a load cycle, $W_{out} = W_{in} \times R$, where R is the resilience of elastin. Finally, the energy storage capacity can be converted into the units of J kg^{-1} by dividing W_{out} by the density of elastin. As indicated in Table 2.3, the elastic energy storage capacity of elastin is about 95 J kg^{-1}. The table also gives the energy storage capacity of spring steel as

Table 2.3. Elastic Energy Storage Capacities for Elastin, Collagen, and Spring Steel Have Been Calculated from Data Given in This Table, as Described in the Text

Material	Modulus (GPa)	Stress-in-Use (MPa)	Density (kg m^{-3})	Resilience	Energy Storage Capacity (J kg^{-1})
Elastin	0.0011	0.55	1300	0.90	95
Collagen	0.12	60	1300	0.90	1000
Spring steel	200	600	7800	0.99	115

115 J kg^{-1}. It may seem remarkable that a material as soft and as weak as elastin (see Table 2.2) has essentially the same elastic energy storage capacity as spring steel. This is indeed the case, and it is a good indicator that elastin is a high-quality spring material. Table 2.3 also gives the spring energy storage capacity of collagen that is about 10 times greater than that for elastin or steel, and this leads to a completely different class of elastic protein.

THE FUNCTIONAL DESIGN OF COLLAGEN

Collagen fibres, as seen in tendons, can hardly be described as stretchy, because their extensibility, ε_{max}, is only about 0.13. Neither is collagen soft, because its modulus is about 1000 times greater than that of elastin or resilin. It is also much stronger and somewhat tougher than elastin or resilin (see Table 2.2). So why consider collagen an elastic protein? Figure 2.5 shows a typical stress-strain curve to failure for tendon collagen, with a load cycle to about 50% of breaking stress superimposed upon it. Note that tendon, and virtually all other

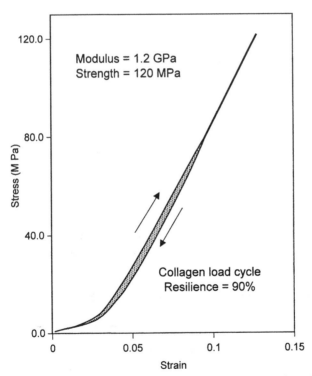

Figure 2.5: Typical stress-strain curve for tendon collagen, based on data from Ker (1981) and Shadwick (1990). The modulus given is for the linear portion of the stress-strain curve. The load cycle overlying the stress-strain curve indicates that collagen has a resilience of about 90%.

collagen-containing connective tissues, have a "J-shaped" stress-strain curve. At low strains, the slope is low; but, as extension proceeds, it rises gradually and becomes constant when the collagen fibres become aligned and then finally stretched. These data, and particularly the load cycle, show that collagen is definitely capable of reversible deformation, and it is this aspect of collagen that establishes it as an elastic protein. Indeed, the resilience of tendon collagen is about 90% (Ker, 1981; Shadwick, 1990). Despite this, collagen does not always function as an elastic protein in animals. Rather, collagen is frequently arranged in parallel with elastin fibres to form connective tissues and, in these circumstances, the "stretchy" elasticity of the tissue is due primarily to the elastin. Collagen provides a network of wavy, reinforcing fibres that become aligned in the direction of stretch. When taut, this network limits tissue deformation and prevents the rupture of the softer and weaker elastin fibres. Some tendons, however, do function as elastic devices, and they have a remarkable energy storage capacity.

Ker et al. (1988) found that there are two distinct classes of tendons, based on the stresses these tendons experience in vivo. Some tendons seem to be over-strong, with safety factors between 8 and 15. Other tendons, particularly those found in the ankle and digital extensor muscles in the limbs of cursorial animals, have much lower safety factors, typically between 1 and 2. In this case, safety factor is defined as the tendon's breaking force divided by the maximum isometric force generated by its attached muscle.

Referring to the stress-strain curve in Figure 2.5, we can see the functional consequences of these differences. The stress-strain curve in this figure is typical of both the over-strong and the low safety-factor tendons. Both classes of tendons have essentially identical stiffness in the linear zone and tensile strength (Pollock & Shadwick, 1994), and this means that the strain levels seen in life by these two tendon types must be very different. In the over-strong tendons, tendon strain should never exceed 0.02 (i.e., 2% extension), even when the attached muscles are maximally activated. Under submaximal loading the strain in these tendons will be even smaller. Functionally, the over-strong tendons provide rigid links between muscle and bone. On the other hand, during locomotion, the low safety-factor tendons experience much larger strains in their normal function, strains that approach the failure strain. This exceptional difference in strain level is a direct reflection of the design of low safety-factor tendons to function as springs that conserve energy in running locomotion (Alexander, 1988). Indeed, these low safety-factor tendons could be called spring tendons, and they are likely the pre-eminent strain energy storage devices in animal skeletons.

The elastic energy storage capacity of tendon collagen can be calculated from the data in Figure 2.5, with some additional information on the fatigue behaviour of tendon collagen. These data are presented in Figure 2.6, which have been redrawn from Ker et al. (2000). Figure 2.6(a) shows the static fatigue

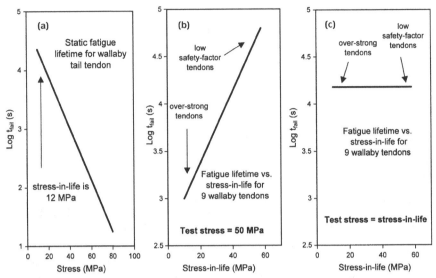

Figure 2.6: Fatigue lifetime of tendon collagen. (a) Static fatigue lifetime (t_{fail}) of wallaby tail tendon at a range of test stress levels. Vertical arrow indicates the maximal stress-in-life for this tendon. (b) Static fatigue lifetime (t_{fail}) obtained for nine different types of wallaby tendons, all taken at a test stress of 50 MPa, are plotted at the stress-in-life for each tendon. (c) Static fatigue lifetime (t_{fail}) for the nine wallaby tendons, obtained at the stress-in-life for each tendon, are plotted against the stress-in-life. Data plotted here are replotted from regression lines in Figures 2, 4, and 5 of Ker et al. (2000). Specific information on the identity and properties of the different tendons can be found in that article.

behaviour of wallaby tail tendon, an over-strong tendon with a safety factor of about ten. Fatigue lifetime in seconds is plotted as a function of static stress level on a log-log plot, and the regression line shown was based on more than 90 fatigue tests. There is a strong correlation between stress and fatigue lifetime, with fatigue lifetimes ranging from about 15 s at the highest stresses to over 10^4 s at low stress. The tail tendon sees a 'stress-in-life' of 12 MPa, as determined from the maximal isometric force of the tail muscle. The vertical arrow indicates the intersection of the stress-in-life with the regression line, and this intersection point indicates that the fatigue lifetime of wallaby tail tendon at its stress-in-life is about 2×10^4 s, or 5.5 hr. It is interesting that the static fatigue lifetime of tendon collagen is many orders of magnitude shorter than that of arterial elastin, but perhaps this is because the stress levels experienced by tendons are so much higher. It certainly confirms that tendons must remodel and repair themselves throughout the life of an animal.

Figure 2.6(b) shows the regression line obtained when the fatigue lifetime for nine different wallaby tendons, all determined at a fixed test stress of 50 MPa, is plotted as a function of the stress-in-life for each type of tendon. The range of tendons used includes several that are over-strong and have low stress-in-life, such as the tail tendon and digital flexor tendon. At the other extreme are limb extensor tendons, the plantaris and digital extensor tendons, which are low safety-factor tendons and have a high stress-in-life. Interestingly, there seems to be a large variation in fatigue lifetimes that correlates with the stress-in-life seen by each type of tendon. This correlation indicates that those tendons that have evolved to function at high stress (the spring tendons) have also developed a better fatigue resistance than the over-strong, link tendons. Indeed, it suggests that if the fatigue lifetime for all tendons is measured at the stress-in-life, then all tendons may have very similar fatigue lifetimes. Figure 2.6(c) shows that this is indeed the case. All tendons have essentially the same fatigue lifetime, approximately 5 hr at the stress-in-life.

With this information, we can estimate the elastic energy storage capacity of the collagen in spring tendons. We will assume that spring tendons evolved to function at stresses up to their stress-in-life and thus that these tendons can be loaded to 60 MPa repeatedly in energy-storing load cycles. Elastic energy storage capacity was calculated as described for elastin, and the values are shown in Table 2.3. As noted previously, collagen has an elastic energy storage capacity that is roughly 10 times greater than that of elastin or spring steel. It is important to note that this storage capacity is not the highest possible, as the limb extensor tendons in animals larger than the wallaby experience considerably higher maximal stress-in-life. In reality, the upper limit of elastic energy storage capacity in collagen is likely closer to 2,000 J kg^{-1}.

Finally, the spring energy storage potential for collagen needs to be considered in light of the other properties that have been determined for collagen. Collagen does function in circumstances where its strength and toughness are crucial, but as listed in Table 2.2, collagen is not particularly stiff, strong, or tough when compared with other materials. For example, the spider silks are 3 to 8 times stronger and 20 times tougher than tendon collagen, and high-performance man-made fibres can be stronger still. Thus, collagen can be described as an exceptional design for elastic energy storage and a modest design for strength and toughness. To find elastic proteins with exceptional design for strength and toughness, we will have to look at the remaining elastic proteins in our list: mussel byssal fibres and spider silks.

THE FUNCTIONAL DESIGN OF MUSSEL BYSSAL FIBRES

The fibres of the mussel byssus are used to attach the animal to rocks in the wave-swept marine intertidal. The fibres look like good candidates for elastic proteins by the 'stretchy' criterion, as shown the force extension curve for whole byssus fibre in Figure 2.7. Whole byssal fibres stretch by about 100% before breaking, but the origins of the whole-fibre behaviour are complex because the fibres are

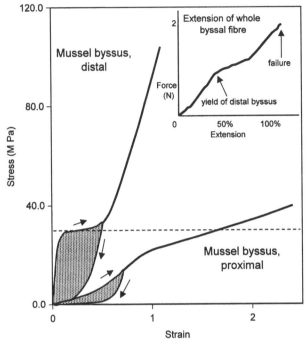

Figure 2.7: Mechanical properties of mussel byssal fibres. The stress-strain plot shows the material properties of proximal and distal portions of byssal fibres from the mussel, *Mytilus californianus*. Dashed line indicates that the yield level for the distal fibre occurs below the failure stress for the proximal fibre. Inset graph shows typical force elongation for a whole byssus fibre. Graphs redrawn from Bell and Gosline (1996).

composed of two segments: a distal segment that makes up about 80% of the fibre and a short, proximal segment that makes up the remainder of the fibre. The material properties of these two segments from *Mytilus californianus* are shown as stress-strain curves in Figure 2.7 (Bell and Gosline, 1996). The proximal segment is extremely stretchy, with an extensibility of about 2.0, suggesting rubber-like behaviour, but its stiffness and strength are about an order of magnitude greater than those for elastin and resilin (Table 2.2). The distal region is somewhat less extensible, with an ε_{max} of about 1.0, but its stiffness and strength approach those of collagen. Because of the combination of high strength and extensibility, the toughness of both proximal and distal byssal threads is about an order of magnitude greater than that of resilin or elastin and about six times greater than that of tendon collagen. The toughness of byssal fibres is quite comparable with that of Kevlar and carbon fibres (Table 2.2), and this high toughness is certainly key to the survival of intertidal marine mussels.

Interestingly, byssal fibres would not be considered as elastic materials if judged by the criterion of reversible elasticity. First, the large, open-load cycles of the stress-strain curves in Figure 2.7 indicate a low resilience. Second, it is not clear if the deformation of these fibres is fully reversible, in that the fibres may not immediately return to their original dimensions following load cycles to high, but submaximal strain levels. The resilience values in Table 2.2 are for the first-cycle behaviour of fibres from *Mytilus galloprovincialis*, taken from Waite et al. (2001) in this volume. Subsequent load cycles show increased resilience, so the mechanical properties are quite complex; but, it is clear that these fibres do not function as efficient springs (Waite et al., 2001). Rather, their strength and toughness, properties that allow mussels to resist dislodgment by breaking waves or hungry starfish, are likely enhanced by their low resilience.

Polymeric materials that dissipate mechanical energy through molecular friction, and hence have low resilience, usually have enhanced toughness. This correlation between low resilience and high toughness in polymeric materials can be seen clearly for elastin in Figure 2.4(a). In fully hydrated elastin, where resilience is high (zone 1), stiffness, strength, and toughness are low (Table 2.2). As water content and temperature are lowered and strain rate increased, elastin enters its glass transition (zone 2). Here, it exhibits higher stiffness and extensibility, but lower resilience, leading to a toughness that is at least an order of magnitude higher. It is worth noting that, because elastin must be dehydrated or cooled to exhibit zone 2 behaviour at moderate strain rates, it may never function in this region of its viscoelastic curve under natural conditions. However, mussel byssal fibres achieve similar toughness levels when fully hydrated in seawater, and so it may be fruitful to discover the molecular origins of this enhanced toughness.

Our analysis to this point indicates that byssal fibres are exceptional for their stretchiness and for their toughness. It is not clear, however, if these properties alone are key to the function of the byssus apparatus. It is possible that the yield behaviour, seen particularly clearly in the distal byssus, might also play a role in its function. Mussels build their byssus apparatus to anchor themselves to rocks in the intertidal, where the waves can strike them from a variety of directions. As a consequence, they create a multidirectional array of fibres, with only a fraction of the fibres oriented to resist forces coming from any particular direction. However, as these fibres are elongated by a breaking wave, other fibres reorient and are recruited into the load-bearing set of fibres. Clearly, the stretchiness of the fibres facilitates this process, but in addition, it appears that a significant yield in their force deformation behaviour has the same effect (Bell and Gosline, 1996). Unfortunately, we know little about the stress-in-life or the fatigue behaviour of these fibres, and it is clear that more research is needed before we truly understand the design of these interesting elastic proteins.

THE FUNCTIONAL DESIGN OF SPIDER SILKS

Spiders produce a variety of structural polymers, called silks, that have evolved to function in air rather than in aqueous media, and whose mechanical properties range from rubber-like to extremely rigid. Given this range of properties, it is not possible to categorize all silks with a single set of criteria. We will therefore focus on the two silks that are best studied and whose properties likely span the range of properties that exist in spider silks. These are the viscid silk, which forms the glue-covered catching spiral of the orb web, and dragline silk, the rigid silk used as a safety line and for the frame of the orb web. Stress-strain curves to failure with load cycles overlaid are shown in Figure 2.8 for these two silks.

Viscid silk easily meets the 'stretchy' criterion for an elastic protein because, like elastin and resilin, it has a low initial stiffness, $E_{init} = 3$ MPa, and high extensibility, with $\varepsilon_{max} = 2.5$. However, it is a great deal stronger than elastin or resilin, and its tensile strength of 450 MPa makes it the strongest rubber-like material known. The rubber-like behaviour of viscid silk is unexpected for an elastic protein because viscid silk fibres function in air, and other rubber-like elastic proteins are brittle polymeric glasses when air-dried. The difference for spider's viscid silk is that molecular mobility in its protein network is maintained by the presence of molar concentrations of low molecular weight organic compounds in the glue (Vollrath et al., 1990). These glue compounds are hygroscopic and draw water out of air (Townley et al., 1991). Some of the glue compounds penetrate the silk network where they, along with their associated water molecules, plasticize the silk proteins and maintain molecular mobility. In addition, water

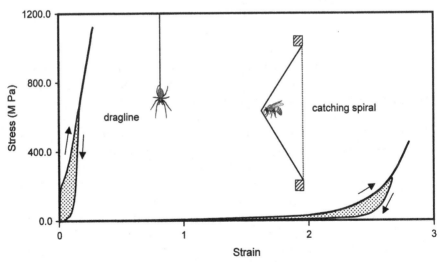

Figure 2.8: Stress-strain curves and overlaid stress-strain cycles for dragline and viscid silks from the spider *Araneus diadematus*. After Gosline et al. (1999).

absorbed into the glue layer keeps the glue sticky for its role in prey capture. If viscid silk is dried over P_2O_5 to remove all water (Vollrath and Edmonds, 1989), or if the glue is washed off and the fibres allowed to air-dry at ambient humidity (Gosline et al., 1995), the fibres become rigid, as expected for an unplasticized protein polymer. Because rubber-like elasticity in viscid silk is maintained by complex mechanisms, long-range elasticity is likely an important property for the silk's function in the web.

Dragline silk, in contrast, is not very stretchy. Its initial modulus at 10 GPa and tensile strength at 1.1 GPa is more like a rigid superfibre, such as Kevlar. This behaviour is a direct reflection of the air-dried, non-plasticised condition of the proteins in these fibres. The extensibility of dragline is about 0.3, and although this is about an order of magnitude greater than that of Kevlar or carbon fibre (see Table 2.2), it is less stretchy than most elastic proteins. It is not clear, therefore, if the label of elastic protein is appropriate for dragline silk. Perhaps the criterion of reversible elasticity will clarify the issue.

The load cycles in Figure 2.8 show typical data for viscid and dragline silks from the spider *Araneus diadematus*. Denny (1976) studied both the viscid and dragline silks of *Araneus sericatus* and observed that both silks failed to return to their initial dimensions at the end of an initial load cycle, but that subsequent cycles had consistent, reversible behaviour. Interestingly, the fibres did return to their initial length and showed the first-cycle behaviour if left slack for about 10 minutes. Characterisation of the shape of the load cycles reveals that both viscid and dragline silks have very low resilience, with a first-cycle resilience of 30–40% for both materials. Shao and Vollrath (1999) obtained similar results in load-cycle tests on dragline silk from *A. diadematus* and from *Nephila edulis*; the silks failed to return to initial length and had very low resilience.

These properties obviously make the web silks poor candidates for use in elastic energy storage devices, but then the orb web functions as an energy-dissipating device. As shown in Figure 2.8, a flying insect hitting one of the stretchy viscid silk strands will deflect the strand as it absorbs the kinetic energy of the insect. If the energy of this impact is stored as elastic strain energy, then elastic recoil of the silk would likely catapult the insect back out of the web. To minimise recoil, absorbed energy must be dissipated as heat through molecular friction. The situation for dragline silk is similar. Its function in the orb web is to support the catching spiral formed by the viscid silk, and its ability to dissipate the energy of impact is equally important to the prey-capture process. In addition, its function as a safety line is to absorb and dissipate the energy of a falling spider. Thus, a key function for both dragline and viscid silks is to dissipate energy in impact loading, which can be achieved because polymers with low resilience have enhanced toughness. The values for toughness in Table 2.2 indicate that these silks are tougher by a factor of 3–4 than the other

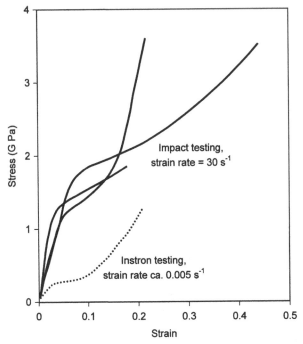

Figure 2.9: Effect of strain rate on the mechanical properties of dragline silk from *Araneus diadematus*. Redrawn from Gosline et al. (1999).

materials listed, including the high-tech superfibres. This, of course, is the key property for the web silks. They are energy-absorbing devices that function only once in the capture of a prey item or in the fall of a spider. Because of this, they are essentially disposable items for which concepts like fatigue are largely irrelevant.

However, the mechanical environment in which they function may strongly affect their properties. Figures 2.3 and 2.4 showed how strain rate, water content, and temperature strongly affected the mechanical behaviour of elastin. These parameters will also have a strong effect on the behaviour of silks. Dragline silk shows dramatic changes in strength and toughness when strain rate is increased to mimic those that likely occur in prey capture, and when dragline is immersed in water it swells and is transformed into a rubber-like elastic material. These two transformations are illustrated in Figures 2.9 and 2.10.

Figure 2.9 shows the effect of increasing strain rate from a low value of $0.005\,s^{-1}$ to about $30\,s^{-1}$, a shift that decreases the time to failure from about 60 s to about 10^{-2} s. The high-strain rate impact tests were designed to mimic an insect flying into a silk strand in a web, and the change in mechanical properties

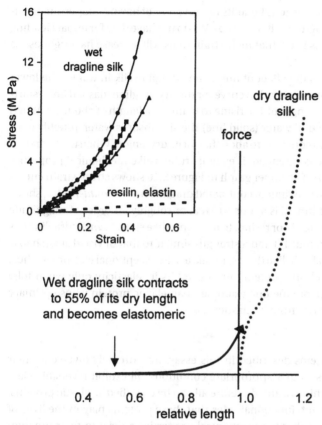

Figure 2.10: Effect of water on the mechanical properties of dragline silk. Lower diagram shows the change in the force elongation behaviour of dragline silk that occurs when it is immersed in water. Inset stress-strain curves compare the material properties of wet dragline silk and the rubber-like proteins elastin and resilin. Initial stiffness for wet dragline silk is about 10 MPa.

are impressive (Gosline et al., 1999). There is a large increase in initial modulus, rising from about 10 MPa to more than 30 MPa, and there is a similar rise in tensile strength. Most test samples fail at stresses of about 2 GPa; but, in some instances (as shown by the two extreme curves in this figure), strength values in the range of 3–4 GPa are seen. That is, strength may approach that of the polymeric superfibres, but the silk's extensibility remains high. This combination brings the toughness to exceptionally high levels. At present, we do not know the maximum toughness possible or the conditions needed to achieve the maximum, but we estimate that the toughness of spider dragline may rise to 500 MJ m^{-3}. This would make spider dragline silk the toughest material known to man, and there are obvious benefits to the spider of having a supertough material for the manufacture of its orb webs. We do not have data for impact loading of viscid silk, but it is likely that high-strain rates will increase its toughness as well.

Figure 2.10 shows the effect of immersing dragline silk in water. The lower diagram shows a force elongation curve for the dry dragline (dashed line), starting from a reference length of 1.0, rising to failure at a length of about 1.3. When dragline is immersed in water (solid line) the silk absorbs water, roughly doubling its volume, and contracts to about half of its dry length. When the hydrated, contracted dragline is extended, it exhibits rubber-like mechanical properties (Gosline et al., 1984). The upper graph in Figure 2.10 shows stress-strain curves for several samples of hydrated, contracted dragline from *A. diadematus*, showing that its initial stiffness is about 10 MPa, or roughly an order of magnitude greater than that of elastin or resilin. In its rubber-like state, *Araneus* dragline has an extensibility of about 1.3 and a strength similar to that of viscid silk (Shao & Vollrath, 1999), making it, like the viscid silk, a truly exceptional elastomer. Thus, dragline silk is an elastic protein, but it exhibits its elasticity only when fully hydrated. It is therefore the one example of an elastic protein whose primary function does not arise from its elasticity.

CONCLUSIONS

The structural proteins described in this essay are extremely diverse in their material properties. Under appropriate conditions, all exhibit reversible elasticity and/or stretchiness, and therefore all can be classified as elastic proteins. When we consider the functional roles that these proteins play in the lives of the animals, we find that the mechanical properties crucial to their function usually, but not always, encompass some aspect of their elasticity. Not all of the elastic proteins are exceptional in their material properties, but all represent good designs because the properties that define their function are well matched to the environmental conditions. Elastin and resilin work well as strain energy storage devices at low frequencies and long times, but they have material properties that could likely be equalled by commonly available synthetic materials. Collagen fibres and spider silks, on the other hand, are at the material limits for their respective classes of polymeric materials, and they are truly exceptional materials. Collagen has unmatched capacity for the storage of elastic strain energy, and spider silks have unmatched toughness. In addition, when we test the natural elastic proteins under environmental conditions beyond those seen in their normal function, we discover a much broader range of material properties. It is this full range of properties that is available for exploitation through biotechnology. For example, dragline silk becomes rubber-like when hydrated, but it retains essentially all of its remarkable strength, making it a candidate as high-performance elastomeric material for bioimplantation. Mussel byssal fibres normally function in seawater, but when dry or perhaps when plasticised by other solvents, they may achieve greater strength with little compromise in extensibility. Clearly, there are interesting and useful molecular mechanisms in elastic proteins that could be incorporated into the design of novel man-made materials.

ACKNOWLEDGEMENTS

This paper was supported by grants from Natural Sciences and Engineering Research Council (NSERC) of Canada and from the Heart and Stroke Foundation of British Columbia and the Yukon (to J.G.) and by the National Science Foundation Grant 0082605 (to E.C.).

REFERENCES

Aaron, B. B., and Gosline, J. M. (1981) Elastin as a random-network elastomer: A mechanical and optical analysis of single elastin fibers. *Biopolymers* **20,** 1247–60.

Alexander, R. M. (1988) *Elastic Mechanisms in Animal Movement.* Cambridge, NY: Cambridge University Press, 141 pp.

Andersen, S. O., and Weis-Fogh, T. (1964) Resilin. In: *Advances in Insect Physiology*, vol. 2, J. W. L. Beament, J. E. Treherne, and V. B. Wigglesworth (eds.). London, UK: Academic Press, pp. 1–65.

Bell, E. C., and Gosline, J. M. (1996) Mechanical design of mussel byssus: Material yield enhances attachment strength. *J. Exp. Biol.* **199,** 1005–17.

Chai, P., Harrykissoon, R., and Dudley, R. (1996) Hummingbird hovering performance in hyperoxic heliox: Effects of body mass and sex. *J. Exp. Biol.* **199,** 2745–55.

Davis, E. C. (1993) Stability of elastin in the developing mouse aorta – A quantitative autoradiographic study. *Histochemistry* **100,** 17–26.

Denny, M. (1976) The physical properties of spider's silk and their role in the design of orb-webs. *J. Exp. Biol.* **65,** 483–506.

Gordon, J. E. (1988) *The Science of Structures and Materials.* New York, NY: Scientific American Library, 217 pp.

Gosline, J. M. (1980) The elastic properties of rubber-like proteins and highly extensible tissues. In: *Mechanical Properties of Biological Materials*, J. F. V. Vincent and J. D. Currey (eds.). London, UK: Society for Experimental Biology, pp. 331–57.

Gosline, J. M., Denny, M.W., and DeMont, M. E. (1984) Spider silk as rubber. *Nature* (Lond.) **309,** 551–52.

Gosline, J. M., and French, C. J. (1979) Dynamic mechanical properties of elastin. *Biopolymers* **18,** 2091–2103.

Gosline, J. M., Guerette, P. A., Ortlepp, C. S., and Savage, K. N. (1999) The mechanical design of spider silks: From fibroin sequence to mechanical function. *J. Exp. Biol.* **202,** 3295–3303.

Gosline, J. M., Nichols, C., Guerette, P., Cheng, A., and Katz, S. (1995) The macromolecular design of spiders' silk. In: *Biomimetics, Design and Processing of Materials*, M. Sarikaya and I. A. Aksay (eds.). Woodbury, NY: A.I.P Press, pp. 237–61.

Jensen, M., and Weis-Fogh, T. (1962) Biology and physics of locust flight V. Strength and elasticity of locust cuticle. *Philos. Trans. R. Soc. Lond.* **254,** 137–69.

Josephson, R. K., Malamud, J. G., and Stokes, D. R. (2000) Asynchronous muscle: A primer. *J. Exp. Biol.* **203,** 2723–35.

Ker, R. F. (1981) Dynamic tensile properties of plantaris tendon of sheep (*Ovis aries*). *J. Exp. Biol.* **93,** 282–302.

Ker, R. F., Alexander, R. M., and Bennett, M. B. (1988) Why are mammalian tendons so thick? *J. Zool.* (Lond.) **216,** 309–24.

Ker, R. F., Wang, X. T., and Pike, A. V. (2000) Fatigue quality of mammalian tendons. *J. Exp. Biol.* **203,** 1317–27.

Lillie, M. A., and Gosline, J. M. (1990) The effects of hydration on the dynamic mechanical properties of elastin. *Biopolymers* **29,** 1147–60.

Pollock, C. M., and Shadwick, R. E. (1994) Relationship between body mass and biomechanical properties of limb tendons of adult mammals. *Regul. Integrative Comp. Physiol.* **35,** R1016–21.

Shadwick, R. E. (1990) Elastic energy storage in tendons: Mechanical differences related to function and age. *J. Appl. Physiol.* **68,** 1022–40.

Shapiro, S. D., Endicott, S. K., Province, M. A., Pierce, J. A., and Campbell, E. J. (1991) Marked longevity of human lung parenchymal elastic fibers deduced from prevalence of d-aspartate and nuclear weapons related radiocarbon. *J. Clin. Invest.* **87,** 1828–34.

Shao, Z., and Vollrath, F. (1999) The effect of solvents on the contraction and mechanical properties of spider silk. *Polymer* **40,** 1799–1806.

Townley, M. A., Bernstein, D. T., Gallagher, K. S., and Tillinghast, E. K. (1991) Comparative study of orb web hygroscopicity and adhesive spiral composition in three araneid spiders. *J. Exp. Zool.* **259,** 154–65.

Vollrath, F., and Edmonds, D. T. (1989) Modulation of the mechanical properties of spider silk by coating with water. *Nature* (Lond.) **340,** 305–7.

Vollrath, F., Fairbrother, W. J., Williams, R. J. P., Tillinghast, E. K., Bernstein, D. T., and Townley, M. A. (1990) Compounds in the droplets of the orb spider's viscid thread. *Nature* (Lond.) **345,** 526–28.

Waite, J. H., Vaccaro, E., Sun, C., and Lucas, J. M. (2001) Elastomeric gradients: A hedge against stress concentration in marine holdfasts? *Philos. Trans. R. Soc. Lond. B* **357,** 143–54.

Weis-Fogh, T. (1961) Molecular interpretation of the elasticity of resilin, a rubber-like protein. *J. Mol. Biol.* **3,** 648–67.

Weis-Fogh, T. (1972) Energetics of hovering flight in hummingbirds and in drosophila. *J. Exp. Biol.* **56,** 79–104.

Weis-Fogh, T. (1960) A rubber-like protein in insect cuticle. *J. Exp. Biol.* **37,** 887–907.

Wells, D. J. (1993a) Muscle performance in hovering hummingbirds. *J. Exp. Biol.* **178,** 39–57.

Wells, D. J. (1993b) Ecological correlates of hovering flight in hummingbirds. *J. Exp. Biol.* **178,** 59–70.

THREE

Elastin as a Self-Assembling Biomaterial

C. M. Bellingham and F. W. Keeley

INTRODUCTION

Elastin is the major extracellular matrix protein in tissues – such as the large arterial blood vessels, lung parenchyma, elastic ligaments, and skin – where it is accepted to be principally responsible for the physical properties of extensibility and elastic recoil which are particularly important for the function of these tissues. Elastin is also the major matrix protein in some cartilaginous tissues – such as ear cartilage – where the functional role of this protein is less evident.

Elastic fibres are made up of two major components: an elastin component and a microfibrillar component. The microfibrillar component, consisting of 10–12 nm filaments, is made up of at least five distinct proteins but thought to be predominantly composed of fibrillin, a 350 kDa protein which exists in two forms: fibrillin-1 and fibrillin-2 (Sakai et al., 1986; Zhang et al., 1994) (see also chapter 5 by Kielty et al., in this volume). The microfibrils also include two microfibril-associated glycoproteins (MAGP-1 and MAGP-2) (Gibson et al., 1991, 1996). Elastic fibres may also include a number of other components, including a 67 kDa chaperone protein known as the elastin binding protein (EBP) (Hinek et al., 1988; Hinek and Rabinovitch, 1994), and lysyl oxidase, the enzyme that catalyzes oxidative deamination of lysine residues in preparation for cross-link formation (Kagan and Trackman, 1991).

Like most structural proteins, elastin is synthesised as a monomer, tropoelastin, a protein with a molecular weight of approximately 70 kDa, which is subsequently assembled into a stable, polymeric structure in the extracellular matrix (Mecham and Davis, 1994; Vrhovski and Weiss, 1998). This polymer is stabilised by covalent cross-links formed through interactions between side chains of lysine residues after oxidative deamination by lysyl oxidase. Polymeric elastin has the unusual property of being essentially insoluble in all reagents except those that break polypeptide bonds, and this persistent insolubility of elastin has been both a benefit in the isolation and purification of the protein, and an impediment to its structural characterisation. In addition, with the notable exception of the uterus (Starcher and Percival, 1985), elastin, once laid down in its insoluble polymeric form in the extracellular matrix of tissues, does not turn over at any appreciable rate under normal circumstances (Davis, 1993). In effect, this means that the elastin present in the aorta of an older human is the elastin that was laid down during aortic development. The fact that this aortic elastin has literally gone through thousands of millions of cycles of extension and recoil without mechanical failure is a testament to the remarkable properties of elastin as a biomaterial.

For more than three decades, there has been a continuing lively debate over the basis for the elastomeric properties of this protein. Attempts to account for the elastomeric properties of elastin through structural models have a long history, from the earlier globular models of Partridge (1966), to the 'oiled coil' model of Gray et al.(1973), to the 'librational motion' model of Urry (1988). However, as alluded to earlier, one of the major impediments to detailed structural analysis of elastin is the insolubility of the protein, limiting at least some of the most powerful approaches such as X-ray crystallography and solution nuclear magnetic resonance. The more recent availability of monomeric elastin has not been of particular assistance in this respect because of the strong tendency of this monomer for self-aggregation, of which more will be said later. Many structural studies have used circular dichroism and other physical techniques either on hydrolytic fragments of insoluble elastin or on synthetic model peptides based on elastin sequences (Urry et al., 1969; Jamieson et al., 1972; Urry, 1988; Tamburro et al., 1992; Castiglione-Morelli et al., 1993; Reiersen et al., 1998; Debelle et al., 1998; Bochicchio et al., 2001). Alternatively, structural conclusions have been drawn from various techniques used to image the polymeric elastin matrix (Ross and Bornstein, 1969; Pasquali Ronchetti et al., 1979; Bressan et al., 1986; Mecham and Heuser, 1991). More recently, computational methods have also been used to predict the structure of specific domains of elastin (Debelle et al., 1998).

Leaving the structural and biomechanical arguments to others, there can be agreement on some fundamentals. First, as data are becoming available on more elastomeric proteins, it is clear there is no obvious single protein sequence which confers properties of elastic recoil on these proteins, although similarities in protein design may be very informative. Second, the way in which the monomers are organized into polymeric structures must be crucial for their elastomeric properties. It is this question of the organisation of elastin monomers into polymeric elastin that we will deal with here.

TROPOELASTIN

Tropoelastin is synthesised from a single gene of approximately 45 kb in size. There are 34 exons in the human gene, each generally encoding either a hydrophobic or a cross-linking domain, with the exception of the first exon that contains the signal peptide and the final exon that contains the C-terminal domain and the 3'-untranslated region (Bashir et al., 1989). Although there is a single gene for elastin, a number of mRNA splice variants have been reported that result in several isoforms of tropoelastin (Indik et al., 1987, 1989; Yeh et al., 1987; Boyd et al., 1991). However, the role of these isoforms is not clear.

Tropoelastin has a repeating domain structure, alternating between hydrophobic and cross-linking regions. In general, tropoelastin is extremely nonpolar, with approximately 80% of the total residues made up of proline (P), valine (V), glycine (G), leucine (L), isoleucine (I), and alanine (A). One-third of the amino acid residues are glycine. These non-polar amino acid residues are found predominantly in the hydrophobic regions of tropoelastin. Typical sequences found in these domains, often present in tandem repeats, include PGGV, PGVGV, PGVGVA, and GGLGV. These types of repetitive motifs are found in elastin of all species, although the specific sequences may vary. For example, the sequence encoded by exon 24 contains the hydrophobic peptide motif PGVGVA repeated seven times in tandem in human and baboon elastin (Szabo et al., 1999). In comparison, exon 24 of bovine elastin contains the sequence PGVGV repeating four times in tandem. In chicken elastin, the PGVGV sequence is repeated 10 times in this exon (Figure 3.1).

The cross-linking regions are rich in alanine and also contain the lysine (K) residues from which the cross-links of elastin are formed. Alanines and lysines are typically found in sequences, such as AAAKAAKAA or AAAKAAAKAA, where the lysine residues occur in alanine-rich sequences and are usually separated by two or three alanine residues (Gerber and Anwar, 1974). In human elastin, lysine makes up 4% of the total amino acid residues with typical cross-linking regions containing either two or three lysine residues. Aromatic residues (tyrosine or phenylalanine) are found C-terminal to some of these lysine residues, and have been suggested to play a role in determining the nature of the cross-links formed (Gerber and Anwar, 1975; Baig et al., 1980).

IN VIVO ASSEMBLY AND CROSS-LINKING

Unlike collagen, tropoelastin does not have 'registration' or propeptides at the N- or C-terminals of the polypeptide that aid in the alignment of monomers for polymer formation. However, several other components of the elastic fibre have been suggested to contribute to the organization of the elastin polymer. Tropoelastin is believed to be escorted through intracellular compartments and presented on the cell surface for orderly incorporation into the elastic fibres by the EBP (Hinek et al., 1988; Hinek and Rabinovitch, 1994). Acting as a chaperone or companion protein, the EBP is believed to prevent both intracellular aggregation and degradation of tropoelastin. The EBP can also bind galactosugars

ELASTIN (exon 24)

Bovine	PGVGV	PGVGV	PGVGV	PGVGV						
Chicken	PGVGV	PGVGV	PGVGV	PGVGV	PGVGV	PGVGV	PGVGV	PGVGV	PGVGV	PGVGV
Human	PGVGVA	PGVGVA	PGVGVA	PGVGLA	PGVGVA	PGVGVA	PGVGVA			

ELASTIN (exon 30)

Rat	GLGGA	GGLGA	GGLGA	GGLGA	GGLGA	GGLGA	GGLGA	GGVI.	.PGAVGLG
Chicken	GVPGA	GVPGV	GGIP.	GGLGV	GGLGV	GGLGA	GGLGA	.....	.....GVG
Human	GLVGA	AGL..	GGLGV	GGLGV	.....	.....	.....	.....	.PGVGGLG

LAMPRIN
(sea lamprey)

LGHPV GGLGY GGLGY GGLGY GGLGA AGLGY GGLGY

CHORION CLASS B

(domestic silkmoth)

IGCGRGC GGRGY GGLGY GGLGY GGLGY GGLGG GCGRG

OOTHECIN
(American cockroach)

YGGY GGLGY GGLGY GGLGY GGLGY GGLGY GGLGY GGLGY GGLGY

SPIDROIN 1
(spider)

GQGGY GGLGS QGAGR GGLGG QGA

PRE-COL-D
(mussel byssus)

PGVGP GGLGG LGGLG AGGLG GGLGG GLGGL GGAGG LGGGL GGLGG

Figure 3.1: Sequence similarities among a number of self-assembling matrix proteins. Exon 24 of chicken tropoelastin contains the repetitive sequence PGVGV ten times in tandem. A similar sequence, PGVGVA, repeats seven times in tandem in human tropoelastin at the same site. Elastin also contains the GGLGV/A repeat sequence, a sequence similar to the GGLGY sequence that is present in other self-assembling proteins, including lamprin, the major matrix protein of lamprey cartilage.

through a galactolectin domain and, once at the cell surface, is suggested to interact with galactosugars of fibrillin and fibrillin-associated chondroitin sulfate proteoglycans, releasing the tropoelastin onto the microfibrillar network (Hinek, 1996; Kielty et al., 1996).

Additional contributions to the formation of an organised polymeric matrix may come from the microfibrillar proteins. Based on early observations that during development of elastic tissues newly synthesised elastin appears to coalesce on a preformed bed of microfibrils in the extracellular matrix, it has been suggested that microfibrils act as a scaffold for elastin deposition (Ross and Bornstein, 1969; Ross et al., 1977). The C-terminal of tropoelastin, containing a conserved tetrabasic sequence, may bind to the N-terminal region of MAGP-1, helping to align the tropoelastin molecules (Rosenbloom et al., 1993; Brown-Augsburger et al., 1996). Recent evidence suggests that both the C-terminus and the region between domains 25 and 26 of tropoelastin, a kallikrein-binding region, may be required for binding of tropoelastin to MAGP-1 (Jensen et al., 2001, 2000). In addition to acting as a scaffold, the microfibrils themselves have some intrinsic properties of extensibility and elastic recoil (McConnell et al., 1996; Baldock et al., 2001).

Once tropoelastin is present in the extracellular matrix, lysyl oxidase catalyses the oxidative deamination of lysine residues found in the cross-linking regions, allowing intermolecular cross-linking into a stable elastin matrix (Eyre et al., 1984; Kagan and Trackman, 1991). Lysyl oxidase is a copper-dependent amine oxidase using a tyrosine-derived cofactor, lysyl tyrosylquinone (Stites et al., 2000). Animals with a dietary deficiency of copper show incomplete cross-linking of tropoelastin to insoluble elastin and an increase in soluble tropoelastin in tissues (Sandberg et al., 1969). Lysyl oxidase may also have an organising role in elastin polymer formation above and beyond simply oxidatively deaminating lysine side chains in preparation for cross-linking.

The mechanisms of elastin cross-linking have not been fully elucidated. However, it has been well-established that cross-links occur through the lysine residues, ultimately forming desmosine and isodesmosine, the predominant cross-links of elastin. In formation of desmosine and isodesmosine cross-links, lysyl oxidase oxidatively deaminates three of the four lysine residues involved, converting them to allysine, an α-aminoadipic-δ-semialdehyde. It has been suggested that the presence of a tyrosine or phenylalanine adjacent to a lysine residue in elastin may prevent deamination of that lysine (Foster et al., 1974; Baig et al., 1980; Reiser et al., 1992), thereby providing the nitrogen for the pyridinium ring of desmosine or isodesmosine. Deamination is followed by spontaneous condensation with other modified and unmodified lysines to form the cross-links of elastin.

There are also a small number of additional cross-links present in insoluble elastin, including lysinonorleucine (formed by the condensation of allysine and lysine side chains), allysine aldol (formed by the condensation of two allysine side chains), and merodesmosine (formed by the addition of a lysine to allysine aldol). Desmosine formation has been suggested to proceed either by the spontaneous condensation of dehydrolysinonorleucine and allysine aldol, or by the addition of an allysine to merodesmosine.

Although it is believed that cross-links usually form between two tropoelastin molecules (Foster et al., 1974), the possibility of cross-links involving three tropoelastin molecules has been suggested. At least two cross-linking domains contain three lysine residues, theoretically permitting formation of a desmosine cross-link between two tropoelastin molecules, and leaving the third lysine in each molecule to form a cross-link with a third tropoelastin molecule (Brown-Augsburger et al., 1995). Formation of cross-links involving three tropoelastin molecules would have important implications for the assembly of elastin.

Once cross-linked, tropoelastin cannot be solubilized from the matrix even under harsh extraction conditions, such as boiling in 0.1 M NaOH for 45 minutes (Lansing et al., 1952). Treatment with oxalic acid or potassium hydroxide does not release soluble tropoelastin from insoluble elastin, but rather disrupts peptide bonds, giving soluble fragments of elastin (α-elastin and β-elastin, or κ-elastin, respectively) and leaving the cross-links intact (Partridge et al., 1955; Jacob and Robert, 1989).

SELF-AGGREGATION OF ELASTIN

One of the striking features of elastin is its ability to self-aggregate. This property of elastin has generally been investigated in vitro using measurements of the ability of elastin to coacervate, a phenomenon in which hydrophobic proteins come out of solution as a second phase with an increase in solution temperature. Coacervation of elastin was originally shown with heterogeneous mixtures of hydrolytically solubilized insoluble elastin (α-elastin) (Partridge et al., 1955; Urry et al., 1969; Jamieson et al., 1972). Subsequently, it has been shown to also take place with synthetic peptides based on hydrophobic domains in elastin (Urry, 1988; Tamburro et al., 1992; Castiglione-Morelli et al., 1993; Reiersen et al., 1998), as well as in the full-length recombinant tropoelastin molecule (Vrhovski et al., 1997). The kinetics of coacervation of elastin are similar to those of a nucleation process.

Unlike most proteins that undergo denaturation with increased temperature, becoming less ordered, elastin appears to become more ordered through the process of coacervation. For this reason, Urry (1995) has termed the coacervation process for elastin an inverse temperature transition. The temperature at which coacervation takes place is affected by a number of factors: pH, ionic strength, protein concentration, and relative hydrophobicity.

Coacervation of tropoelastin, α-elastin, and hydrophobic polypeptides based on the sequences found in the hydrophobic domains of elastin resulted in coacervates with fibrillar structures resembling those of native, insoluble elastin when viewed using electron microscopy (Cox et al., 1973, 1974; Urry et al., 1974; Cleary and Cliff, 1978; Bochicchio et al., 2001). Optical diffraction of coacervates of tropoelastin and α-elastin indicated a diameter of 5 nm for the parallel filaments making up the coacervate fibre (Volpin et al., 1976). Bressan and coworkers (1983) showed fibrillar structures formed from coacervates of tropoelastin. Their studies also suggested that, prior to coacervation, tropoelastin exists in solution as globular monomers and small oligomers. Immediately after coacervation, a network of filaments is seen, transforming into a more ordered fibrillar structure after overnight incubation at the coacervation temperature (Bressan et al., 1986).

Other hydrophobic matrix proteins also show the ability to self-aggregate, including lamprin (the major matrix protein of lamprey cartilage), silkmoth chorion proteins, mussel byssus threads, and spidroin 1 (a spider dragline silk protein) (Hamodrakas et al., 1985; Xu and Lewis, 1990; Robson et al., 1993; Qin et al., 1997). These matrix proteins all contain repetitive sequence motifs similar to those present in elastin, including repeats of GGLGY (Figure 3.1). These sequences have been suggested to adopt short β-sheet/β-turn or β-spiral structures, and it has been proposed that these sequences may promote self-organisation by interdigitation or stacking of the hydrophobic amino acid side chains in a 'lego'-like manner (Robson et al., 1993). In this way, such hydrophobic motifs may have a more general role in promoting self-assembly of proteins into organised, polymeric matrices.

SELF-ASSEMBLY OF RECOMBINANT HUMAN ELASTIN POLYPEPTIDES

In recent years, there has been renewed interest in mechanisms of assembly of elastin, spurred particularly by technology for expression of recombinant proteins. In our laboratory, we have used a 'minimalist' approach to understanding the assembly of elastin and the relationship between the organisation of elastin polymers and their structural and mechanical properties. We have attempted to establish the minimal sequence requirements for self-organisation and assembly of elastin monomers into polymeric matrices with elastin-like properties. Although this work is still in its early stages, data clearly indicate that even rather small polypeptides based on elastin domains not only have the ability for directed and organised self-assembly, but also result in matrices with physical properties similar to elastin.

The recombinant human elastin polypeptides used by us were designed based on hydrophobic exons 20 and 24 and cross-linking exons 21 and 23 (Bellingham et al., 2001). Elastin polypeptide (EP) 20–24 is made up of exons 20-21-23-24, maintaining the alternating domain structure of tropoelastin. Exon 24 contains the repetitive motif PGVGVA, repeating seven times in tandem, the most striking tandem repeat found in human elastin. Exons 21 and 23 have sequences typical of cross-linking exons, including two lysines in each exon separated by either two or three alanine residues. A second polypeptide, EP20-24-24 contains one additional hydrophobic and cross-linking domain (exons 20-21-23-24-21-23-24), and, finally, EP21–23 contains only the cross-linking exons 21 and 23 (Figure 3.2).

As expected, with the exception of EP21–23, which contained no hydrophobic domains, these recombinant elastin polypeptides had the ability to coacervate, confirming the requirement for hydrophobic domains. As had previously been

EP 20-24

EP 20-24-24

EP 21-23

| 20 | FPGFGVGVGGIPGVAGVPGVGGVPGVGGVPGVGIP |

(21) (23) EAQAAAAAKAAKYGVGTPAAAAAKAAAKAAQFG

| 24 | LVPGVGVAPGVGVAPGVGVAPGVGLAPGVGVAPGVGVAPGVGVAPAIG |

Figure 3.2: Recombinant polypeptides based on sequences in human elastin. Polypeptides were produced as glutathione *S*-transferase fusion proteins, using a methionine residue placed just upstream of the first amino acid of the elastin sequence as a site for cleavage with cyanogen bromide. The released elastin polypeptides were purified by ion exchange and gel filtration chromatography. Polypeptides contained exons 20, 21, 23, and 24 in combinations shown. Sequences of these exons are given herein. Exons 20 and 24 are hydrophobic domains with tandem repeats. Exons 21 and 23 together form a classic cross-linking domain containing two pairs of lysine residues separated by either two or three alanine residues.

reported for tropoelastin (Vrhovski et al., 1997), the temperature of coacervation was inversely related both to the concentration of the polypeptides and to the ionic strength of the coacervation solution. Equimolar concentrations of EP20-24-24 coacervated at a significantly lower temperature (approximately 29°C), compared with EP20-24 (approximately 41°C). These results confirmed that polypeptides containing both cross-linking and hydrophobic domains, but representing as little as 15% of the entire tropoelastin monomer, possessed the ability to self-aggregate.

Data have suggested that the propensity of these polypeptides to self-assemble, as measured by coacervation temperature, is related not only to their hydropathy and molecular weight, but also to the specific hydrophobic sequences that they contain (Bellingham et al., 2001). We are presently investigating details of the relationship among specific hydrophobic sequences, tandem sequence repeats, and capacity of these polypeptides for self-aggregation.

AGGREGATION AS AN ORDERING PROCESS

The ability of these polypeptides to aggregate is, in itself, a relatively uninteresting property unless that self-association results in an alignment of these monomers which contributes to the structural organisation of the polymer. Several years ago, Bressan et al. (1986) suggested that the inherent ability of elastin to self-aggregate might be an important factor contributing to the polymeric

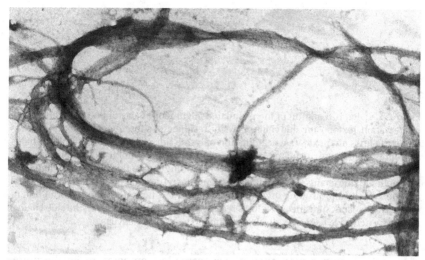

Figure 3.3: Spontaneous formation of fibrillar matrices by EP20-24-24 after coacervation. The polypeptide was dissolved in coacervation buffer (25 μM), and the solution temperature slowly increased until coacervation took place. Solution temperature was maintained at 39°C for 16 hr, after which time the coacervate was layered onto prewarmed copper grids and stained with 2% aqueous uranyl acetate for transmission electron microscopy.

organisation of elastin. That is, the formation of these aggregates played an organisational role in the alignment of tropoelastin monomers for polymeric assembly and cross-link formation.

Evidence for the formation of oriented, fibrillar structures on coacervation of tropoelastin (Cox et al., 1973; Urry et al., 1974; Bressan et al., 1986) had supported the view that self-aggregation involved a specific spatial organization of the elastin monomers. Similarly, electron microscopy of coacervates of EP20–24 and EP20-24-24 has demonstrated that both of these polypeptides were capable of formation of fibrillar structures on self-aggregation similar to those reported for tropoelastin (Figure 3.3).

Formation of fibrillar structures on self-aggregation, either of these polypeptides representative of tropoelastin or of the whole tropoelastin molecule, only implies a rather limited level of structural organisation of these monomers during self-assembly. A much more significant test of the self-organisational ability of these monomers is whether, as a consequence of the association of the hydrophobic domains, lysines in the cross-linking domains of these polypeptides are aligned such that cross-linking can take place (Figure 3.4). Narayanan and coworkers (1978) reported that formation of desmosine cross-links between tropoelastin molecules in vitro, after oxidation with lysyl oxidase, takes

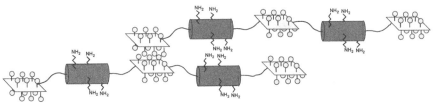

Figure 3.4: Representation of the postulated alignment of elastin polypeptides during coacervation, juxtaposing side chains of lysine residues for subsequent formation of covalent cross-links after oxidative deamination of lysine side chains. Hydrophobic domains are represented by square planar structures with large hydrophobic side chains protruding above and below the plane ('lego motifs'), allowing interactions by interdigitation of hydrophobic side chains. Cross-linking domains are represented by cylinders, with pairs of lysine residues protruding on the same side of the α-helical structure. Formation of desmosine or isodesmosine cross-links requires condensation of four lysine side chains, three of which have been oxidatively deaminated to lysine aldehydes.

place only at temperatures above the coacervation temperature. Furthermore, Bedell-Hogan and coworkers (1993) had shown that oxidative deamination of recombinant tropoelastin using lysyl oxidase resulted in the formation of covalent, lysine-derived cross-links.

We have used a pyrroloquinoline quinone/copper redox system to oxidize lysine residues in the polypeptides (Shah et al., 1992). Amino acid analysis confirmed oxidative deamination of approximately 75% of the lysine residues, with no other amino acid affected. In the absence of coacervation, although lysine residues in the polypeptides were oxidatively deaminated, no significant cross-linking of the polypeptides took place. In the absence of oxidation of lysines, aggregates of polypeptides were formed by coacervation, but these aggregates could be completely solubilized in solutions of 2% sodium dodecyl sulphate/4 M urea. In contrast, coacervation of polypeptides in which lysines were oxidatively deaminated formed aggregates that were insoluble in 2% sodium dodecyl sulphate/4 M urea. Moreover, lysine-derived cross-links, including both desmosine and isodesmosine, the major naturally occurring covalent cross-links in insoluble elastin, were identified in these aggregates. These results demonstrated that the process of self-aggregation of these polypeptides involved the organisational alignment of the monomers such that zero-length cross-linking involving oxidized lysines could take place.

Fabrication of these covalently cross-linked aggregates of recombinant elastin polypeptides into membrane-like matrices has allowed further assessment of their solubility characteristics and physical properties. Like insoluble elastin, these matrices are resistant both to treatment with cyanogen bromide and to extraction with hot 0.1 M NaOH. Preliminary mechanical testing of these matrices indicates physical properties that are remarkably similar to those of

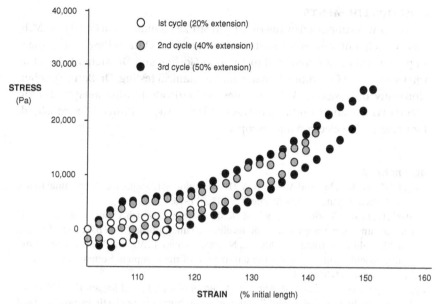

Figure 3.5: Stress-strain curves for membrane structures fabricated from EP20-24-24 after coacervation, followed by oxidative deamination of lysine side chains with pyrrolo-quinoline quinone. These stress-strain curves represent three cycles of loading and un-loading, with the length of extension progressively increased. Breaking strain is at approx-imately 100% of the initial length. Young's modulus of these membranes is approximately 0.2 MPa.

native insoluble elastin, including comparable elastic moduli, extensibility, and elastic recoil characteristics (Figure 3.5), and similar strains to failure.

These data strongly support the view that monomers of elastin possess the inherent ability to organise themselves into polymeric structures, aligning lysine residues for subsequent cross-link formation. Furthermore, this ability for self-alignment can be mimicked in elastin polypeptides containing small numbers of hydrophobic and cross-linking domains. Thus, although the presence of the microfibrillar scaffolding, the chaperoning role of the elastin binding protein, and the action of lysyl oxidase are all, without doubt, important for establishing the appropriate architecture of insoluble elastin in tissues, it appears that the fundamental requirements for organisation of monomeric elastin into a matrix with properties of extensibility and elastic recoil can be provided by the intrinsic self-organisational ability conferred by the presence of a few domains of the elastin monomer and the presence of an oxidising agent. Such recombinant polypeptides, therefore, provide a useful tool for the investigation of the role of polypeptide sequence and domain organisation in determining the structural organisation and material properties of insoluble elastin.

ACKNOWLEDGMENTS

This work was supported by the Heart and Stroke Foundation of Ontario. C.M.B. was the recipient of a Heart and Stroke Foundation of Ontario Research Traineeship. The authors are indebted to Dr. John Gosline and Dr. Margo Lillie, University of British Columbia, Canada, for mechanical testing; Dr. Barry Starcher, University of Texas, USA, for the desmosine/isodesmosine assays; and Dr. Glenda Wright and Dorota Wadowska of University of Prince Edward Island, for transmission electron microscopy.

REFERENCES

Baig, K. M., Vlaovic, M., and Anwar, R. A. (1980) Amino acid sequences C-terminal to the cross-links in bovine elastin. *Biochem. J.* **185,** 611–16.

Baldock, C., Koster, A., Ziese, U., Rock, M., Kadler, K., Shuttleworth, C., and Kielty, C. (2001) The supramolecular organization of fibrillin-rich microfibrils. *J. Cell Biol.* **152,** 1045–56.

Bashir, M., Indik, Z., Ornstein-Goldstein, N., Rosenbloom, J., Abrams, W., Fazio, M., Uitto, J., and Rosenbloom, J. (1989) Characterization of the complete human elastin gene. *J. Biol. Chem.* **264,** 8887–91.

Bedell-Hogan, D., Trackman, P., Abrams, W., Rosenbloom, J., and Kagan, H. (1993) Oxidation, crosslinking and insolubilization of recombinant tropoelastin by purified lysyl oxidase. *J. Biol. Chem.* **268,** 10345–50.

Bellingham, C. M., Woodhouse, K. A., Robson, P., Rothstein, S., and Keeley, F. W. (2001) Self-aggregation characteristics of recombinantly expressed human elastin polypeptides. *Biochim. Biophys. Acta* **1550,** 6–19.

Bochicchio, B., Pepe, A., and Tamburro, A. M. (2001) On (GGLGY) synthetic repeating sequences of lamprin and analogous sequences. *Matrix Biol.* **20,** 243–50.

Boyd, C. D., Christiano, A. M., Pierce, R. A., Stolle, C. A., and Deak, S. B. (1991) Mammalian tropoelastin: Multiple domains of the protein define an evolutionarily divergent amino acid sequence. *Matrix* **11,** 235–41.

Bressan, G., Castellani, I., Giro, M., Volpin, D., Fornieri, C., and Ronchetti, I. (1983) Banded fibers in tropoelastin coacervates at physiological temperatures. *J. Ultrastruct. Res.* **82,** 335–40.

Bressan, G., Pasquali-Ronchetti, I., Fornieri, C., Mattioli, F., Castellani, I., and Volpin, D. (1986) Relevance of aggregation properties of tropoelastin to the assembly and structure of elastic fibers. *J. Ultrastruct. Res.* **94,** 209–16.

Brown-Augsburger, P., Broekelmann, T., Rosenbloom, J., and Mecham, R. P. (1996) Functional domains on elastin and microfibril-associated glycoprotein involved in elastic fibre assembly. *Biochem. J.* **318,** 149–55.

Brown-Augsburger, P., Tisdale, C., Broekelmann, T., Sloan, C., and Mecham, R. (1995) Identification of an elastin cross-linking domain that joins three peptide chains. *J. Biol. Chem.* **270,** 17778–83.

Castiglione-Morelli, M., DeBiasi, M., DeStradis, A., and Tamburro, A. (1993) An aggregating elastin-like pentapeptide. *J. Biomol. Struct. Dynamics* **11,** 181–90.

Cleary, E. G., and Cliff, W. J. (1978) The substructure of elastin. *Exp. Mol. Biol.* **28,** 227–46.

Cox, B. A., Starcher, B. C., and Urry, D. W. (1973). Coacervation of alpha-elastin results in fiber formation. *Biochim. Biophys. Acta* **317,** 209–13.

Cox, B. A., Starcher, B. C., and Urry, D. W. (1974) Coacervation of tropoelastin results in fiber formation. *J. Biol. Chem.* **249,** 997–98.

Davis, E. (1993) Stability of elastin in the developing mouse aorta: A quantitative radioautographic study. *J. Histochem.* **100,** 17–26.

Debelle, L., Alix, A., Wei, S., Jacob, M., Huvenne, J., Berjot, M., and Legrand, P. (1998) The secondary structure and architecture of human elastin. *Eur. J. Biochem.* **258,** 533–39.

Eyre, D., Paz, M., and Gallop, P. (1984) Cross-linking in collagen and elastin. *Ann. Rev. Biochem.* **53,** 717–48.

Foster, J. A., Rubin, L., Kagan, H. M., Franzblau, C., Bruenger, E., and Sandberg, L. B. (1974) Isolation and characterization of crosslinked peptides from elastin. *J. Biol. Chem.* **249,** 6191–96.

Gerber, G. E., and Anwar, R. A. (1974) Structural studies on cross-linked regions of elastin. *J. Biol. Chem.* **249,** 5200–7.

Gerber, G. E., and Anwar, R. A. (1975) Comparative studies of the cross-linked regions of elastin from bovine ligamentum nuchae and bovine, porcine and human aorta. *Biochem. J.* **149,** 685–95.

Gibson, M. A., Hatzinikolas, G., Kumaratilake, J. S., Sandberg, L. B., Nicholl, J. K., Sutherland, G. R., and Cleary, E. G. (1996) Further characterization of proteins associated with elastic fibre microfibrils including the molecular cloning of MAGP-2 (MP25). *J. Biol. Chem.* **271,** 1096–1103.

Gibson, M. A., Sandberg, L. B., Grosso, L. E., and Cleary, E. G. (1991) Complementary DNA cloning establishes microfibril-associated glycoprotein (MAGP) to be a discrete component of the elastin-associated microfibrils. *J. Biol. Chem.* **266,** 7596–601.

Gray, W., Sandberg, L., and Foster, J. (1973) Molecular model for elastin structure and function. *Nature* **246,** 461–66.

Hamodrakas, S., Etmektzoglou, T., and Kafatos, F. (1985) Amino acid periodicities and their structural implications for the evolutionarily conservative central domain of some silkmoth chorion proteins. *J. Mol. Biol.* **186,** 583–89.

Hinek, A. H. (1996) Biological roles of the non-integrin elastin/laminin receptor. *J. Biol. Chem.* **377,** 471–80.

Hinek, A. H., and Rabinovitch, M. (1994) 67-kD elastin-binding protein is a protective "companion" of extracellular insoluble elastin and intracellular tropoelastin. *J. Cell Biol.* **126,** 563–74.

Hinek, A. H., Wrenn, D., Mecham, R., and Barondes, S. (1988) The elastin receptor: A galactoside-binding protein. *Science* **239,** 1539–41.

Indik, Z., Yeh, H., Ornstein-Goldstein, N., Kucich, U., Abrams, W., Rosenbloom, J. C., and Rosenbloom, J. (1989) Structure of the elastin gene and alternative splicing of elastin mRNA: Implications for human disease. *Am. J. Med. Genet.* **34,** 81–90.

Indik, Z., Yeh, H., Ornstein-Goldstein, N., Sheppard, P., Anderson, N., Rosenbloom, J. C., Peltonen, L., and Rosenbloom, J. (1987) Alternative splicing of human elastin mRNA indicated by sequence analysis of cloned genomic and complementary DNA. *Proc. Nat. Acad. Sci. U.S.A.* **84,** 5680–84.

Jacob, M., and Robert, L. (1989) Isolation, characterization, and biochemical properties of elastin. In: *Elastin and Elastases,* vol. I, L. Robert and W. Hornebeck (eds.). Boca Raton, FL: CRC Press, pp. 49–65.

Jamieson, A., Downs, C., and Walton, A. (1972) Studies of elastin coacervation by quasi-elastic light scattering. *Biochim. Biophys. Acta* **271,** 34–47.

Jensen, S. A., Reinhardt, D. P., Gibson, M. A., and Weiss, A. S. (2001). MAGP-1: Protein interaction studies with tropoelastin and fibrillin-1. *J. Biol. Chem.* **276,** 39661–66.

Jensen, S. A., Vrhovski, B., and Weiss, A. S. (2000) Domain 26 of tropoelastin plays a dominant role in association by coacervation. *J. Biol. Chem.* **275,** 28449–54.

Kagan, H. M., and Trackman, P. C. (1991) Properties and function of lysyl oxidase. *Am. J. Respir. Cell. Mol. Biol.* **5,** 206–10.

Kielty, C. M., Whittaker, S. P., and Shuttleworth, C. A. (1996) Fibrillin: Evidence that chondroitin sulphate proteoglycans are components of microfibrils and associate with newly sythesized monomers. *FEBS Lett.* **386,** 169–73.

Kielty, C. M., Baldock, C., Sherratt, M. J., Rock, M. J., Lee, D., and Shuttleworth, C. A. Fibrillin: From microfibril assembly to biomechanical function. Chapter 5 in this volume.

Lansing, A., Rosenthal, T. B., Alex, M., and Dempsey, E. W. (1952) The structure and chemical characterization of elastic fibers as revealed by elastase and by electron microscopy. *Anat. Record* **114,** 555–75.

McConnell, C. J., Wright, G. M., and DeMont, M. E. (1996) The modulus of elasticity of lobster aorta microfibrils. *Experientia* **52,** 918–21.

Mecham, R. P., and Davis, E. (1994) Elastic fiber structure and assembly. In: *Extracellular Matrix Assembly and Structure,* P. Yurchenco, D. Birk, and R. Mecham (eds.). San Diego, CA: Academic Press, pp. 281–314.

Mecham, R. P., and Heuser, J. E. (1991) The elastic fiber. In: *Cell Biology of Extracellular Matrix,* 2nd ed., E. D. Hay (ed.). New York: Plenum Press, pp. 79–109.

Narayanan, A. S., Page, R. C., Kuzan, F., and Cooper, C. G. (1978) Elastin cross-linking in vitro. Studies on factors influencing the formation of desmosines by lysyl oxidase action on tropoelastin. *Biochem. J.* **173,** 857–62.

Partridge, S. M. (1966) Biosynthesis and nature of elastin structures. *Proc. Fed. Am. Soc. Exp. Biol.* **25,** 1023.

Partridge, S. M., Davis, H., and Adair, G. (1955) The chemistry of connective tissues. 2. Soluble proteins derived from partial hydrolysis of elastin. *Biochem. J.* **61,** 11–21.

Pasquali Ronchetti, I., Fornieri, C., Baccarani-Contri, M., and Volpin, D. (1979) The ultrastructure of elastin revealed by freeze-fracture electron microscopy. *Micron* **10,** 89–99.

Qin, X., Coyne, K., and Waite, J. (1997) Tough tendons: Mussel byssus has collagen with silk-like domains. *J. Biol. Chem.* **272,** 32623–27.

Reiersen, H., Clarke, A., and Rees, A. (1998). Short elastin-like peptides exhibit the same temperature-induced structural transitions as elastin polymers: Implications for protein engineering. *J. Mol. Biol.* **283,** 255–64.

Reiser, K., McCormick, R., and Rucker, R. (1992) Enzymatic and nonenzymatic cross-linking of collagen and elastin. *FASEB J.* **6,** 2439–49.

Robson, P., Wright, G., Sitarz, E., Maiti, A., Rawat, M., Youson, J., and Keeley, F. W. (1993) Characterization of lamprin, an unusual matrix protein from lamprey cartilage. *J. Biol. Chem.* **268,** 1440–47.

Rosenbloom, J., Abrams, W., and Mecham, R. (1993) Extracellular matrix 4: The elastic fiber. *FASEB J.* **7,** 1208–18.

Ross, R., and Bornstein, P. (1969) The elastic fiber. I. The separation and partial characterization of its macromolecular components. *J. Cell Biol.* **40,** 366–81.

Ross, R., Fialkow, P., and Altman, K. (1977) The morphogenesis of elastic fibres. *Adv. Exp. Med. Biol.* **79,** 7–17.

Sakai, L. Y., Keene, D. R., and Engvall, E. (1986) Fibrillin, a new 350-kD glycoprotein is a component of extracellular microfibrils. *J. Cell Biol.* **103,** 2499–509.

Sandberg, L., Hackett, T. N., and Carnes, W. H. (1969) The solubilization of an elastin-like protein from copper-deficient porcine aorta. *Biochim. Biophys. Acta* **181,** 201–7.

Shah, M. A., Bergethon, P. R., Boak, A. M., Gallop, P. M., and Kagan, H. M. (1992) Oxidation of peptidyl lysine by copper complexes of pyrroloquinoline quinone and other quinones. A model for oxidative pathochemistry. *Biochim. Biophys. Acta* **1159,** 311–18.

Starcher, B., and Percival, S. (1985) Elastin turnover in the rat uterus. *Connect. Tissue Res.* **13,** 207–15.

Stites, T. E., Mitchell, A. E., and Rucker, R. B. (2000) Physiological importance of quinoenzymes and the o-quinone family of cofactors. *J. Nutr.* **130,** 719–27.

Szabo, Z., Levi-Minzi, S., Christiano, A., Struminger, C., Stoneking, M., Batzer, M., and Boyd, C. (1999) Sequential loss of two neighboring exons of the tropoelastin gene during primate evolution. *J. Mol. Biol.* **49,** 664–71.

Tamburro, A., Guantieri, V., and Gordini, D. (1992) Synthesis and structural studies of a pentapeptide sequence of elastin. Poly(Val-Gly-Gly-Leu-Gly). *J. Biomol. Struct. Dyn.* **10,** 441–54.

Urry, D. W. (1988) Entropic elastic processes in protein mechanisms. I. Elastin structure due to an inverse temperature transition and elasticity due to internal chain dynamics. *J. Protein Chem.* **7,** 1–34.

Urry, D. W. (1995) Elastic biomolecular machines. *Sci. Am.* **272,** 44–49.

Urry, D. W., Long, M. M., Cox, B. A., Ohnishi, T., Mitchell, L. W., and Jacobs, M. (1974) The synthetic polypentapeptide of elastin coacervates and forms filamentous aggregates. *Biochim. Biophys. Acta* **371,** 597–602.

Urry, D. W., Starcher, B., and Partridge, S. M. (1969) Coacervation of solubilized elastin effects a notable conformational change. *Nature* **222,** 795–96.

Volpin, D., Urry, D., Cox, B., and Gotte, L. (1976) Optical diffraction of tropoelastin and α-elastin coacervates. *Biochim. Biophys. Acta* **439,** 253–58.

Vrhovski, B., Jensen, S., and Weiss, A. (1997) Coacervation characteristics of recombinant human tropoelastin. *Eur. J. Biochem.* **250,** 92–98.

Vrhovski, B., and Weiss, A. (1998). Biochemistry of elastin. *Eur. J. Biochem.* **258,** 1–18.

Xu, M., and Lewis, R. (1990). Structure of a protein superfiber: Spider dragline silk. *Proc. Nat. Acad. Sci. U.S.A.* **87,** 7120–24.

Yeh, H., Ornstein-Goldstein, N., Indik, Z., Sheppard, P., Anderson, N., Rosenbloom, J. C., Cicila, G., Yoon, K., and Rosenbloom, J. (1987) Sequence variation of bovine elastin mRNA due to alternative splicing. *Collagen Related Res.* **7,** 235–47.

Zhang, H., Apfelroth, S. D., Hu, W., Davis, E. C., Sanguineti, C., Bonadio, J., Mecham, R. P., and Ramirez, F. (1994) Structure and expression of fibrillin-2, a novel microfibrillar component preferentially located in elastic matrices. *J. Cell Biol.* **124,** 855–63.

FOUR

Ideal Protein Elasticity: The Elastin Models

D. W. Urry, T. Hugel, M. Seitz, H. Gaub, L. Sheiba, J. Dea,
J. Xu, L. Hayes, F. Prochazka, and T. Parker

INTRODUCTION

Definition of Ideal or Perfect Elasticity

Ideal elasticity is the property whereby the energy expended in deformation of the elastomer is completely recovered on removal of the deforming force. Because the energy expended in deformation is given by the area under the force, f, versus increase in length, ΔL, curve, a perfectly reversible force-extension curve means complete recovery on relaxation of the energy expended on deformation. Therefore, ideal elastomers exhibit perfectly reversible force-extension curves.

Perhaps our earliest perspective of the mechanism underlying ideal elasticity comes from a fundamental observation concerning rubber elasticity. In the mid-nineteenth century, Joule and Thomson noted a quantitative correlation between the increase in temperature of the elastomer due to stretching and the increase in force due to increasing the temperature (Flory, 1968). Thermodynamics provides for the analysis underlying this correlation, and the Boltzmann relation provides the bridge between experimental thermodynamic quantities and statistical mechanical description of molecular structures.

Continuing qualitatively with the Joule and Thomson correlation, heat produces motion, and the energy represented by heat distributes into the various available degrees of freedom in the chain molecules comprising the elastomer. Accordingly, the release of heat on stretching correlates with a loss of motion. By means of statistical mechanics, the loss of motion is seen as a decrease in entropy on extension. In addition, should solvent be essential for elasticity, this requires explicit consideration.

But there is more to elastomeric force than entropy changes arising from changes in mobility. There is also internal energy, and a proper understanding of elasticity requires delineation of internal energy and entropy components of the elastomeric force. As the internal energy component of the elastic force increases, irreversible processes (such as chain breakage) become more probable.

So, greater durability results as elastomers become more dominantly entropic and durability becomes a feature of an ideal or more perfect elastomer.

Delineation of Internal Energy and Entropy Components of Elastomeric Force

Delineation of internal energy and entropy components of elastomeric force can begin with the definition of the Helmholtz free energy, A, also referred to as the maximal work function,

$$A \equiv E - TS, \tag{4.1}$$

where E is the internal energy, T is the absolute temperature ($°$K), and S is entropy for the system. The differential of the maximum work function, dA, can be written to include the work done on an elastomer by application of a force, f, over the change in length, dL, i.e.,

$$dA = -PdV - SdT + fdL, \tag{4.2}$$

where P is pressure, and V is volume. Writing the partial differential of A in Eq. (4.1) with respect to length at constant V, T, and composition, n, gives

$$(\partial A/\partial L)_{V,T,n} = (\partial E/\partial L)_{V,T,n} - T(\partial S/\partial L)_{V,T,n}, \tag{4.3}$$

but by Eq. (4.2), $(\partial A/\partial L)_{V,T,n} = f$, such that

$$f = (\partial E/\partial L)_{V,T,n} - T(\partial S/\partial L)_{V,T,n}. \tag{4.4}$$

Accordingly, the force is seen to be comprised of two components, an internal energy component, f_E, and an entropy component, f_S, i.e.,

$$f = f_E + f_S. \tag{4.5}$$

Now it becomes useful to find an expression that will allow experimental estimation of the relative magnitude of the internal energy and entropy components of the force. Following Flory et al. (1960), when the functions exist and are continuous, the order of a partial differential does not matter, and it can be shown that $(\partial S/\partial L)_{V,T} = (\partial f/\partial T)_{V,L}$, such that,

$$f = (\partial E/\partial L)_{V,T,n} + T(\partial f/\partial T)_{V,L,n}. \tag{4.6}$$

Interestingly, this can be rewritten as

$$f_E/f = -T(\partial \ln[f/T]/\partial T)_{V,L,n}, \tag{4.7}$$

which allows for an experimental estimate of the f_E/f ratio from the slope of a plot of $\ln(f/T)$ versus temperature under conditions of constant V, L, and composition, n.

By means of an approximate correction term, the f_E/f ratio can be estimated under more usual experimental conditions of constant P, L, and at equilibrium (e.g., with surrounding solvent), i.e.,

$$f_E/f = -T(\partial \ln[f/T]/\partial T)_{P,L,\text{eq}}, -\beta_{\text{eq}}T/[\alpha^3(V_i/V)-1], \tag{4.8}$$

where $\beta_{\text{eq}} = (\partial \ln V/\partial T)_{P,L,\text{eq}}$ is the thermal expansion coefficient; α is the fractional increase in length, L/L_i, with the subscript i indicating initial length; and V_i and V are the volumes of the elastomer before and after elongation, respectively (Hoeve and Flory, 1962; Dorrington and McCrum, 1977). The correction term was derived under the assumption of random chain networks with a Gaussian distribution of end-to-end chain lengths.

As we will see, a random chain network comprised of a Gaussian distribution of end-to-end chain lengths between cross-links is not an accurate structural description of the elastin protein nor of its simpler models. Also, there are important solvation changes that can occur under experimental conditions. Nonetheless, insight into the magnitude of the entropic component of the elastomeric force provided by Eq. (4.8), with experimental estimates of the quantities in the correction term or when neglecting the correction term altogether, is informative, but only when the temperature range for determination of slope is judiciously chosen.

Basic Statistical Mechanical Expression for Entropy

The Boltzmann Relation

The starting point for the statistical mechanical expressions of entropy is the Boltzmann relation,

$$S = R\ln W, \tag{4.9}$$

where R (1.987 cal/degree-mol) is the gas constant; $R = Nk$, with N being Avogadro's number (6.02×10^{23}/mol) and k the Boltzmann constant (1.38×10^{-16} erg/$^\circ$K); and W is the number of a priori equally probable states accessible to the system (Eyring et al., 1964). W is the volume in phase space occupied by a particular state of a molecular system. In practice, W becomes the product of partition functions for each of the degrees of freedom of the molecular system. There are $3n - k$ degrees of freedom, where n is the number of atoms in the molecule and k is the holonomic constraint on the system. In general, there are three translational and three rotational degrees of freedom, and the remaining are vibrational degrees of freedom. For large enough molecules, the vibrational degrees of freedom also include motions of rotation (or torsional oscillations) about bonds. Due to the constraints of the elasticity measurement, k is 6; the ends of the molecules are effectively fixed in space such that there are neither translational nor whole molecule rotational degrees of freedom.

Fundamental Expression for the Change in Entropy on Extension

Fortunately, when calculating entropic contributions to the elasticity, interest centers on the change in entropy on extension from a relaxed, S^r, to an extended, S^e, state of the elastomer,

$$\Delta S = (S^e - S^r) = R \ln(W^e / W^r). \tag{4.10}$$

Because the change in entropy is a ratio, it becomes relatively straightforward to calculate the contribution to the change in entropy of a particular expression of motion accessible to the representation of the molecular structure to be examined in its relaxed and extended states. For our purposes, the β-spiral structure of the (Gly-Val-Gly-Val-Pro)$_n$, or poly(GVGVP), model of elastin is used, and two different means of representing that structure are considered. The first is the enumeration of states in configuration (φ-ψ torsion angle) space within a given energy cut-off for relaxed and extended states, and the second is to use the molecular dynamics approach to determine the change in root mean square (rms) torsional oscillations that occur on extension by the same amount as used in the enumeration of states approach. The magnitudes of the calculated entropy changes on extension obtained by each approach can then be compared.

Historical Notes of Proposed Mechanisms for Protein Entropic Elasticity

Classical (Random Chain Network) Theory of Rubber Elasticity

During the last half of the last century, identification of ideal (dominantly entropic) protein elasticity required, by the recognized authorities of the period, the conclusion that the ideal elastomer was comprised of random chain networks with a Gaussian distribution of end-to-end chain lengths between cross-links. For elastin, the protein of interest here, this point of view began with a paper by Hoeve and Flory (1958). This perspective became entrenched in the minds of the interested scientific community by the award in the Fall of 1974 of the Nobel Prize to Paul Flory following reaffirmation of the random chain network conclusion (Hoeve and Flory, 1974). The message of the paper is unmistakable. In the Synopsis, it is stated that "*A network of random chains within the elastic fibers, like that in a typical rubber, is clearly indicated.*" Furthermore, in Figure 1 of their paper is a structural representation of the chains between cross-links with a statement in the figure legend that, "*Configurations of chains between cross-linkages are much more tortuous and irregular than shown.*"

Decrease in Solvent Entropy on Extension

The purpose of the 1974 Hoeve and Flory paper, which presented no additional data, was to refute an earlier publication by Weis-Fogh and Andersen (1970), who had suggested an alternate mechanism for the elasticity of elastin. The alternate mechanism proposed that hydration of hydrophobic side chains of the protein that become exposed to solvent on extension would be responsible for the stretch-induced decrease in entropy (i.e., by this proposal, changes in solvent entropy became the source of the entropic component of the elastomeric force).

Damping of Internal Chain Dynamics on Extension

Just over a decade later, yet another perspective was put forward. Considering structural studies and molecular mechanics calculations of the most prominent repeating sequence of bovine elastin, (Gly-Val-Gly-Val-Pro)$_{11}$, entropic elasticity was described as arising from a decrease in available configuration space on extension (Urry et al., 1982). Equivalently, treating the experimentally and computationally derived regularly repeating structure from the perspective of molecular dynamics, the damping of internal chain dynamics on extension described the same decrease in entropy for the same degree of extension (Chang and Urry, 1989).

In the following Discussion section, these three mechanisms will be considered in more detail, after consideration of new experimental data on models of elastin, along with previously published relevant data.

Inverse Temperature Transition Behaviour of Elastin and Its Models

Increase in Order with an Increase in Temperature

Filament Formation on Raising the Temperature. When the temperature of an aqueous solution of the precursor protein, tropoelastin, is increased, the protein aggregates to form a more dense viscoelastic phase called a coacervate. When a droplet of the aqueous suspension of incipient aggregates is placed on a carbon-coated grid, negatively stained with uranyl acetate and oxalic acid at the appropriate pH, and examined in the transmission electron microscope, fibrils comprised of parallel-aligned filaments are observed with a 5 nm periodicity (Cox et al., 1974). Similar results are obtained for α-elastin, a 70,000 Da fragment of the elastic fiber (Cox et al., 1973) and for high polymers, poly(GVGVP) and poly(GVGVAP), of repeating sequences of elastin (Volpin et al., 1976). Fibrils of similar dimensions have been reported for fibrous elastin itself using similar techniques (Gotte et al., 1974).

Crystallization of Cyclic Analogues on Raising the Temperature of the Inverse Temperature Transition. Most strikingly, cyclic analogues of repeating sequences of elastin crystallize on raising the temperature (Urry et al., 1978; Cook et al., 1980) and redissolve on lowering the temperature. Without ambiguity, these polymers increase order with an increase in temperature through a transition temperature range.

We refer to this phase transition as an *inverse temperature transition.* Whereas the peptide component of this water-peptide system increases in order on increasing the temperature, ordered water molecules surrounding hydrophobic residues become less ordered bulk water as hydrophobic groups separate from solution. Accordingly, the overall change effected by the phase transition is to less order in keeping with the second law of thermodynamics.

Indeed, the phase diagram for high molecular weight polymers of repeating sequences of elastin is inverted with the soluble phase below, at lower temperature, and the insoluble phase above the binodal or coexistence line and with an inverted coexistence line [i.e., with the curvature of the coexistence line convex to the volume fraction axis (Sciortino et al., 1993; Manno et al., 2001)]. On the other hand, the usual circumstance for polymers is for the coexistence line to be concave to the volume fraction axis and for the polymer to be insoluble below and soluble above the coexistence line (Flory, 1953).

Composition of Poly(GVGVP) in Water as a Function of Temperature

The temperature dependence of composition of the poly(GVGVP)-water system provides a clear visualization of a water-containing, structured state at intermediate temperatures (Urry et al., 1985b). A plot of the percentage of water by weight as a function of temperature is schematically shown in Figure 4.1. As the polymer is miscible with water in all proportions below 20°C, the plot arbitrarily starts at about 90% water by weight near 0°C. On raising the temperature from 20° to 30°C, there occurs a reversible phase separation to form a state that is 63% water by weight. The percentage of water by weight of this intermediate water-retaining state remains essentially unchanged until the temperature is raised above 60°C. Then, a very slow irreversible transition occurs on raising the temperature from 60° to 80°C to form a state that is 32% water by weight. Dialysis of the polymer against 100,000 Da cutoff membranes, after prolonged heating at 80°C, resulted in less than 0.1% loss of polymer (i.e., prolonged heating did not result in chain breakage).

The results on poly(GVGVP) are compared in Figure 4.1 to those of Grinberg et al. (1999) on the petroleum-based polymer, poly(N-isopropylacrylamide) (PNIPAM). This polymer shows an inverted phase diagram like that of the elastin-based polymers. Though with a composition more hydrophobic than that of poly(GVGVP), the transition for this lower critical solution temperature

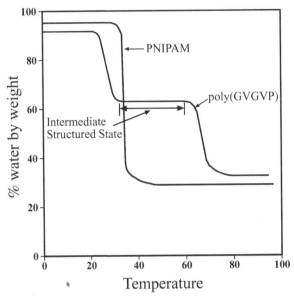

Figure 4.1: Plot of the percentage of water by weight as a function of temperature for poly(GVGVP) showing a stable intermediate structured state of 63% water by weight between 30° and 60°C, a state that slowly denatures above 60°C to form a state containing 32% water by weight. Plotted from the data of Urry et al. (1985b). Amphiphilic (lower critical solution temperature) polymers – such as PNIPAM, poly(*N*-isopropyl acrylamide) – exhibit only a single, sharper transition, and it is to what is generally considered a disordered state with about 30% water by weight. Plotted from the data of Grinberg et al. (1999).

polymer occurs at a higher temperature (34°C), is much sharper, and directly results in the formation of the state with approximately 30% water by weight. This PNIPAM state of 30% water by weight may reasonably be described as a random chain network. It is clear that poly(GVGVP) forms an intermediate, water-containing state that holds a fixed amount of water until irreversible denaturation occurs at high temperature. This state of 63% water by weight is an elementary property that simply is not consistent with the recently computed 'compact amorphous globule' for $(GVGVP)_{18}$ due to Li et al. (2000).

A Structured State at Intermediate Temperatures
for Elastin-Based Systems

Both the elastin fiber and α-elastin at intermediate temperatures form similar water-containing states of about 50% water by weight (Partridge et al., 1955; Partridge and Davis, 1955; Partridge, 1966, 1967). It will also be shown below (see Figure 4.9) that both cross-linked poly(GVGVP) and fibrous elastin exhibit a slow, irreversible loss of elastic force when at fixed extension after prolonged exposure to temperatures above 60°C. These are further demonstrations of the slow irreversible denaturation observed in Figure 4.1 above 60°C for poly(GVGVP). Furthermore, detailed dielectric relaxation studies on the elastin-based systems – fibrous elastin, α-elastin, and poly(GVGVP) – demonstrate molecular motions, relaxations, limited to localized frequency ranges (see Figures 4.6, 4.7, and 4.8), and these could only be the result of non-random, regularly structured, albeit dynamic, conformational states. These experimental results are consistent with neither random chain networks nor compact amorphous globules.

MATERIALS

Preparation of Elastin Models
In its native state, elastin, a protein of about 70,000 Da, occurs as the overwhelming dominant component (90%+ by weight) of the mammalian elastic fiber, which resides within the extracellular matrix, for example, as a structure several microns in diameter in the vascular wall. Non-elastin components are trapped enzymatic protein involved in the cross-linking and a fine fibrillar coating of some hundreds of nanometer thickness. The elastic fiber is a substantially cross-linked, insoluble material of the extracellular matrix. Efforts to isolate it for purposes of defining its properties require draconian methods, for example, refluxing in sodium hydroxide, to remove all of the surrounding matrix components. This product is called purified fibrous elastin. Alternatively, model approaches are used, such as isolating and characterizing (1) the precursor protein, tropoelastin, which may be obtained in ill-defined intermediate states of cross-linking; (2) the pure precursor protein prepared by microbial biosynthesis; (3) chemical degradation products of purified fibrous elastin; and (4) chemically and microbially synthesized repeating sequences found within the protein sequence, as well as interesting analogues.

Synthesis of Model Systems
Chemical Synthesis. Initially, several thousand protein-based polymers were synthesized to determine the conformation and function of these model proteins. This involved the synthesis of oligomers and high polymers of repeating tetra-, penta-, hexa-, nona-, and even decapeptides reported in the sequence of elastin. By means of chemical synthesis, many compositions could be prepared and characterized, and axioms were developed for the function of these elastic model proteins in energy conversion and for their use as biomaterials for medical and non-medical applications.

Microbial Biosynthesis. Once a specific composition was identified as being of sufficient interest, it was prepared by means of recombinant DNA technology, in which hundreds of grams could be prepared in a single fermentation that by means of bench chemical synthesis would require nearly a year to prepare.

Preparation of Natural Materials

Isolation of the Precursor Protein, Tropoelastin. Initially, the precursor protein was prepared from animals made copper-deficient and treated with inhibitors of enzymatic cross-linking. The product was meager in amount, and a mixture of partially cross-linked and oxidized lysyl side chains was used. It was nonetheless helpful in determining many of the fundamental physical properties relating to structure and elastogenesis.

Preparation of Tropoelastin by Microbial Biosynthesis. In an important development, Weiss and coworkers used recombinant DNA technology to prepare the precursor protein, tropoelastin, in *Escherichia coli*, which allowed preparation of pure protein in large quantities (Vrhovski et al., 1997; Wu et al., 1999). Important aspects deduced by model systems and by preparations of model structures derived from the elastin and from knowledge of sequence have been confirmed and extended by recombinantly prepared tropoelastin.

Preparation of α-Elastin. A model system useful in stepping systematically from polymers of model-repeating sequences to the natural fibrous elastin has been a chemical fragmentation product from fibrous elastin called α-elastin. It is a 70,000 molecular weight fragment comprised of 16 cross-linked chains (Partridge, 1966, 1967; Partridge et al., 1955; Partridge and Davis, 1955).

Purification Using Phase Transitional Behaviour

In all cases, these protein-based polymers are soluble in water at a lower temperature and phase separate on raising the temperature above that for the onset of a hydrophobic folding and assembly transition. This phase separation process has been called coacervation. It is a fundamental property of these polymers with the correct balance of apolar and more polar residues.

Cross-Linking of Elastin Models

General Cross-Linking Procedure

Cobalt 60 γ-irradiation has been used to cross-link the phase-separated, coacervated, and structured state of the elastic protein-based polymers with interpenetrating polymer chains. The effective cross-linking doses have ranged from 6 to 30 Mrads (million radiation absorbed dose), with 20 Mrads chosen as the usual circumstance for the elastic model proteins.

Efforts to Cross-Link α-Elastin

Even though α-elastin undergoes phase separation, it does not form an elastic matrix on γ-irradiation. As a cluster of some 16 cross-linked chains of a total of 70 kDa molecular weight, it does not cross-link because there are no chains of sufficient length to interpenetrate.

Efforts to Cross-Link Heat-Denatured Poly(GVGVP)

Although the phase-separated state of poly(GVGVP) with 63% water by weight (Figure 4.1) cross-links very effectively, the heat-denatured state obtained on prolonged heating above 60°C does not. Even though the density of chains is twice as great for the denatured state formed above 60°C (a state of 32% water by weight as seen in Figure 4.1), the chains no longer form the intermediate structured state of Figure 4.1, and the chains no longer sufficiently interpenetrate to result in an elastic matrix on γ-irradiation. This is simply another demonstration that the intermediate between 30° and 60°C is a structured state of interpenetrating chains and filaments.

SPECIALIZED METHODOLOGIES AND ANALYSIS OF RESULTS

Atomic Force Microscopy (AFM) in the Force-Extension Mode

Preparation of the Sample for AFM

Glass microscope slides were cleaned and coated with about 30 nm of gold in a home-built machine. Si_3N_4 AFM tips were used (Microlevers, Park Scientific Instruments, Sunnyvale, CA) and also coated with gold. Quantities of 1 mg of the polypentapeptides and 0.5 mg methoxy-PEG-thiol (M_W 5,000) were dissolved in 1 ml Milli-Q water unless otherwise noted. Twenty microliters of this solution were incubated on the gold-coated slide for about 30 minutes at 3°C and then rinsed with Milli-Q water.

The AFM Instrument

A detailed description of the AFM force spectroscopic technique and details of the instrumental set-up have been given elsewhere (Oesterhelt et al., 1999; Clausen-Schaumann et al., 2000). Briefly, the tip of a cantilever is brought into contact with molecules on the surface and then retracted, its deflection – and therefore the force – is detected with a laser by optical lever detection (Cleveland et al., 1993), whereas the z-distance is controlled by a strain gauge (Figure 4.2). The nominal spring constants of cantilevers used in the experiments were 10 mN/m. Before the first approach of the AFM tip to the surface, the spring constants of each lever were individually calibrated by measuring the amplitude

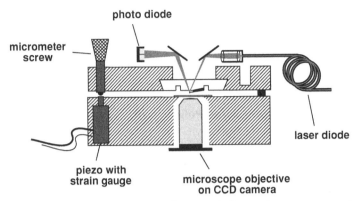

Figure 4.2: Schematic diagram of an atomic force microscope apparatus adapted for varying the z-direction for obtaining force-extension curves of single chains.

of its thermal oscillations (Butt and Jaschke, 1995). The sensitivity of the optical lever detection was measured by indenting the AFM tip into a hard surface. All experiments were conducted in Milli-Q water at room temperature (21°C), unless otherwise specified.

Obtaining the Single-Chain Force-Extension Curve

For characterization of elastic properties, the gold-coated AFM tip was brought into contact with the polypentapeptides on the microscope slide by manual control. The AFM tip and the polymer layer were kept in contact under a contact force of several nanoNewton (nN) for approximately 30 s to allow for an attachment of the chain to the tip. Usually, the chain adsorbed non-covalently at some position on the tip, maybe sometimes a chemical reaction between the gold-coated cantilever and the cysteine on the end of a polypentapeptide allowed for really high rupture forces. Upon retraction of the cantilever, individual polypentapeptides were stretched between the surface and the AFM tip.

Note that the first force-distance profile recorded after tip-substrate contact can be rather complex, consisting of contributions from stretching several polypentapeptides, desorption from the substrate and/or cantilever, bond rupture of short strands, as well as interchain aggregation and entanglements. Therefore, in each measurement, the cantilever was first retracted from the substrate to a distance at which unspecific adhesion was no longer observed. Then, in successive retraction-approaching cycles the distance range is increased while trying to avoid contact between the tip and additional polypentapeptide strands at the substrate surface, until only one polymer strand remained between tip and substrate. The force-distance profile of this strand was then measured repeatedly until rupture.

Analysis of the Results

The experimental traces were fit by an extended worm-like chain (WLC) model, including linear elastic contributions arising from the stretching of bond angles and covalent bonds:

$$F \cdot \frac{L_P}{k_B T} = \frac{R_z}{L} - \frac{F}{K_o} + \frac{1}{4(1 - R_z/L + F/K_o)^2} - \frac{1}{4}. \tag{4.11}$$

In this expression, R_z is the measured end-to-end-distance at any given force, F; and L is the contour length of the stretched chain (polypentapeptide) under zero force ($F = 0$). The peptides bending rigidity is expressed by the chain's persistence length, L_P. Finally, the chain's extensibility upon stretching is described by the segment elasticity, K_o, which is introduced into Eq. (4.9) as a linear term (hereby, K_o can be understood as the inverse of the normalized compliance of a Hookean spring; the spring constant of the polymer chain is given by K_o/L). (It should be appreciated that the WLC model is not modeled at the molecular structure level but rather is based on more macroscopic considerations.)

This resulted in values for the persistence length L_P of $(GVGVP)_{n \times 251}$: $L_P = 0.4$ nm in the low force regime and $L_P = 0.6$ nm when fitted to higher forces, which are comparable to values measured previously for other polypeptide backbones, e.g., of proteins (Rief et al., 1997). The reversible traces provide evidence for single-chain ideal elasticity (entropic and therefore reversible – most probably from backbone torsional movements (rocking), wherein the contribution to the entropic component of the force increases with decrease in the frequency of the oscillation). The small deviations from perfect reversibility in traces 2, 3, and 5 of Figure 4.3 occurred when the chain was held at least 30 s in the relaxed state. These deviations could have several origins. There could be a *configurational transition* (e.g., an opening of β-turns and/or hydrophobic interaction), a backfolding, between different parts of the chain. The latter is favored by the observed rate dependence of this non-equilibrium contribution.

The persistence length L_P of $(Gly-Val-Gly-Ile-Pro)_{n \times 260}$, $(GVGIP)_{n \times 260}$ is difficult to determine, when intra- and/or interchain aggregation gives rise to a strong hysteresis even when being far away from the surface within two successive cycles (Figure 4.4). It is determined to exhibit an $L_P = 0.7$ nm, which might be a little higher than with $(GVGVP)_{n \times 251}$, but this value lies well within the error range of about 20%. This strong aggregation of $(GVGIP)_{n \times 260}$ happens

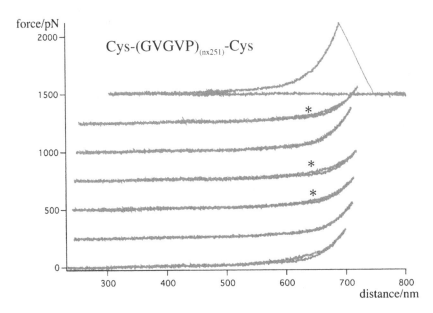

Successive traces from bottom to top, shifted by 250 pN. Before the traces with the *where taken, the chain was at least for 30 s in the relaxed state.

Figure 4.3: Single-chain force-extension curves for Cys-(GVGVP)$_{n \times 251}$-Cys at a temperature below the onset temperature for hydrophobic folding and assembly for this composition. The initial trace is given at the bottom, and subsequent traces of the same chain without intervening detachment are displaced vertically by 250 pN. The second and fifth curves exhibit perfect reversibility. The curves marked with an asterisk and the initial curve were the result of a period of at least 30 s in the relaxed state, where one possibility is that some hydrophobic back-folding occurs. Reproduced with permission from 'Elastin: A Representative Ideal Protein Elastomer', by D. W. Urry et al., 2002, *Philos. Trans. R. Soc. Lond. B* **357,** 169–84. © 2002.

at the same concentration, at which the aggregation for (GVGVP)$_{n \times 251}$ is much smaller. This reflects the increase in hydrophobicity due to an additional CH$_2$ group in GVGIP, which is the only difference with GVGVP. Unfortunately, it is not yet clear whether the aggregation is mainly due to backfolding within a single chain or association between different chains. This is under investigation.

Comparison of AFM Single-Chain and Macroscopic Elastic Moduli
For the characterization of elastomers at the macroscopic scale, force is plotted versus elongation, but instead of being given as actual increase in length as in Figures 4.3 and 4.4, elongation is given in terms of the ratio of the initial length to the extended length. Furthermore, deformation is generally uniform so that the initial cross-sectional area, *A*, is used. So, Young's elastic modulus, *Y*, is the

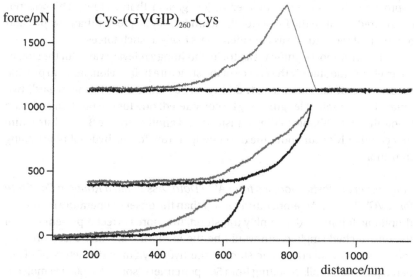

Figure 4.4: Single-chain force-extension curves for Cys-(GVGIP)$_{n \times 260}$-Cys at a temperature above the onset temperature for hydrophobic folding and assembly for this composition. In each case, the energy expended in deformation is greater than that recovered on relaxation. In addition, there appears to be a complex relaxation curve. Reproduced with permission from 'Elastin: A Representative Ideal Protein Elastomer', by D. W. Urry et al., 2002, *Philos. Trans. R. Soc. Lond. B* **357**, 169–84. © 2002.

force, f, times the initial length, L_i, divided by the elongation, L, and the cross-section of the elastomer, that is, $Y = fL_i/AL$. The units are N/m^2 or dynes/cm^2, where N/m$^2 = 10$ dynes/cm^2 and N stands for Newtons. The elastic modulus of 20 Mrad γ-irradiation cross-linked (GVGVP)$_{251}$ [i.e., X^{20}-(GVGVP)$_{251}$] is 1.6×10^5 N/m^2.

To use the data in Figure 4.3 to obtain an estimate of single-chain elastic modulus, estimates of the L_i/L ratio and of A are required. Initially, the total number of repeats being extended must be deduced. In Figure 4.3, the length to rupture is just greater than 700 nm, which we take to be the fully extended chain. This is too long for a chain of 251 pentamers, but appears to be too short for a chain of 753 pentamers. By elimination of chain lengths with 251 and 753 pentamers, this indicates that the structure would be (GVGVP)$_{2 \times 251}$, giving 502 pentamers (i.e., 2,510 residues with a value of 0.29 nm per residue). This is shorter than the usual value of 0.35 nm per residue presumably due to the presence of a proline residue in every pentamer. This is considered to be due to the *cis*-Val–Pro peptide bond, which introduces a kink in the extended backbone for every pentamer. It should be noted, however, that if all of the β-turns were retained at rupture, the 3×251 chain length could apply with approximately 1 nm per pentamer, or just greater than 700 nm. This was not considered because the forces for detachment can be greater than 500 pN, and retention of the β-turn was considered less likely at such forces.

Perhaps the most significant factor in obtaining a relevant value for the single-chain elastic modulus is the choice of the structure before elongation to provide an estimate of an initial length and cross-sectional area. In this regard, two quite different initial lengths may be considered: one for the random coil state below the hydrophobic folding transition and another for the β-spiral structure that could only occur above the onset temperature for the hydrophobic folding transition.

Calculation of Elastic Modulus from AFM Data, Assuming a Random Coil State for $(GVGVP)_{n \times 251}$. At temperatures lower than the onset temperature for the hydrophobic folding and assembly transition, the more correct representation of the unstretched single chain would be the random coil state. Based on photon correlation spectroscopic studies, the hydrodynamic diameters of disordered polymeric coils resulting from 502 pentamers (some 200,000 Da molecular weight) below the transition would be about 18 nm (San Biagio et al., 1988). By way of example, at a distance of 378 nm in Figure 4.3, the L_i/L ratio would be 1/20. With a slope of 0.08 ± 0.03 pN/nm from Figure 4.3, at a distance of 378 nm from the substrate, the increase in force would be 36 pN (i.e., due to elongation from 18 to 378 nm), and the change in length, ΔL, would be a factor of 20 or a slope of 1.44 ± 0.54 pN for a 100% extension. Next, this quantity should be divided by an appropriate cross-sectional area [i.e., π times the hydrodynamic radius squared of the random coil before extension of the single chain; $\pi r^2 = \pi(9 \text{ nm})^2 = 254 \text{ nm}^2$]. Given the foregoing assumptions, the calculated single-chain elastic modulus would be $5.7(\pm 2.1) \times 10^3$ N/m^2. As might have been expected, without taking cross-link density into consideration (Urry et al., 1984), this value is a factor of 30 from the experimental value at $37°$C for X^{20}-$(GVGVP)_{251}$ of 1.6×10^5 N/m^2, but more closely approximates what is expected below the transition temperature or when X^{20}-$(GVGVP)_{251}$ has been thermally denatured, as seen in Figure 4.9B. Concerning polymer density, it may be noted that the density of the low-temperature random-coil state is about one-quarter that of the phase separated state.

Calculation of Elastic Modulus from AFM Data Assuming a Single-Chain β-Spiral Structure for $(GVGVP)_{n \times 251}$. For comparison, it is useful to consider the structure representative of the state above the temperature of the inverse temperature for hydrophobic folding and assembly. This addresses the importance of choosing structure when converting the force-extension curve into a macroscopic elastic modulus. If the single β-spiral structure were assumed before extension with three pentamers per turn and a translation along the axis of 1 nm per turn, the length of a 502 pentamer would be *ca.* 170 nm. This means that an elongation of 100% would occur at 340 nm, and the change in force for extension of the β-spiral from 170 to 340 nm would be 170 nm $\times$ 0.08 $\pm$ 0.03 pN nm^{-1} = 13.6 $\pm$ 5.1 pN per β-spiral.

The next issue is an estimate of the number of β-spirals per m^2 for comparison with the cross-linked matrix from which the experimental macroscopic value of 1.6 $\times$ 10^5 N m^{-2} is obtained. For a macroscopic state of 37% water by weight (see Figure 4.1), an assumed density of 1.33 g cm^{-2} for the polymer, and a cross-sectional area for the β-spiral of 3.2 nm^2, the number would be *ca.* 8.7 $\times$ 10^{16} β-spirals m^{-2}. The single-chain elastic modulus, calculated on the basis of the β-spiral structure, would be (13.6 $\pm$ 5.1 $\times$ 10^{-12} N per β-spiral) $\times$ (8.7 $\times$ 10^{16} β-sprirals m^{-2}) = 1.2 $\times$ 10^6 N m^{-2}.

Thus, by assuming a different structure, instead of being low by a factor of 30 as was the case above for a random-coil structure, which occurs below the transition temperature, the assumption of the β-spiral structure resulted in a value for the single-chain elastic modulus that is seven and a half times larger than the macroscopic value. We note that the orientation of a random coil is not relevant in the argument relating single-chain data to macroscopic data. For the γ-irradiation cross-linked macroscopic matrix, however, the β-spirals would be randomly oriented with respect to the direction of extension. This would decrease the contribution of each β-spiral not aligned in the direction of extension. It will now be particularly interesting to obtain the elastic modulus for the twisted filament structure depicted in Figure 4.10F. Work is in progress for the preparation of multistranded twisted filaments.

The Acoustic Absorption Experiment

Sample Preparation and Experimental Set-up
The polymer is prepared in the phase-separated state (i.e., the intermediate structured state of Figure 4.1, and 20 Mrad γ-irradiation cross-linked in the form of the desired cylindrical shape and with path lengths that differ by a factor of 2, as shown in the lower part of Figure 4.5).

The Naval Research Laboratory at SSC-SD (Spatial Warfare Systems Center-San Diego) has the capability to measure accurately the elastic properties of materials by using a broadband technique developed by Sheiba (1996). The technique is under U.S. Patent Pending. Older techniques are more cumbersome and less reliable. The frequency of measurement ranges from subsonic to

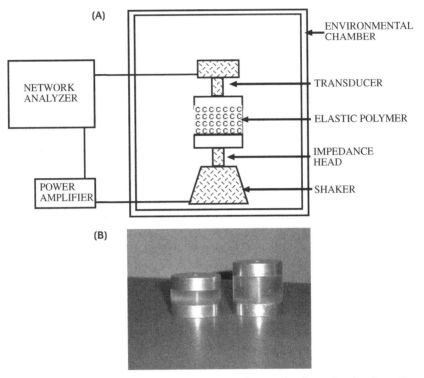

Figure 4.5: (A) Block diagram of the instrumentation used to determine the absorption per unit volume of elastic samples in the acoustic range. (B) Example of the use of two samples that differ in length by a factor of 2, as required for the analysis of Sheiba (1996) to be utilized.

about 15 kHz. The measurements are performed in air inside an environmental chamber. Figure 4.5 also shows the block diagram of the instrumental set-up for these measurements.

The network analyzer sends sweeping signals to the shaker. The transducer (sometimes called an accelerometer) and impedance head record the velocities at the base and at the top of the sample. After recording data during this frequency sweep, a new set of data is then obtained at a new temperature setting. Commonly, temperatures can range from $-30°$ to $+70°$C. The velocity record constitutes the transfer matrix. By mathematically inverting the transfer matrix, it is possible to extract the complex modulus and Poisson ratio. These data provide a complete description of a material's elastic properties. Further analysis yields the shear modulus, μ, the Poisson ratio, ν, and the loss factor, η, each as a function of frequency and temperature. The loss factor, η, is the main parameter of interest. It represents the energy absorption per unit volume at a given frequency. The loss factor versus frequency and temperature were found for several different elastic protein-based polymer compositions. Data for $(GVGIP)_{n \times 260}$ is given in Figure 4.6(A). MATHCAD 7.0 was used to perform the mathematical analysis.

Comparison of the Acoustic Absorptions of (GVGIP) and Natural Rubber

Natural rubber is well characterized as a random chain network. This means that there are no regular repeating conformational features in the polymer chain. The consequence of a random chain is a broad, slowly changing dependence on frequency of physical properties, such as absorption. Simply stated, the absence of structure means that the barrier to rotation about each bond will tend to be different. This is what is observed in Figure 4.6(A) for natural rubber and a synthetic petroleum-based polymer, polyurethane, chosen for its relatively high loss factor in the frequency range reported.

On the other hand, at frequencies lower than those of the infrared spectroscopic range, any material that exhibits a localized absorption band as a function of frequency must do so by virtue of a regular repeating structure. Interestingly, this is seen in Figure 4.6(A) for X^{20}-$(GVGIP)_{260}$ in the range of 200 Hz–7 kHz, and it is seen that the intensity of the absorption per unit volume increases as the temperature is raised through the temperature interval of the inverse temperature transition. Besides demonstrating the remarkable acoustic absorption property of this material, the acoustic absorption data provide additional demonstration of the existence of a regular structure for poly(GVGIP) and for related polymers. Actually, this is simply further confirmation of earlier dielectric relaxation data, which demonstrated the development of intense localized relaxations near 5 MHz as the temperature is raised through the interval of the inverse temperature transition.

Dielectric Relaxation Studies on Elastin-Related Systems

Dielectric Relaxation and Acoustic Absorption Data for (GVGIP)$_n$ Over the Frequency Range, 0.1 to 100 kHz

In the acoustic absorption measurement of Figure 4.6(A), an oscillating compressional wave activates the mechanical relaxation in the vicinity of 1 kHz, and the relaxation requires reorientation of peptide moieties for the polypentapeptide to respond to the acoustical frequency of the vibrational wave. Because peptide moieties have very large dipole moments, poly(GVGIP) could be expected to function as a transducer in which the acoustically activated

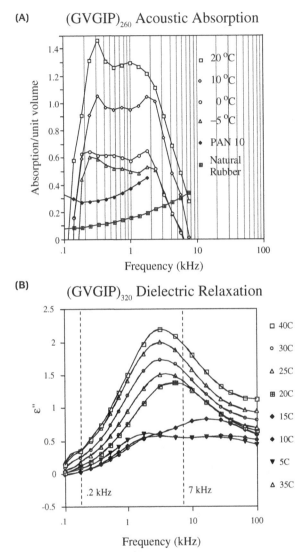

Figure 4.6: (A) Plots of the loss factor (absorption per unit volume) for a series of elastomers. The bottom curve is of natural rubber, and the slightly higher structureless curve is for polyurethane, which is recognized as a rubber with a high loss factor in the acoustic absorption range. The four upper curves are of 20 Mrad cross-linked $(GVGVP)_{260}$. As the temperature is raised above the temperature for the inverse temperature transition of hydrophobic folding and assembly for this composition, the intensity for a localized relaxation grows dramatically. To the best of our knowledge, these model elastic protein-based polymers exhibit the greatest sound absorption for the 100–10,000 Hz frequency range. This demonstrates the formation of a regular structure for $(GVGIP)_n$ in contrast to the lack of a regular structure for the classical rubbers. Reproduced with permission

mechanical relaxation would be transduced into a dielectric relaxation signal of the same frequency. The experimental equipment to demonstrate this free energy transduction has yet to be designed.

The existence of the acoustic absorption, however, indicates that a dielectric relaxation should be observed in a not too distant frequency range. An exact overlap is not expected in the two different experiments, because in the dielectric relaxation experiment that alternates positive and negative charges on a pair of plates directly drives the peptide reorientation, and the particular repeating structure determines the frequency range and intensity of the relaxation. On the other hand, in the acoustically driven relaxation, the peptide reorientations are a secondary result that is required for the repeating structure to respond to the hammering frequency of the acoustic wave. Of course, the fact that a mechanical resonance grows in intensity as the inverse temperature transition progresses requires progressive formation of a regular repeating structural feature on raising the temperature.

It is to be expected, therefore, that a dielectric relaxation be observed in the frequency range of 0.1–100 kHz as reported in Figure 4.6(B), which plots the imaginary (absorption) part of the dielectric permittivity of poly(GVGIP) as a function of temperature. Just as for the acoustic absorption study in Figure 4.6(A), there is an increase in the absorption (imaginary) component of the dielectric relaxation as the temperature is raised through the temperature interval of the inverse temperature transition for hydrophobic folding and assembly. Interestingly, intensity of the relaxation continues to grow above 20°–40°C, suggesting that an acoustic absorption even more intense than reported in Figure 4.6(A) might be expected.

Related Dielectric Relaxation (Mechanical Resonance) Near 5 MHz for Poly(GVGIP)

For poly(GVGIP) in the frequency range of 1–1,000 MHz and over the temperature range of 0°–60°C, the real part of the dielectric permittivity is given in Figure 4.7(A) and the imaginary (absorption) part in Figure 4.7(B). As with the data in the 1 kHz range of Figure 4.6, there is an increase in regular structure as the temperature is raised from below to above the inverse temperature transition.

---

from 'Elastin: A Representative Ideal Protein Elastomer', by D. W. Urry et al., 2002, *Philos. Trans. R. Soc. Lond. B* **357,** 169–84. © 2002. (B) Imaginary (absorptive) part of the dielectric permittivity of poly(GVGIP). The same phenomenon of increased absorption on passing through the temperature range of the inverse temperature transition is observed in the 100–10,000 Hz frequency range as exhibited by the same composition in the acoustic absorption range. In this case, the temperature is increased to 40°C, and the intensity of the relaxation continues to rise.

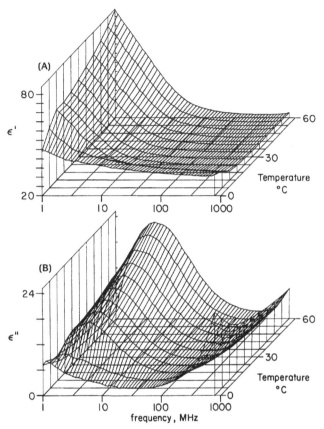

Figure 4.7: (A) Real part of the dielectric permittivity for poly(GVGIP). (B) Imaginary (absorptive) part of the dielectric permittivity of poly(GVGIP). The same phenomenon of increased absorption on passing through the temperature range of the inverse temperature transition is observed in the 1–100 MHz range as exhibited by the same composition in the acoustic absorption range in Figure 4.6. Such low-frequency motions provide a source of entropic elasticity and low energy for the structural state of the elastic protein-based polymers. Reproduced with permission from 'Dielectric Relaxation Studies on Analogs of the Polypentapeptide of Elastin', by R. Buchet et al., 1988, *J. Phys. Chem.* **92,** 511–17. © 1988.

This type of data was used more than a decade ago to demonstrate the formation of regular structure; it is equally valid today, and it is what allowed anticipation of the data recently obtained in Figure 4.6 (i.e., anticipation of the remarkable acoustic absorption properties of these materials). This successful trail of proposed structural regularity, testing of the proposal, and predicting of additional properties continues unbroken for much of two decades.

The predictive success and usefulness of a perspective is the truest test of its validity. The perspectives of disorder and randomness have given rise to no such progress. On the other hand, the perspective of a regular dynamic structure has led to properties such as numerous free energy transductions otherwise unimaginable. Now, data on the elastin models can be directly compared with α-elastin and fibrous elastin itself. Mechanical resonances due to regularly repeating structures are observed both in the vicinities of 1 kHz and 5 MHz.

Comparison of Poly(GVGVP) and α-Elastin

Plotted in Figure 4.8(A) is the real part of the dielectric permittivity as a function of temperature for α-elastin and for poly(GVGVP) in the inset (Urry et al., 1985a). In both cases, there is the development of an intense relaxation near 5 MHz as the protein-based elastomers hydrophobically fold and assemble on increasing the temperature through the range of the inverse temperature transition. Both elastomeric molecular systems exhibit the same development of regular structure on raising the temperature through the range of the inverse temperature transition.

Comparison of Cross-Linked Poly(GVGVP) to Fibrous Elastin

In Figure 4.8(B) is plotted the imaginary (absorption) part of the dielectric relaxation for purified natural fibrous elastin (from bovine ligamentum nuchae) as a function of temperature from 23° to 58°C (Luan et al., 1988). Also, in Figure 4.8(B) is plotted the same data at 58°C for the elastomer, 20 Mrad cross-linked poly(GVGVP). Although there is a small increase in mean frequency for the X^{20}-$(GVGVP)_{251}$ relaxation in Figure 4.8(B) as in Figure 4.8(A), it is unmistakable that there is hydrophobic folding and assembly resulting in the formation of regular, dynamic, non-random structures in the natural fibrous elastin that are similar to those of cross-linked poly(GVGVP) and poly(GVGIP). Incidentally, the decrease in intensity at 1 GHz for fibrous elastin in Figure 4.8(A) for α-elastin and in Figure 4.8(B) for fibrous elastin reflects the loss of hydrophobic hydration as the hydrophobic folding proceeds. This is due to a dielectric relaxation near 5 GHz, which arises from hydrophobic hydration (Urry et al., 1997).

The experimental data clearly reflect dynamic regular structures in the elastic protein-based polymers with related structures and corresponding behaviours in the natural elastin protein.

DISCUSSION

Relevance of Proposed Mechanisms for Protein Entropic (Ideal) Elasticity

The kernel for the classical theory of rubber elasticity originated with K. H. Meyer in 1932, which according to Flory (1968) can be stated as "... *the orientation of the molecules of a piece of rubber when it is stretched would entail a decrease in entropy, and thus could account for the retractive force.*" A common

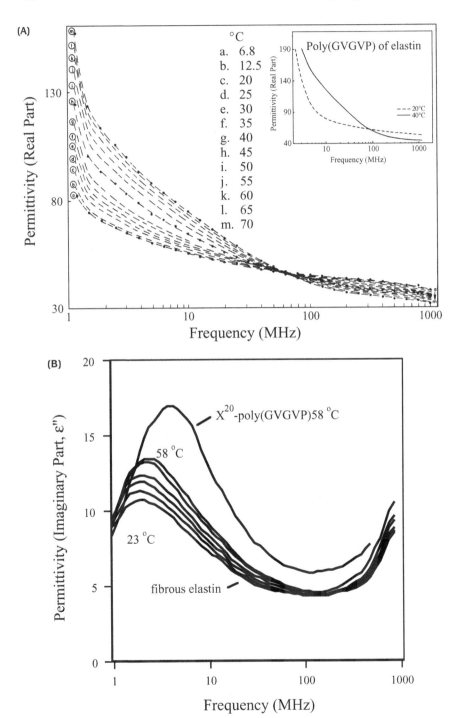

restatement has been that a collection of disordered chains becomes more aligned as the elastomer is stretched and that this alignment of chains constitutes the decrease in entropy on extension. Flory's restatement in 1968 was given in terms of two more fundamental subsidiary hypotheses: (1) the restoring force originates within the chain molecules by distortion of the distribution of their configurations and not from interactions between chains; and (2) the restoring force is due "exclusively to the *entropy* of the chain configuration." For ideal or perfect elastomers, we have no quarrel with these subsidiary hypotheses, but those elastomers exhibiting hysteresis appear not to be consistent with the first hypothesis.

Obviously, with respect to protein elasticity, we do disagree with the derivative dogma (Hoeve and Flory, 1958, 1974; Flory, 1968) that only random chain networks comprised of a Gaussian distribution of end-to-end chain lengths can satisfy the two hypotheses, which has been a tenet of the classical theory of rubber elasticity. Clearly, entropic elasticity exhibited by a single chain limits the source of entropy change to the damping of internal chain dynamics within a single chain and cannot be due to the relative orientation of many chain molecules within a random chain network.

The Flory Random Chain Network Theory of Entropic Elasticity?

Flory's publications describe a Gaussian distribution of end-to-end chain lengths that results from a random chain network, and they assert that the decrease in entropy due to extension or compression arises from distortion of the network from this most probable Gaussian distribution (Flory, 1953, 1968). Hoeve and

---

Figure 4.8: (A) Real part of the dielectric permittivity of α-elastin showing an increasing relaxation near 5 MHz with increasing temperature from below to above the temperature range for the inverse temperature transition. Data for α-elastin are nearly identical to that found for poly(GVGVP) shown in the inset for a single temperature below and one above that of the inverse temperature transition. This demonstrates that the same relaxation occurs in a fragment of the natural elastic fiber as has been so thoroughly characterized in the model elastic proteins, poly(GVGVP) and poly(GVGIP). Reproduced with permission from 'Temperature Dependence of Dielectric Relaxations in α-Elastin Coacervate: Evidence for a Peptide Librational Mode', by D. W. Urry et al., 1985a, *Biochem. Biophys. Res. Commun.* **128,** 1000–6. © 1985. (B) Imaginary (absorptive) part of the dielectric permittivity of natural fibrous elastin showing an increasing relaxation near 5 MHz with increasing temperature from below (23°C) to above (58°C) the temperature range for the inverse temperature transition. The same relaxation is seen at 58°C with a somewhat higher intensity, and a slightly higher frequency is found for the 20 Mrad cross-linked poly(GVGVP). It seems quite evident that dynamic-structured states occur in the natural elastic fiber as demonstrated for the model elastic protein-based polymer. Adapted with permission from 'On the Source of Entropic Elastomeric Force in Polypeptides and Proteins: Backbone Configurational Vs. Side Chain Solvational Entropy', by C.-H. Luan et al., 1989, *Int. J. Quant. Chem.: Quant. Biol. Symp.* **16,** 235–44. © 1989.

Flory (1958, 1974) concluded that the entropic elasticity of elastin arises from such a random chain network. There are four principal reasons why this perspective is to be questioned: (1) a limited potential for cross-links in the fixed sequence of elastin; (2) the above spectroscopic evidence for regular, non-random, dynamic structural features in components of elastin; (3) the above report that single chains of repeating sequences of elastin exhibit entropic elasticity; and (4) irreversible loss of elastic force due to heating above 60°C resulting from thermal denaturation.

Relevance of the Probability of Cross-Link Formation to Structure. In rubber, each repeating unit is a potential cross-link, and, on vulcanization, any pair of repeating units can be made into a cross-link wherever two chain segments are sufficiently proximal. In elastin, only lysine residues form the cross-links, and yet there are only some forty lysine residues for an elastin protein of about 800 residues (i.e., one lysine in every twenty residues). Furthermore, the lysine residues are arranged in an orderly manner along the sequence. Most commonly, four lysine residues combine to form a single cross-link, called a desmosine, and essentially all of the lysine residues in fibrous elastin have been formed into cross-links. Such cross-linking cannot be achieved within a random chain network. Achievement of such rare cross-linkages require elements of order (Urry, 1976).

Low-Frequency Motions Provide Much Entropy to Chain Structures. Motions due to rotations about backbone bonds in a chain molecule that exhibit intense relaxations limited to localized frequency intervals (i.e., intense mechanical resonances) can only result from polymers containing regular, non-random, albeit dynamic structures. For model elastin systems, including purified fibrous elastin, such mechanical relaxations are observed near 1 kHz and 5 MHz (see Figures 4.6, 4.7, and 4.8: Urry et al., 1985a; Buchet et al., 1988; Luan et al., 1988). As will be treated later, these relaxations provide an ample source of entropy to structured elastin systems. These contributions to entropy increase greatly as the frequency of the relaxation is lower, and they are contributions to the free energy of protein structure that have yet to be adequately treated in the current computations of low-energy protein structures. Most importantly, for entropic elasticity, low-frequency motions provide an abundant and ready source of entropy decrease on extension.

Entropic Elastomeric Force Arising from a Single Chain. At a fixed extension in the AFM single-chain force-extension curves of Figure 4.3, there is only one end-to-end chain length. The entropic elastic force measured at a given extension cannot have arisen from displacement of a network of chains from a Gaussian distribution of end-to-end chain lengths. The decrease in entropy that provides the entropic restoring force in the single-chain experiment must arise from a damping of internal chain dynamics of the single chain. There are no other degrees of freedom from which the decrease in chain entropy can arise.

Loss of Elastomeric Force of Elastin-Based Systems Due to Thermal Denaturation. In Figure 4.1, it was noted that poly(GVGVP) exhibits a slow irreversible denaturation when held at temperatures above 60°C. This same phenomenon is apparent in the thermoelasticity studies on fibrous elastin, as originally noted by Dorrington and McCrum (1977). In the thermoelasticity experiments of Figure 4.9 (Urry, 1988a, 1988b), the elastomer is equilibrated at 40°C and at that temperature is extended to 60%. At this fixed length, the temperature is lowered to about 20°C and slowly raised to 80°C. The force dramatically increases as the temperature is raised through the range of the inverse temperature transition from 20° to 40°C. Then, the force/T (°K) tends to remain nearly constant from 40° to 60°C, especially for X^{20}-poly(GVGVP). The value of $\ln f/T$ begins to decrease above 60°C in a time-dependent manner. When using a rate of increase of 4°C per 30 minutes, 8°C/hr, there is a decrease in $\ln f/T$, and, when the temperature is returned to 40°C, there has occurred an irreversible loss of force. The decrease in force above 60°C is more dramatic on raising the temperature more slowly (e.g., when raising the temperature at a rate of 1°C/hr). Irreversible thermal denaturation is most dramatically seen when kept above 60°C for a longer time. Furthermore, a plot of ln(elastic modulus) versus time at 80°C is linear in evidence of a first-order process (Urry, 1988a, 1988b). Quite interestingly, while heating above 60°C decreases force, it also shortens the length at which zero force is obtained, as expected for a filamentous, β-spiral-like structure that became randomized by thermal denaturation.

As discussed in relation to Eqs. (4.7) and (4.8), a slope of zero for plots like those in Figure 4.9 would mean an ideal (entirely entropic) elastomer. As clearly seen for X^{20}-poly(GVGVP) in the 40°–60°C temperature range when data are collected at the rate of 30 minutes per data point, a near zero slope is obtained. Commonly, an f_E/f ratio of 0.1 is obtained for X^{20}-poly(GVGVP) (Luan et al., 1989). Because of the steep positive slope between 20° and 40°C (due to the hydrophobic folding and assembly of the inverse temperature transition) and because of the slow irreversible thermal denaturation above 60°C, the slope – obtained in the thermoelasticity experiment from which the entropic contribution is estimated – depends on the rate at which the data are collected and the temperature interval over which it is collected.

A steep rise in force in the 20° to 40°C temperature interval for fibrous elastin in Figure 4.9(A), though less steep than seen in Figure 4.9(B) for X^{20}-poly(GVGVP), demonstrates the effect of the inverse temperature transition within elastin

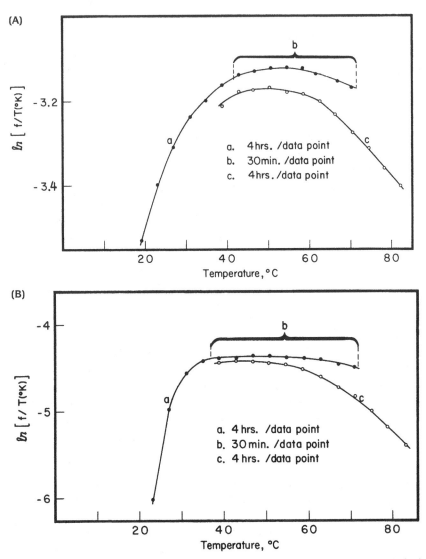

Figure 4.9: Thermoelasticity curves for (A) fibrous elastin and (B) 20 Mrad cross-linked poly(GVGVP). See text for discussion. Whether or not a near zero slope (i.e., an $f_E/f \ll 1$), is obtained depends on the rate at which the temperature is raised. This is due to slow denaturation at temperatures above 60°C. Reproduced with permission from 'Entropic Elastic Processes in Protein Mechanisms. I. Elastic Structure Due to an Inverse Temperature Transition and Elasticity Due to Internal Chain Dynamics', by D. W. Urry, 1988a, *J. Protein Chem.* **7,** 1–34. © 1988; 'Entropic Elastic Processes in Protein Mechanisms. II. Simple (Passive) and Coupled (Active) Development of Elastic Forces', by D. W. Urry, 1988b, *J. Protein Chem.* **7,** 81–114. © 1988.

itself. More dramatic for fibrous elastin is the loss of elastomeric force on raising the temperature above 60°C. This finding of a slow loss of force at elevated temperature would not have been so apparent in the limited time of a single day to perform the experiment (without the advantage of computerized equipment), as was the case when Hoeve and Flory (1958) carried out their original study on fibrous elastin. For the thermoelasticity study of fibrous elastin shown in Figure 4.9, however, the rate of loss of force at elevated temperature is even greater than for X^{20}-poly(GVGVP).

Additionally, the plot of ln(elastic modulus) versus time for fibrous elastin maintained at 80°C, as determined by a stress-strain curve every 24 hr, is a straight line with a half-life of approximately 10 days (Urry, 1988b). Another important feature of the data for determining half-life is that the length at which zero force is obtained decreases with time at 80°C. Remarkably, this means that the loss of force in Figure 4.9 and the progressive loss of elastic modulus exhibited by a series of stress-strain curves at 80°C is not due to structural rearrangement to an extended state under load or for that matter due to slipping in the grip. Instead, extended periods of time at high temperature results in a denaturation, which entails a randomization of linear filamentous structures to denatured, more nearly, globular structures. This has analogy to the loss of percentage of water by weight for time above 60°C in Figure 4.1.

What is clearly being observed in the thermoelasticity data of elastin-based systems in Figure 4.9, in the half-life studies for the elastic modulus at 80°C, and in the structural change above 60°C in Figure 4.1, is a slow thermal denaturation of protein-based elastomers. Because random chain networks do not exhibit thermal denaturation, these elastin-based systems are not properly described as random chain networks. *Thus, the evidence is clear and abundant; elastin-based systems contain regular repeating conformations and those ordered structures denature.*

Solvent Entropy Changes as a Source of Entropic Elastic Force?

Effect of Removal of Solvent Entropy Change during Inverse Temperature Transition. The change in solvent entropy is fundamental to the occurrence of an inverse temperature transition. This transition has been extensively used to perform the family of energy conversions that sustain living organisms (Urry, 1993a, 1993b, 1997), and therefore the inverse temperature transition is extensively characterized. One approach to testing whether the change in solvent entropy contributes to the elastomeric force is to use an approach due to Flory, that is to reduce the heat of the inverse temperature transition to near zero by an appropriate solvent mixture. Hoeve and Flory (1958) had used a mixture of 30% ethylene glycol by weight in water in their thermoelasticity study that deduced a dominant entropic elastic force for fibrous elastin. When Luan et al.

(1989) compared the thermoelasticity-derived f_E/f ratio of X^{20}-poly(GVGVP) in water and in 30% ethylene glycol, the values were the same, 0.1. Based on the approximations of Eqs. (4.7) and (4.8), X^{20}-poly(GVGVP) is a dominantly entropic elastomer in both solvent systems. The effect of adding ethylene glycol is to reduce the heat of the transition to near zero. Now, because entropy of the desolvation process is the heat of the transition divided by the temperature of the transition, $\Delta H_t/T_t$, the solvent entropy change would be approaching zero, and yet the solvent mixture, instead of reducing the elastomeric force, remarkably resulted in a substantial increase in force. Furthermore, the increase in force at fixed length in the $20°–40°C$ temperature range results in an elastomeric force that is 90% entropic (see Figure 4.9) – occurs while the solvent entropy change is positive rather than negative. The conclusion is that there is no experimental basis for believing that solvent entropy change contributes to elastomeric force.

The Polymer Chains Bear the Force; Solvation Changes Alter the Fraction of Chain Segments That Sustain the Force. The force of deformation must be borne by the chains; it cannot be borne by solvent. At fixed length, raising the temperature of a band of elastomer from below to above the temperature of the inverse temperature transition increases the force. Alternatively, increasing chemical potential to lower the transition temperature from above to below the operating temperature at fixed length causes an increase in elastomeric force (Urry et al., 1997, 1988a, 1988b). These results are due to an increase in hydrophobic folding and assembly of the chains. Because this is the case at fixed length, the fraction of a chain length that is not folded must be further extended. Accordingly, to maintain a fixed length, the unfolded segments must be further extended. This provides for the increase in entropic elastic force by means of a decrease in configurational entropy and not directly due to a change in solvent entropy.

Entropic Elasticity Due to Damping of Internal Chain Dynamic on Extension

Molecular Structure of Poly(GVGVP). To understand the concept of the damping of internal chain dynamics on extension, the structure given in Figure 4.10 is helpful. The Pro-Gly containing elastomeric sequential polypeptides exhibit a repeating β-turn structure as schematically shown in Figure 4.10(A), and the β-turn is given in crystallographic detail in Figure 4.10(B). On raising the temperature, the inverse temperature transition results in the optimization of intramolecular and intermolecular hydrophobic association. Intramolecularly, this results in a helical structure called a β-spiral, which is shown schematically as a helix in Figure 4.10(C), in more detail with β-turns schematically shown as

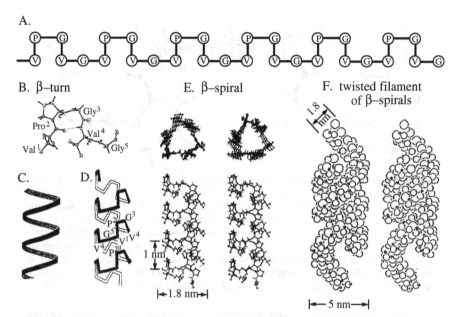

Figure 4.10: Description of the proposed molecular structure of poly(GVGVP). See text for discussion.

spacers between turns of the helix in Figure 4.10(D), and in atomic stereo detail in Figure 4.10(E). Intramolecularly, the β-spirals associate to form multistranded twisted filaments, as depicted in Figure 4.10(F). As there is water within the β-spiral and as a result of the β-turns functioning as spacers, the peptide segments connecting β-turns are free to undergo large amplitude torsional oscillations.

Unrolled Perspective of β-Spiral Showing Suspended Segment between β-Turns. Shown in Figure 4.11 is an unrolled perspective of the polypentapeptide β-spiral showing a suspended segment between β-turns where peptide moieties are free to undergo large amplitude torsional oscillations. In particular, the ψ_4–ϕ_5 pair of torsion angles and the ψ_5–ϕ_1 pair of torsion angles couple in each case to result in rocking motions of the respective peptides (i.e., in peptide librations). The point to be made is that the peptide moieties can undergo large torsional oscillations due to the absence of steric interactions in the hydrophobically folded β-spiral structure. This freedom of motion of the peptide moieties of the suspended segment within the β-spiral structure gives much entropy to the structure and is considered to be a primary source of the mechanical resonances of Figures 4.6, 4.7, and 4.8, as will be treated later.

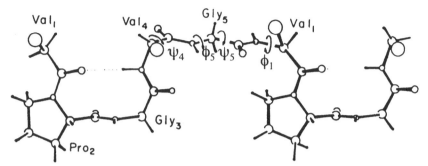

Figure 4.11: Unrolled perspective of one-half turn of a β-spiral showing the suspended segment between β-turns. Due to the near collinearity of the bonds on each side of a peptide moiety, the peptide moiety can undergo rocking motions. In particular, in the indicated suspended segment, there are two peptide moieties with the ψ_4–ϕ_5 pair of torsion angles and the ψ_5–ϕ_1 pair of torsion angles. These pairs of torsion angles couple in each case to result in rocking motions of the respective peptides, referred to as peptide librations. Such motions provide much entropy to the β-spiral structure and to all protein structures that allow for these coupled motions; and it is a decrease in this chain dynamics on extension that is a dominant source of entropic elasticity exhibited by single chains.

Effect of Extension on the Amplitude of the Torsional Oscillations and on Chain Entropy. The magnitude of the torsional (dihedral angle) excursions for a single pentamer sandwiched within a turn of β-spiral (three pentamers) for a 1.5 kcal/mol residue cut-off energy is shown in Figure 4.12(A) when using the ECEPP (empirical conformational energies of peptides and proteins) molecular mechanics calculations of the Scheraga approach. Dihedral angle oscillations of 160° are possible. Lambda oscillational plots of ψ_i–ϕ_{i+1} for the two peptide moieties of Figure 4.11 are given in Figure 4.13 for the relaxed and extended (130%) states of poly(GVGVP), where the correlated motions of the pair of angles is apparent by the data points falling on a line with a slope of -1.

The effect of a 130% extension of the β-spiral pitch, shown structurally in Figure 4.12(B) and in the plots of Figure 4.13, is a dramatic decrease in the amplitude of the torsional oscillations allowed within a 1.5 kcal/mol residue cut-off energy. Using an enumeration of states approach with each change of 5° in any torsion angle counted as a new state, the decrease in numbers of states for a 0.6 kcal/mol residue cut-off energy (i.e., within an RT; where R is the gas constant, 1.987 cal/mol °K and T is the temperature in °K) is from a W^r of 342 to a W^e of 24. Using Eq. (4.10), this gives for ΔS a value of -5.28 cal/mol-deg/pentamer. If the Boltzmann summation over all energies is used, the value is -5.06 cal/mol-deg/pentamer (Urry et al., 1985c).

The plots of Figure 4.13(C) show similar reductions in freedom of motion to the 130% extension on replacement of the Gly_5 residue by a D-Ala_5 residue. An expected reduction in the intensity of the 5 MHz relaxation has been observed, and

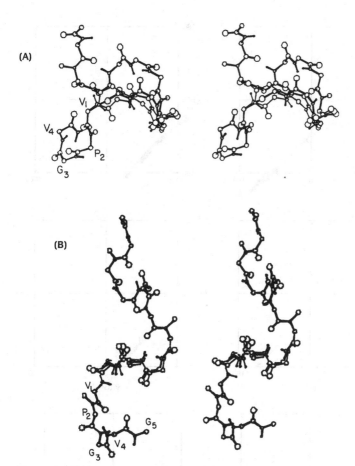

Figure 4.12: Stereoview of the torsional oscillations possible within a 2 kcal/mol residue cut-off energy for the central pentamer of one turn (three pentamers) of the β-spiral structure of poly(GVGVP) in the relaxed state (A) and in the 130% extended state (B). See text for discussion. Reproduced with permission from 'A Librational Entropy Mechanism for Elastomers with Repeating Peptide Sequences in Helical Array', by D. W. Urry and C. M. Venkatachalam, 1883, *Int. J. Quant. Chem.: Quant. Biol. Symp.* **10,** 81–93. © 1983.

both are calculable using the peptide dipole reorientations of the pentapeptide moiety in the β-spiral structure (Venkatachalam and Urry, 1986). Furthermore, replacement of the Gly_5 residue by an L-Ala_5 residue resulted in complete loss of elasticity (Urry et al., 1983).

Alternatively, when using the molecular dynamics approach of Karplus, the expression becomes

$$\Delta S = R \ln \left[\prod_i \Delta \phi_i^e \, \Delta \psi_i^e \Big/ \prod_i \Delta \phi_i^r \, \Delta \psi_i^r \right], \tag{4.12}$$

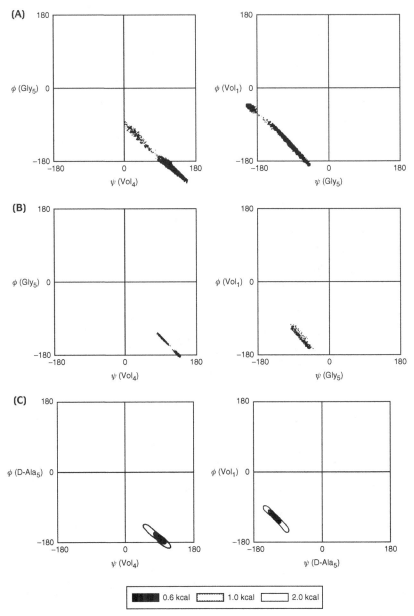

Figure 4.13: Lambda plots of the torsion angles attending the peptide moieties of the suspended segment of poly(GVGVP) and an analogue. Slopes of the negative one demonstrate the correlated motions of the β-spiral structure of poly(GVGVP) in (A) and (B) and of the D-Ala (A′) analog, poly(GVA′VP). Extension to 130% dramatically decreases the

where the product Π_i is over the rms torsion angle fluctuations $\Delta\phi_i^e$ and $\Delta\psi_i^e$ for the extended state divided by the product of the rms torsion angle fluctuations $\Delta\phi_i^r$ and $\Delta\psi_i^r$ for the relaxed state as obtained from the molecular dynamics trajectories for the relaxed and 130% extended states. The set of rms torsion angle fluctuations for the relaxed and extended states are given in Table 4.1, where the decreases in amplitude of the torsional oscillations as the result of extension are apparent. Using Eq. (4.11), the calculated decrease in entropy is −5.5 cal/mol-deg/pentamer for the 130% extension (Chang and Urry, 1989). Whether the calculation of entropy is by means of an enumeration of states in configuration space or the amplitude of the torsional oscillations in a molecular dynamic simulation, the calculated entropy change for a 130% extension is the same.

The $\Delta\phi$–$\Delta\psi$ representation of entropy change on extension of Eq. (4.11) provides an opportunity for a particularly insightful sense of entropy. Figure 4.14 plots in three dimensions the magnitude of the $\Delta\phi$, $\Delta\psi$ values for a chain segment of three torsion angles. As such, the volume is recognized by Eq. (4.11) as being proportional to the configurational entropy of the chain segment. The decrease in magnitude of the torsion angle oscillations on extension is seen as a decrease in volume, which is proportional to the decrease in the configurational entropy of the chain segment due to extension. Volume in configuration space provides a representation of chain entropy and a visualization of the decrease in entropy on extension. Obviously, for a single chain of N-residues fixed at its ends, entropy of the chain can be represented as a volume in 2 N-dimensional configurational space, and extension decreases the volume in N-dimensional configuration space, in a manner represented in Figure 4.14 for three dimensions. In particular, the values in Table 4.1 are actually the rms $\Delta\phi$ and $\Delta\psi$ for the relaxed state and the extended state. Especially for the suspended segment, the decrease in rms $\Delta\phi$ and $\Delta\psi$ values are substantial on extension, leading to the calculated decrease in entropy on extension.

---

number of states of the polymer that are available within a given energy cut-off as does the substitution of the D-residue. Decrease in the number of states, as discussed in the text, represents a decease in entropy of the structure. Reproduced with permission from 'Dynamic β-Spirals and a Librational Entropy Mechanism of Elasticity', by D. W. Urry et al. In: *Conformation in Biology*, by R. Srinivasan and R. H. Sarma. Guilderland, NY: Adenine Press, 1982, pp. 11–27. © 1982; 'A Librational Entropy Mechanism for Elastomers with Repeating Peptide Sequences in Helical Array', by D. W. Urray and C. M. Venkatachalam, 1983, *Int. J. Quant. Chem.: Quant. Biol. Symp.* **10**, 81–93. © 1983; 'Calculation of Dipole Moment Changes Due to Peptide Librations in the Dynamic β-Spiral of the Poly-pentapeptide of Elastin', by C. M. Venkatachalam and D. W. Urray, 1986, *Int. J. Quant. Chem.: Quant. Biol. Symp.* **12**, 15–24. © 1986.

Table 4.1. Listing of the φ and ψ Torsion Angles for the Central Six Pentamers of (VPGVG)$_{11}$ and the Root Mean Square (RMS) Fluctuations of Those Torsion Angles, (i.e., the $\Delta\phi_i$ and $\Delta\psi_j$) in Columns for the Relaxed and 130% Extended States Arising from 45 Picosecond Equilibration Time and 80 Picosecond of Molecular Dynamics Simulation for Each State

	Angle	Relaxed	Extended	Angle	Relaxed	Extended	Angle	Relaxed	Extended
	ψ_{16}	10.87	14.17	ψ_{26}	27.33	07.64	ψ_{36}	20.19	14.84
	ϕ_{17}	09.86	15.18	ϕ_{27}	11.71	08.33	ϕ_{37}	08.99	10.47
β-Turns	ψ_{17}	47.59	46.68	ψ_{27}	11.70	13.51	ψ_{37}	21.53	32.96
	ϕ_{18}	61.70	47.41	ϕ_{28}	08.61	10.36	ϕ_{38}	11.15	10.66
	ψ_{18}	09.37	16.05	ψ_{28}	09.33	08.16	ψ_{38}	11.09	27.29
	ϕ_{19}	14.25	08.67	ϕ_{29}	09.70	07.31	ϕ_{39}	12.70	10.24
Suspended	ψ_{19}	44.09	10.99	ψ_{29}	47.32	10.48	ψ_{39}	52.00	12.50
segments	ϕ_{20}	41.94	09.29	ϕ_{30}	48.57	11.39	ϕ_{40}	55.88	08.37
	ψ_{20}	14.50	11.15	ψ_{30}	42.56	10.62	ψ_{40}	40.67	11.08
	ϕ_{21}	27.13	24.17	ϕ_{31}	11.43	11.38	ϕ_{41}	36.44	19.06
	ψ_{21}	09.39	22.73	ψ_{31}	12.17	09.21	ψ_{41}	12.97	14.80
	ϕ_{22}	09.94	08.00	ϕ_{32}	09.90	08.93	ϕ_{42}	11.59	07.33
β-Turns	ψ_{22}	11.58	16.13	ψ_{32}	15.30	10.80	ψ_{42}	11.34	13.17
	ϕ_{23}	16.37	09.33	ϕ_{33}	09.60	07.62	ϕ_{43}	09.23	09.76
	ψ_{23}	14.33	14.25	ψ_{33}	09.88	09.43	ψ_{43}	10.60	12.53
	ϕ_{24}	11.39	29.20	ϕ_{34}	11.86	09.71	ϕ_{44}	11.06	12.82
Suspended	ψ_{24}	19.53	37.87	ψ_{34}	63.80	08.36	ψ_{44}	41.89	35.22
segments	ϕ_{25}	25.02	23.06	ϕ_{35}	91.70	10.20	ϕ_{45}	48.98	31.31
	ψ_{25}	49.32	32.10	ψ_{35}	15.03	11.51	ψ_{45}	42.05	56.89
	ϕ_{26}	31.43	27.24	ϕ_{36}	21.49	18.66	ϕ_{46}	21.55	30.33

Reproduced with permission. (Chang and Urry, 1989.)

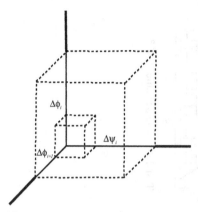

Figure 4.14: Representation of entropy as a volume in a $\Delta\phi_i$–$\Delta\psi_i$ configuration space. See text for further discussion.

Calculation of Contribution of Force Due to a Decrease in Freedom of Motion on Extension. From Eqs. (4.4) and (4.5), we have that

$$f_S = -T(\partial S/\partial L)_{V,T}. \tag{4.13}$$

At physiological temperatures, $310°K$, and for an extension from 0.35 to 0.8 nm for a pentamer within the β-spiral structure of Figure 4.10(E) with a calculated decrease in entropy of 5.5 cal/mol-deg/pentamer, the calculated value for the entropic contribution to the elastomeric force is 24 pN for a single chain of β-spiral.

To put this in terms of an elastic modulus in units of N/m^2, it is necessary to estimate the number of β-spirals per m^2. From Figure 4.1, the functional state is 37% peptide by weight. Taking the density of the peptide to be 1.3 gm/cm³, there would be some 10^{17} β-spirals/m^2. The result would be 24×10^5 N/m^2. The calculated result is an order of magnitude higher than the experimental value of 1.6×10^5 N/m^2 for 20 Mrad cross-linked (GVGVP)$_{251}$. Several factors would be expected to result in a lower experimental elastic modulus for the matrix. Firstly, in the cross-linked matrix, the β-spirals would be randomly oriented with respect to direction of extension. Secondly, the β-spirals occur as part of a multistranded twisted filament as shown in Figure 4.10(F), which would some-what restrict the freedom of motion of the torsional oscillations and result in a lower entropy change on extension. *What has been demonstrated, however, is that the damping of internal chain dynamics on extension provides an abundant source of entropic-restoring force in elastin-based systems.* Furthermore, given the entropic elasticity of single chains demonstrated by the AFM single-chain force-extension curves for many different polymers, damping of internal chain

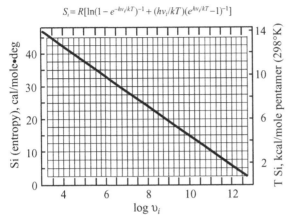

$$S_i = R[\ln(1 - e^{-hv_i/kT})^{-1} + (hv_i/kT)(e^{hv_i/kT} - 1)^{-1}]$$

Figure 4.15: Plot of the equation (see top of figure) for entropy as a function of log(oscillator frequency) using the partition function of a harmonic oscillator. On the left-hand ordinate is plotted the entropy in cal/mol-deg (EU, entropy units). On the right-hand ordinate is plotted the Gibbs free energy in kcal/mol at 298°K. This illustrates the increasing contribution of low-frequency oscillations to the entropy of a polypeptide chain.

dynamics on extension would seem to be a fundamental source for the development of entropic elastic force in chain molecules.

Dependence of Entropy and Structural Free Energy on Oscillator Frequency

One final point should be noted. It is the significance of the occurrence of intense low-frequency mechanical motions in elastin-based systems. The insight comes from the expression for the dependence of entropy on frequency of a harmonic oscillator as given in Figure 4.15. From the expression for the partition function of the harmonic oscillator (Dauber et al., 1981), it is seen that the slope for the frequency dependence of entropy is -4.6 EU/logv_i. The extrapolation of the harmonic oscillator treatment to low-frequency motions is surely not quantitative, but it is nonetheless informative. Although the contribution of a classical vibration is the order of 5 cal/mol-deg (EU), giving a Gibbs free energy of 1.5 kcal/mol at 298°K, the contribution of the 5 MHz relaxation, seen in Figures 4.7 and 4.8, would be some 30 cal/mol-deg or 9 kcal/mol, and that for the acoustically activated mechanical resonance near 1 kHz, reported in Figure 4.6(A), would approach 47 cal/mol-deg or 14 kcal/mol. From this it should be apparent that the mechanical relaxations of Figures 4.6, 4.7, and 4.8, arising from torsional oscillations, are fundamental in providing for the entropic elasticity of elastin-based systems. On a more general note, computations of protein structure that do not include these important low-frequency contributions to the free energy would seem to be limited in their treatment of the structure/function issue.

ACKNOWLEDGMENTS

This work was supported in part by the Office of Naval Research under Contracts N00014-00-C-0178 and N00014-00-C-0404.

REFERENCES

Buchet, R., Luan, C.-H., Prasad, K. U., Harris, R. D., and Urry, D. W. (1988) Dielectric relaxation studies on analogs of the polypentapeptide of elastin. *J. Phys. Chem.* **92,** 511–17.

Butt, H. J., and Jaschke, M. (1995) Calculation of thermal noise in atomic force microscopy. *Nanotechnology* **6,** 1–7.

Chang, D. K., and Urry, D. W. (1989) Polypentapeptide of elastin: Damping of internal chain dynamics on extension. *J. Comput. Chem.* **10,** 850–55.

Clausen-Schaumann, H., Rief, M., Tolksdorf, C., and Gaub, H. E. (2000) Mechanical stability of single DNA molecules. *Biophys. J.* **78,** 1997–2007.

Cleveland, J. P., Manne, S., Bocek, D., and Hansma, P. K. (1993) A nondestructive method for determining the spring constant of cantilevers for scanning force microscopy. *Rev. Sci. Instrum.* **64,** 403–5.

Cook, W. J., Einspahr, H. M., Trapane, T. L., Urry, D. W., and Bugg, C. E. (1980) Crystal structure and conformation of the cyclic trimer of a repeat pentapeptide of elastin, cyclo-(L-valyl-L-prolylglycyl-L-valylglycyl)$_3$. *J. Am. Chem. Soc.* **102,** 5502–5.

Cox, B. A., Starcher, B. C., and Urry, D. W. (1973) Coacervation of α-elastin results in fiber formation. *Biochim. Biophys. Acta* **317,** 209–13.

Cox, B. A., Starcher, B. C., and Urry, D. W. (1974) Coacervation of tropoelastin results in fiber formation. *J. Biol. Chem.* **249,** 997–98.

Dauber, P., Goodman, M., Hagler, A. T., Osguthorpe, D., Sharon, R., and Stern, P. (1981) In: *ACS Symposium Series No. 173, Supercomputers in Chemistry,* P. Lykos and I. Shavitt (eds.). Washington, DC: American Chemical Society, pp. 161–91.

Dorrington, K. L., and McCrum, N. G. (1977) Elastin as a rubber. *Biopolymers* **16,** 1201–22.

Eyring, H., Henderson, D., Stover, B. J., and Eyring, E. M. (1964) *Statistical Mechanics and Dynamics.* New York: John Wiley & Sons Inc., 92 pp.

Flory, P. J. (1953) *Principles of Polymer Chemistry.* Ithaca, NY: Cornell University Press.

Flory, P. J. (1968) Molecular interpretation of rubber elasticity. *Rubber Chem. Tech.* **41,** G41–G48.

Flory, P. J., Ciferri, A., and Hoeve, C. A. J. (1960) The thermodynamic analysis of thermoelastic measurements on high elastic materials. *J. Polymer Sci.* **XLV,** 235–36.

Gotte, L., Giro, G., Volpin, D., and Horne, R. W. (1974) The ultrastructural organization of elastin. *J. Ultrastruct. Res.* **46,** 23–33.

Grinberg, N. V., Dubovik, A. S., Grinberg, V. Y., Kuznetsov, D. V., Makhaeva, E. E., Grosberg, A. U., and Tanaka, T. (1999) Studies of the thermal volume transition of poly(N-isopropylacrylamide) hydrogels by high sensitivity differential scanning microcalorimetry. 1. Dynamic effects. *Macromolecules* **32,** 1471–75.

Hoeve, C. A. J., and Flory, P. J. (1958) The elastic properties of elastin. *J. Am. Chem. Soc.* **80,** 6523–26.

Hoeve, C. A. J., and Flory, P. J. (1962) Elasticity of crosslinked amorphous polymers in swelling equilibrium and diluents. *J. Polymer Sci.* **60,** 155–64.

Hoeve, C. A. J., and Flory, P. J. (1974) Elastic properties of elastin. *Biopolymers* **13,** 677–86.

Li, B., Alonso, D. O. V., and Daggett, V. (2000) The molecular basis for the inverse temperature transition of elastin. *J. Mol. Biol.* **305,** 581–92.

Luan, C.-H., Harris, R. D., and Urry, D. W. (1988) Dielectric relaxation studies on bovine ligamentum nuchae. *Biopolymers* **27,** 1787–93.

Luan, C.-H., Jaggard, J., Harris, R. D., and Urry, D. W. (1989) On the source of entropic elastomeric force in polypeptides and proteins: Backbone configurational vs. side chain solvational entropy. *Int. J. Quant. Chem.: Quant. Biol. Symp.* **16,** 235–44.

Manno, M., Emanuele, A., Martorana, V., San Biagio, P. L., Bulone, D., Palma-Vitorelli, M. B., McPherson, D. T., Xu, J., Parker, T. M., and Urry, D. W. (2001) Interaction of processes on different time scales in a bioelastomer capable of performing energy conversion. *Biopolymers* **59,** 51–64.

Oesterhelt, F., Rief, M., and Gaub, H. E. (1999) Single molecule force spectroscopy by AFM indicates helical structure of poly(ethylene-glycol) in water. *N. J. Phys.* **1,** 6.1–6.11.

Partridge, S. M., Davis, H. F., and Adair, G. S. (1955) The chemistry of connective tissues. I. Soluble proteins derived from partial hydrolysis of elastin. *Biochem. J.* **61,** 11–21.

Partridge, S. M., and Davis, H. F. (1955) The chemistry of connective tissues. II. Composition of soluble proteins derived from elastin. *Biochem. J.* **61,** 21–30.

Partridge, S. M. (1966) Biosynthesis and nature of elastin structures. *Fed. Proc.* **25,** 1023–29.

Partridge, S. M. (1967) Diffusion of solutes in elastin fibers. *Biochim. Biophys. Acta* **140,** 132–41.

Rief, M., Oesterhelt, F., Heymann, B., and Gaub, H. E. (1997) Single molecule force spectroscopy on polysaccharides by atomic force microscopy science. *Science* **275,** 1295–97.

San Biagio, P. L., Madonia, F., Trapane, T. L., and Urry, D. W. (1988) Overlap of elastomeric polypeptide coils in solution required for single-phase initiation of elastogenesis. *Chem. Phys. Letters* **145,** 571–74.

Sciortino, F. K., Prasad, K. U., Urry, D. W., and Palma, M. U. (1993) Self-assembly of bioelastomeric structures from solutions: Mean field critical behavior and Flory-Huggins free-energy of interaction. *Biopolymers* **33,** 743–52.

Sheiba, L. (14 May 1996) Broad band method and apparatus for the precise measurement of the complex modulus of viscoelastic materials. U.S. Patent 08/645,878, Pending.

Urry, D. W. (1976) On the molecular mechanisms of elastin coacervation and coacervate calcification. *Faraday Discussions Chem. Soc.* **61,** 205–12.

Urry, D. W. (1988a) Entropic elastic processes in protein mechanisms. I. Elastic structure due to an inverse temperature transition and elasticity due to internal chain dynamics. *J. Protein Chem.* **7,** 1–34.

Urry, D. W. (1988b) Entropic elastic processes in protein mechanisms. II. Simple (passive) and coupled (active) development of elastic forces. *J. Protein Chem.* **7,** 81–114.

Urry, D. W. (1993a) Molecular machines: How motion and other functions of living organisms can result from reversible chemical changes. *Angew. Chem.* (German) **105,** 859–83.

Urry, D. W. (1993b) Molecular machines: How motion and other functions of living organisms can result from reversible chemical changes. *Angew. Chem. Int. Ed. Engl.* **32,** 819–41.

Urry, D. W. (1997) Physical chemistry of biological free energy transduction as demonstrated by elastic protein-based polymers. *J. Phys. Chem. B* **101,** 11007–28.

Urry, D. W., Long, M. M., and Sugano, H. (1978) Cyclic analog of elastin polyhexapeptide exhibits an inverse temperature transition leading to crystallization. *J. Biol. Chem.* **253,** 6301–2.

Urry, D. W., Venkatachalam, C. M., Long, M. M., and Prasad, K. U. (1982) Dynamic β-spirals and a librational entropy mechanism of elasticity. In: *Conformation in Biology* (G. N. Ramachandran Festschrift Volume), R. Srinivasan and R. H. Sarma (eds.). Guilderland, NY: Adenine Press, pp. 11–27.

Urry, D. W., and Venkatachalam, C. M. (1983) A librational entropy mechanism for elastomers with repeating peptide sequences in helical array. *Int. J. Quant. Chem.: Quant. Biol. Symp.* **10**, 81–93.

Urry, D. W., Wood, S. A., Harris, R. D., and Prasad, K. U. (1984) Polypentapeptide of elastin as an elastomeric biomaterial. In: *Polymers as biomaterials*, S. W. Shalaby, T. Horbett, A. S. Hoffman, and B. Ratner (eds.). New York: Plenum Publishing Corporation, pp. 17–32.

Urry, D. W., Henze, R., Redington, P., Long M. M., and Prasad, K. U. (1985a) Temperature dependence of dielectric relaxations in α-elastin coacervate: Evidence for a peptide librational mode. *Biochem. Biophys. Res. Commun.* **128**, 1000–6.

Urry, D. W., Trapane, T. L., and Prasad, K. U. (1985b) Phase-structure transitions of the elastin polypentapeptide-water system within the framework of composition-temperature studies. *Biopolymers* **24**, 2345–56.

Urry, D. W., Venkatachalam, C. M., Wood, S. A., and Prasad, K. U. (1985c) Molecular structures and librational processes in sequential polypeptides: From ion channel mechanisms to bioelastomers. In: *Structure and motion: Membranes, Nucleic Acids and Proteins*, E. Clementi, G. Corongiu, M. H. Sarma, and R. H. Sarma (eds.). Guilderland, NY: Adenine Press, pp. 185–203.

Urry, D. W., Harris, R. D., and Prasad, K. U. (1988a) Chemical potential driven contraction and relaxation by ionic strength modulation of an inverse temperature transition. *J. Am. Chem. Soc.* **110**, 3303–5.

Urry, D. W., Haynes, B., Zhang, H., Harris, R. D., and Prasad, K. U. (1988b) Mechanochemical coupling in synthetic polypeptides by modulation of an inverse temperature transition. *Proc. Natl. Acad. Sci. U.S.A.* **85**, 3407–11.

Urry, D. W., Peng, S. Q., Xu, J., and McPherson, D. T. (1997) Characterization of waters of hydrophobic hydration by microwave dielectric relaxation. *J. Am. Chem. Soc.* **119**, 1161–62.

Urry, D. W., Hugel, T., Seitz, M., Gaub, H., Sheiba, L., Dea, J., Xu, J., and Parker, T. (2002) Elastin: A representative ideal protein elastomer. *Philos. Trans. R. Soc. Lond. B* **357**, 169–84.

Venkatachalam, C. M., and Urry, D. W. (1986) Calculation of dipole moment changes due to peptide librations in the dynamic β-spiral of the polypentapeptide of elastin. *Int. J. Quant. Chem.: Quant. Biol. Symp.* **12**, 15–24.

Volpin, D., Urry, D. W., Pasquali-Ronchetti, I., and Gotte, L. (1976) Studies by electron microscopy on the structure of coacervates of synthetic polypeptides of tropoelastin. *Micron.* **7**, 193–98.

Vrhovski, B., Jensen, S., and Weiss, A. S. (1997) Coacervation characteristics of recombinant human tropoelastin. *Eur. J. Biochem.* **250**, 92–98.

Weis-Fogh, T., and Andersen, S. O. (1970) New molecular model for the long-range elasticity of elastin. *Nature* **227**, 718–21.

Wu, W. J., Vrhovski, B., and Weiss, A. S. (1999) Glycosaminoglycans mediate the coacervation of human tropoelastin through dominant charge interactions involving lysine side chains. *J. Biol. Chem.* **274**, 21719–24.

FIVE

Fibrillin: From Microfibril Assembly to Biomechanical Function

Cay M. Kielty, Clair Baldock, Michael J. Sherratt, Matthew J. Rock, David Lee, and C. Adrian Shuttleworth

INTRODUCTION

Fibrillin-rich microfibrils are thin filamentous connective tissue assemblies present in virtually all dynamic connective tissues (Fig. 5.1) (Kielty and Shuttleworth, 1995; Handford et al., 2000; Sherratt et al., 2000). In elastic tissues – such as aorta, lung, skin, and elastic cartilage – preformed bundles of microfibrils form a template for tropoelastin deposition during elastic fibre formation and are retained as an outer mantle of mature elastic fibres (Mecham and Heuser, 1991). Microfibril arrays are often abundant in tissues which do not express elastin, such as the ciliary zonules of the eye which hold the lens in dynamic equilibrium (Ashworth et al., 2000).

Several recent studies, based on whole tissues and isolated fibrillin-rich microfibrils, have highlighted their unique elastic properties that are critical to their biological function alone or in association with elastin. The extensibility of lobster aorta was accounted for by microfibril arrays that intersperse medial smooth muscle cells (McConnell et al., 1996). Extracted sea cucumber microfibrils exhibited long-range elastomeric properties (Thurmond and Trotter, 1996). X-ray diffraction, tensile testing, and stress-relaxation tests demonstrated that hydrated mammalian ciliary zonules and microfibril bundles are reversibly extensible in the presence or absence of calcium (Wess et al., 1998a; Wright et al., 1999; Eriksen et al., 2001). Isolated human microfibrils tangled in debris during preparation for electron microscopy can become stretched to periodicities up to ~165 nm (Keene et al., 1991). Extended microfibrils, often observed in matrix metalloproteinase-treated preparations, may reflect pathological loss of elastic recoil (Ashworth et al., 1999a).

Early microscopy studies highlighted distinctive staining characteristics of microfibrils that indicated a strongly anionic nature (Cleary and Gibson, 1983). Rotary shadowing and negative staining, extensively used to examine isolated microfibrils, have since revealed a 'beads-on-a-string' arrangement, with an average periodicity of ~56 nm and an apparent diameter of ~10–14 nm (Fig. 5.2)

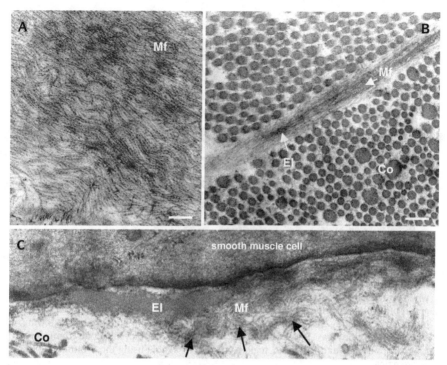

Figure 5.1: Vertebrate and invertebrate fibrillin-rich microfibrils. (A) Ultrathin fixed section of a lobster aorta showing fibrillin-rich microfibrils in longitudinal section, after staining with uranyl acetate and lead citrate. (B) Ultrathin fixed section of a murine skin biopsy stained with uranyl acetate and lead citrate. (C) Ultrathin section of murine aorta, showing an assembling elastic fibre adjacent to a smooth muscle cell. In (B) and (C), the microfibrillar mantle (Mf), elastin core (El), and transversely sectioned collagen fibrils (Co) are indicated. Scale bars = 400 nm.

(Wright and Mayne, 1988; Fleischmajer et al., 1991; Keene et al., 1991; Kielty et al., 1991; Ren et al., 1991; Wallace et al., 1991). Scanning transmission electron microscopy (STEM) analysis of microfibril mass and its axial distribution has revealed that beads correspond to mass peaks, whereas interbeads comprise a mass shoulder to one side of the bead, then a trough (Sherratt et al., 1997). The molecular composition of this class of microfibrils remained unresolved until ~15 years ago when antibodies raised to a human amnion extract were found to label the microfibrils and recognise a novel glycoprotein, designated fibrillin (Sakai et al., 1986). Molecular cloning studies rapidly followed, and it soon became clear that there were two highly homologous isoforms: fibrillin-1 encoded by a gene on chromosome 15 and fibrillin-2 encoded by a gene on chromosome 5 (Lee et al., 1991; Maslen et al., 1991; Pereira et al., 1993; Zhang et al., 1994). Linkage of mutations in the gene encoding fibrillin-1 to Marfan syndrome, and

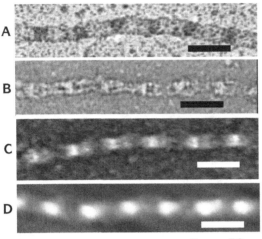

Bars =56 nm

Figure 5.2: Ultrastructural appearance of isolated fibrillin-rich microfibrils. Isolated untensioned microfibrils have a 'beads-on-a-string' appearance and ~56 nm periodicity
when visualised by a range of microscopy techniques. (A) Transmission electron micrograph of a rotary shadowed microfibril. (B) Transmission electron micrograph following negative staining. (C) STEM micrograph in the absence of staining or shadowing,
showing microfibril mass. (D) Surface topography revealed by tapping mode atomic force
microscopy. Beads and interbead regions are clearly visible using these ultrastructural
approaches.

in the gene encoding fibrillin-2 to the related disorder congenital contractural
arachnodactyly, confirmed both the key role of fibrillins in microfibril formation and the importance of microfibrils to connective tissue integrity (Robinson
and Godfrey, 2000). Numerous biochemical and antibody mapping studies have
since confirmed that fibrillin molecules do, indeed, assemble to form the
microfibril scaffold.

Both fibrillin isoforms are large glycoproteins (~350 kDa) with multidomain
structures dominated by 43 calcium-binding consensus sequences (calcium-
binding epidermal growth factor-like (cbEGF) domains) (Fig. 5.3a) (Pereira et al.,
1993; Zhang et al., 1994; Downing et al., 1996). These domains are interspersed
with 8-cysteine domains, also known as TB modules because they are homologous to transforming growth factor-β binding 8-cysteine containing motifs
found in latent transforming growth factor-β binding proteins (Yuan et al., 1997;
Sinha et al., 1998). Fibrillin-1 has a 58 amino acid proline-rich region towards the
amino-terminus which may act as a 'hinge-like' region which, in fibrillin-2, is
replaced by a glycine-rich sequence. As predicted by the many cbEGF domains,
fibrillin molecules bind calcium, which results in an extended rod-like conformation (Reinhardt et al., 1997). Ultrastructural and X-ray diffraction studies

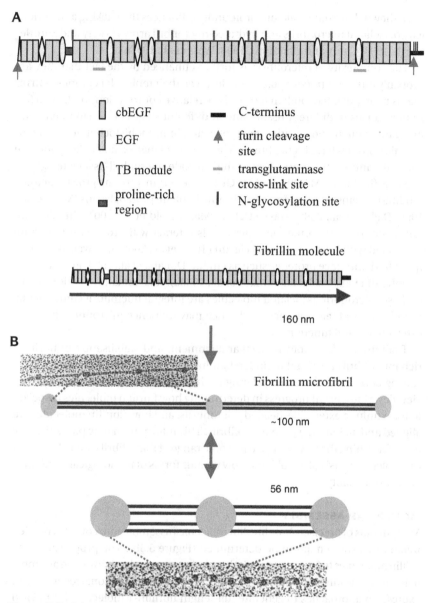

Figure 5.3: Schematic representation of the fibrillin-1 molecule and its assembly into extensible microfibrils. (A) Domain structure of fibrillin, highlighting how the contiguous arrays of calcium-binding epidermal growth factor (cbEGF)-like domains are interspersed with TB modules, and the locations of the proline-rich region, N-glycosylation consensus sequences, and putative transglutaminase cross-link sites. EGF, epidermal growth factor. (B) Diagram highlighting how fibrillin-1 molecules (∼160 nm) assemble to form beaded microfibrils with variable periodicity (∼56–100 nm). (See color section.)

have shown that bound calcium profoundly influences the packing and periodicity of isolated microfibrils and hydrated microfibril arrays (Kielty and Shuttleworth, 1993; Cardy and Handford, 1998; Wess et al., 1998a).

The complexity of microfibrils is further enhanced by the fact that they are probably multicomponent polymers. However, their molecular composition remains poorly defined, both in terms of the relative contributions of the two fibrillin isoforms in different tissues and at different stages of development, and in regard to interactions with other molecules. Immunohistochemical and biochemical studies have highlighted that a number of matrix molecules colocalise with microfibrils in various tissues; they include microfibril-associated glycoproteins (including MAGP 1 and MAGP 2), latent transforming growth factor-β binding proteins (LTBPs), and chondroitin sulphate proteoglycans (Kielty et al., 1996; Dallas et al., 2000; Trask et al., 2000a; Unsold et al., 2001). In addition, during elastic fibre formation, microfibrils interact with tropoelastin and are associated with lysyl oxidase and elastin (Trask et al., 2000b), as well as elastin–microfibril interface proteins, such as emilin (Doliana et al., 1999) and fibulin-2 (Reinhardt et al., 1996a; Raghunath et al., 1999a). It is not known which, if any, of these microfibril-associated molecules are integral microfibril components, or indeed how their presence or absence may influence microfibril structure, extensibility, and function.

Defining the assembly, molecular alignment, and stabilisation of fibrillin-rich microfibrils is central to understanding their structural properties and resolving current controversies relating to their structure and function. This review describes recent progress in determining how fibrillin molecules associate to form unique extensible-beaded microfibrils, and how fibrillin molecules are aligned and stabilised within microfibrils. This information now paves the way for determining the physiological periodic range of microfibril extensibility and the molecular basis of 'recoil', and how various forms of pathological insult may affect extensibility.

MOLECULAR ASSEMBLY

Many details of precisely how fibrillin molecules assemble into complex beaded microfibrils still remain to be determined (Figure 5.3b). The propensity of fibrillin molecules to form disulphide-bonded aggregates has hampered attempts to identify true intermediates of assembly. Moreover, conventional recombinant expression approaches to generate full-length fibrillin-1 molecules for in vitro assembly studies have, to date, proved ineffective. However, new recombinant approaches have at last begun to provide insights into the key fibrillin interactions that drive assembly, the defined intermediates of assembly, and the role of processing in polymerisation.

The N-terminal region of fibrillin-1 has been shown by several groups readily to aggregate in vitro, forming extensive disulphide-bonded aggregates (Ashworth et al., 1999b; Reinhardt et al., 2000). Specific N-terminal interactions are predicted to drive this aggregation which can then become stabilised with covalent disulphide linkages; an unpaired cysteine residue in the first hybrid TB module has been implicated in intermolecular cross-linking (Reinhardt et al., 2000). Distinct, but overlapping, fibrillin N-terminal peptides have been shown by two groups using different mammalian recombinant systems to form dimers (Ashworth et al., 1999b; Trask et al., 1999); however, it remains uncertain whether dimerisation occurs intracellularly or upon secretion. We have now obtained evidence that recombinant full-length human fibrillin-1 can form dimers within the secretory pathway in a time-dependent manner (Figure 5.4). These studies

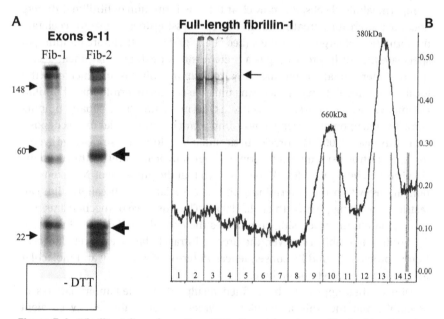

Figure 5.4: Fibrillin-1 dimer formation. (A) Proline-rich region and its flanking domains (exons 9–11) form dimers and higher assemblies. (B) Full-length fibrillin-1 molecule, purified by heparin affinity chromatography (see inset, sodium dodecyl sulphate-polyacrylamide gel electrophoresis in reducing conditions), resolves into monomers and dimers by size fractionation in native non-reducing conditions. The X-axis represents column fractions (volume), and the Y-axis is OD_{280}. These recombinant molecules were expressed in an in vitro transcription/translation system supplemented with semipermeabilised cells as source of secretory organelles (Ashworth et al., 1999b; Lee et al., 2001). Fib, fibrillin; DTT, dithiothreitol. Panel A is reproduced with permission from The Company of Biologists, UK.

used an effective well-characterised in vitro transcription/translation system supplemented with semipermeabilised cells, in which purified fibrillin-1 mRNA is translated as a [^{35}S]-cysteine-labelled protein which can readily and rapidly be monitored in sodium dodecyl sulphate-polyacrylamide gel electrophoresis and size fractionation studies. This system, which provided a powerful means of examining the fate of newly translocated fibrillin-1 molecules within the endoplasmic reticulum, has shown, for the first time, that full-length fibrillin-1 spontaneously formed disulphide-bonded dimers, then larger non-reducible intermediates in a time-dependent manner.

Cleavage sites for furin/paired acidic convertase enzymes (PACE) are present close to the N- and C-termini of each isoform, and C-terminal processing is known to be a prerequisite for extracellular deposition (Raghunath et al., 1999b; Ritty et al., 1999). Using the in vitro transcription/translation system with semipermeabilised cells, it is now clear that the formation of fibrillin-1 dimers does not require furin treatment, although processing does appear to accelerate the formation of larger aggregates (Lee et al., 2001). It is therefore tempting to speculate that furin processing is a prerequisite for extracellular linear accretion of dimers, or larger intermediates, to form microfibrils, in which case the location and regulation of processing must be crucial determinants of fibrillin assembly. A recent report suggests that a C-terminal fibrillin-1 heparin binding site activated upon processing may be important in extracellular matrix deposition (Ritty et al., 2000). The precise pericellular location of these cleavage events remains controversial, although furins can be catalytically active within the trans Golgi and secretory vesicles. We have shown that the presence of N-linked carbohydrates within the C-terminus, and its association with calreticulin, can restrict processing (Ashworth et al., 1999c). The existence of mechanisms that can regulate furin processing emphasises the crucial significance of this step for fibrillin assembly. However, the precise intracellular or extracellular location of processing, and its consequences in terms of assembly and molecular interactions, remain to be defined.

Dimers thus appear to be defined disulphide-bonded intermediates of assembly, and the fully assembled polymer is also stabilised by covalent cross-links, both disulphide bonds and transglutaminase-derived peptide linkages. Indeed, fibrillin-1 homotypic transglutaminase cross-links have been identified in peptides starting at residues 580 and 2312 (Qian and Glanville, 1997). Tissue transglutaminase activity is likely to be an extracellular event that could stabilise newly assembled linear arrays of fibrillin. In our in vitro assembly studies, the presence of tissue transglutaminase activity rapidly generates fibrillin aggregates. Whether this reflects in vivo cross-linking events is unclear, but it is certain that fibrillin is an extremely attractive substrate for this enzyme. Thus, the physiological role of transglutaminase cross-linking in the assembly process

and/or in subsequent stabilisation of microfibrils in different tissues remains to
be defined.

FIBRILLIN MOLECULE ALIGNMENT IN MICROFIBRILS

The intriguing regular beaded periodicity of microfibrils and detailed information on the molecular dimensions of fibrillin molecules has led to several
proposed models of fibrillin alignment in microfibrils. A model based on antibody epitope mapping and measured molecular dimensions suggested a parallel head-to-tail alignment of unstaggered fibrillin monomers with amino- and
carboxy-termini at, or close to, the beads (Reinhardt et al., 1996b). STEM showed
that isolated untensioned microfibrils have an asymmetric repeating bead–
interbead organisation, implying that microfibrils are directional and that fibrillin molecules are parallel (Sherratt et al., 1997). Staggered arrangements
based on extrapolation of molecular length from cbEGF-like domain dimensions and untensioned microfibril periodicity (approximately one-half stagger)
(Downing et al., 1996), or on alignment of the fibrillin-1 homotypic transglutaminase cross-links in approximately one-third stagger, were also proposed (Qian
and Glanville, 1997). None of these models provides a molecular explanation for
how assembled fibrillin forms extensible beaded microfibrils with variable periodicity, nor are they compatible with the axial mass distribution of untensioned
microfibrils revealed by our STEM studies.

NEW MODEL OF FIBRILLIN ALIGNMENT IN
EXTENSIBLE MICROFIBRILS

In an attempt to clarify outstanding anomalies in microfibril structure, we recently used automated electron tomography (AET) to develop the first three-
dimensional reconstructions of microfibrils (Baldock et al., 2001) (Figure 5.5.).
We also localised fibrillin antibody and gold-binding epitopes in direction-
ally oriented untensioned microfibrils, and mapped bead and interbead mass
changes on extension. Together, these data provide compelling evidence that,
whereas in extended configurations fibrillin molecules are strung out in head-
to-tail arrays, in the untensioned state the molecules clearly adopt a complex
folded arrangement. Such folding, perhaps driven by favourable hydrophobic
and electrostatic interactions, could form the molecular engine of microfibril
extensibility.

The data, summarised below, formed the basis of a new model of fibrillin
alignment which is consistent with observed microfibril ultrastructure, indicates that fibrillin polymers undergo conformational maturation to a reversibly
extensible beaded polymer, and provides a molecular explanation of microfibril extensibility. Our model (described following the structural evidence) predicts maturation from a parallel head-to-tail alignment to an approximately

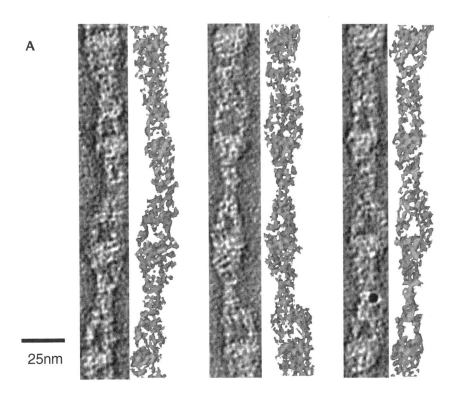

25nm

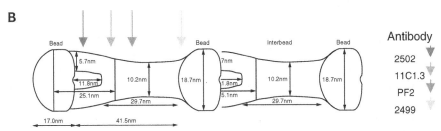

Figure 5.5: Automated electron tomography images of negatively stained untensioned bovine zonular microfibrils. (A) Three Z-slices shown in black and white, compared with the same region of microfibril three-dimensional rendered (green) using AVS Express (Advanced Visual Systems, Inc.). (B) Schematic diagram of two microfibril repeating units with the mean measurements taken from a representative data set. The binding positions of four anti–fibrillin-1 antibodies are shown with coloured arrows. Reproduced with permission from 'The Supramolecular Organization of Fibrillin-Rich Microfibrils', by C. Baldock et al., 2001, *J. Cell Biol.* **152,** 1045–56. © 2001 The Rockefeller University Press. (See color section.)

one-third stagger (Figure 5.6), accounts for observed structural features, suggests that intramolecular interactions drive conformation changes to form extensible microfibrils, and predicts that there are eight molecules in cross-section.

Untensioned Microfibrils

Three-dimensional reconstructions of microfibrils to 18.6Å resolution revealed new structural features and dimensions (Figure 5.5). Beads appeared as dense masses that were more 'heart-shaped' than spherical, with undulating surfaces and resolution at domain level. Bead diameter varied axially between 14.8 and 18.7 nm. Some bead morphological variability occurred in all data sets. In most repeats, two prominent arms emerged from the broader bead face, meeting at a fixed position of ~43% of bead-to-bead distance (14.7 nm from bead edge, ~29% of the interbead). They appeared as stain-excluding regions that bow out between the bead, between which is stain-accessible space. In some repeats, the arms were less clearly defined, appearing as a number of fine filaments. The point where the arms terminated within the interbead probably corresponded to an interbead 'striation' detected by rotary shadowing. The remaining interbead had a much less open structure, was generally narrower than the bead and the arms, extended for the remaining 71% of the interbead region, then merged into the following bead. The repeating units often looked symmetrical about the longitudinal axis of the microfibril, although some sample variability and noise were apparent. Twisting along isolated microfibrils within the interbead may occur [Figure 5.5(b,c)]. In many repeats, the two interbead arms appeared to cross-over between consecutive beads, which emphasised stain pooling between them and gave a 'bow-tie' appearance. Twisting supports the concept that the isolated microfibrils are in an untensioned, relaxed conformation.

Antibody binding studies on isolated microfibrils have been successful in locating fibrillin epitopes as clues to the alignment of fibrillin molecules within a microfibril. A study of the binding sites of three monoclonal antifibrillin-1 antibodies on partially extended microfibrils showed that N- and C-terminal epitopes were or either side of the bead and suggested that fibrillin molecules were assembled in a head-to-tail arrangement (Reinhardt et al., 1996b). More recently, we mapped the epitopes of four fibrillin-1 antibodies accessible on native untensioned microfibrils (Figure 5.5) (Baldock et al., 2001). The binding site for monoclonal antibody 2052, which recognises an N-terminal fibrillin-1 sequence within residues 45–450, was close to the bead on the side opposite the arms (15.9% of bead-to-bead distance). Monoclonal antibody 11C1.3 (epitope within fibrillin-1 residues 654–909) bound to microfibrils at the interbead striation where the arms terminate. Incubation with this antibody generated numerous extensive parallel microfibril arrays with beads in register and the antibody binding at 41.1% of bead-to-bead distance. PF2 polyclonal antibody,

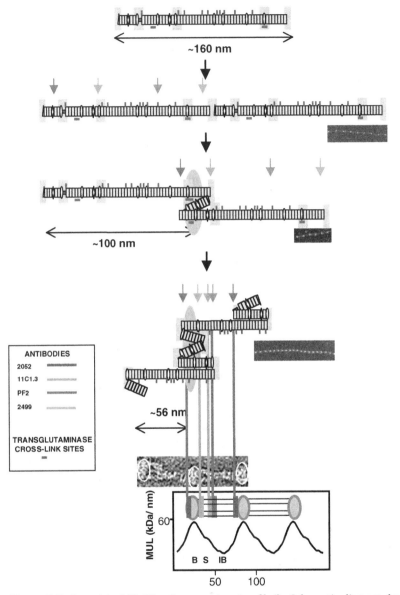

Figure 5.6: A model of fibrillin alignment in microfibrils. Schematic diagram depicting a possible folding arrangement of fibrillin molecules in a beaded microfibril. Two N- and C-terminally processed molecules associate head-to-tail to give ~160 nm periodicity. Subsequent molecular folding events could generate ~100 nm periodicity, then ~56 nm periodicity. The model is supported by automated electron tomography, STEM data, antibody localisations, and microfibril extension studies (Baldock et al., 2001). While the predicted fold sites are shown (orange boxes), it should be noted that the precise packing

which recognises pepsin fragment PF2 (Maddox et al., 1989) (sequences encoded by exons 41–45), bound close to the centre of the interbead (47.1% or 52.9% of bead-to-bead distance). The binding site for monoclonal antibody 2499, which recognises a C-terminal fibrillin-1 sequence within residues 2093–2732 (assuming furin cleavage), was on the arms close to the bead (20.2% of bead-to-bead distance).

Extended Microfibrils

Several experimental approaches have been investigated to extend microfibrils in vitro. When microfibril preparations were centrifuged at 60,000g for 1 hr, a small proportion of microfibrils appeared extended in the range of 70–110 nm, but most retained untensioned periodicity. When microfibril preparations were repeatedly drawn through narrow bore needles, the majority of microfibrils retained untensioned (~56 nm) periodicity, although a few were extended to ~70 nm. Surface tensional forces did provide a means of trapping many microfibrils in the extended state (also 70–110 nm) directly from sample drop surfaces onto carbon-coated grids; interbead morphology of these microfibrils often appeared diffuse, suggesting major conformational changes. In addition, when antibody-treated (11C1.3) human microfibrils were extended in this way, antibody-banded microfibril arrays were detected that were partially extended to ~70 nm, with the antibody position remaining at the end of the interbead arms (23 nm from bead centre). This corresponds to 41.1% of bead-to-bead distance for untensioned 56 nm microfibrils, but 32% of bead-to-bead distance for 70 nm microfibrils. Thus, microfibrils can extend at least 10 nm before this epitope has to move. At higher extensions, all antibody banding was lost.

Colloidal gold particles bind proteins through charge, hydrophobic interaction or dative binding to sulphur-containing groups (Hermanson, 1996). They associate periodically, and often pairwise, with untensioned microfibrils at the ends of the interbead arms. But, on extended microfibrils, the gold bound periodically only at the beads, confirming major molecular rearrangements on extension.

Unstretched (~56 nm) and extended microfibrils isolated from canine zonules were examined by STEM to investigate how mass distribution changes on extension (Baldock et al., 2001). Within any single microfibril, many repeats

---

arrangement of folded segments contributing to the bead remains unresolved. Inset: STEM images of microfibrils with periodicities corresponding to those predicted. Observed antibody binding sites are overlaid on the ~56 nm periodic folding arrangement. The axial mass distribution of ~56 nm microfibrils (shown at the bottom) correlates well with the corresponding predicted molecular folding. B, bead; IB, interbead; MUL, mass per unit length; S, shoulder region (See color section.)

were ~56 nm, but at stretched regions, there was always a sharp periodicity transition from <95 nm to >110 nm (Figure 5.7). Total mass per repeat remained unchanged, irrespective of microfibril periodicity, but there was a progressive reduction in interbead mass per unit length and loss of the shoulder at periodicities from ~56 to 100 nm, then a rapid reduction in bead mass at higher periodicities. This is evidence that interbead unfolding accounts for extension between ~56 and 100 nm, and that bead unravelling occurs at periodicities >100 nm. Interestingly, such highly extended arrays appear to be irreversibly stretched since they are stable in solution.

Molecular Alignment and Reversible Extensibility

1. We have proposed that the first step in microfibril assembly is N- and C-terminal association at the cell surface, perhaps regulated by terminal processing [Figure 5.6(a)]. This would generate linear head-to-tail polymers with a repeating periodicity corresponding to a molecular length of ~160 nm.

2. Molecular association between internal sequences (within exons 12–15 and 50–64, after processing) driven by folding at the termini and proline-rich region, would reconfigure consecutive molecules into an approximately one-third stagger with N- and C-terminal sequences 'looped out' [Figure 5.6(b,c)]. This association would align fibrillin-1 transglutaminase cross-link sequences (exons 14–16 and 56–58) and allow cross-link stabilisation of this folded arrangement. The repeating periodic distance of these polymers would be ~100 nm (32 cbEGFs, 1 hybrid motif, and 5 TB modules), and the 'looped out' termini could contribute to the beads. This arrangement may provide the structural context for reversible extension between ~100 to a more energetically stable 56 nm.

3. Microfibril 'recoil' could involve intramolecular folding at potentially flexible TB-cbEGF linkages (Yuan et al., 1997, 1998) (calcium-bound cbEGF arrays form rod-like arrays) [Figure 5.6(b,c)]. Since five TB-cbEGF junctions exist between the cross-link sites, a number of potential folding arrangements could be adopted in untensioned microfibrils. However, two lines of evidence suggest that major hingeing may occur at the TB3-cbEGF junction such that the central 12 cbEGF array folds back, juxtaposing the centre of this array with the core bead, thereby enhancing its mass and reducing periodicity to ~56 nm (18 cbEGFs, 3 TB modules). TB3, which precedes the central 12-cbEGF array, has the longest linker region (19 residues) and may be particularly flexible. More specifically, only this folding arrangement fits the observed antibody binding pattern and formation of interbead arms of observed dimensions (14.7 nm or ~6 domains). N-glycosylation sites would be accommodated within the

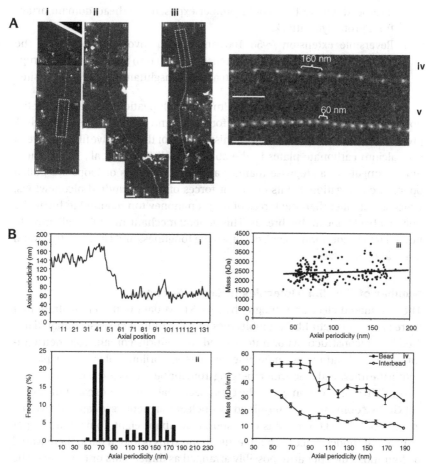

Figure 5.7: STEM mass mapping of untensioned and stretched microfibrils. (A) Dark-field STEM composite of a highly stretched fibrillin-rich microfibril isolated from canine zonules (i, ii, and iii). High magnification (iv, v) of regions indicated by boxes in (I) and (iii). Microfibril periodicity varied between 160 nm (iv) and 56 nm (v), with a short, sharp transition zone. Scale bars = 200 nm. (B) Analysis of microfibril mass and its distribution in untensioned and extended microfibril regions. (i and ii) Bead-to-bead periodicity showing a sharp transition from untensioned to extended state. (iii) Total mass per repeat (kDa) was invariant with changes in periodicity. (iv) Central bead and central interbead mass measurements. Interbead mass falls as periodicity increases from ~56 nm to ~100 nm, but bead mass remains constant until ~90–100 nm, then falls rapidly. Periodicities between ~56–100 nm may be reversible, whereas those more than ~100 nm may be irreversible. Reproduced with permission from 'The Supramolecular Organization of Fibrillin-Rich Microfibrils', by C. Baldock et al., 2001, *J. Cell Biol.* **152**, 1045–56. © 2001 The Rockefeller University Press.

interbead, where they could protect exposed interbead domain arrays from proteolytic attack.

4. Reversible extension (∼56–100 nm) would involve unfolding of the TB3-cbEGF fold. Irreversible extension (>100 nm) would involve disruption of the molecular association at the transglutaminase cross-link site.

In summary, this model predicts a 'modular' elongation mechanism for the fibrillin-rich microfibrils, based on the formation and opening of defined folds. A precedent for a broadly similar mechanism exists for the adhesive fibres between the calcium carbonate plates in the abalone shell (Smith et al., 1999). These fibres elongate in a stepwise manner as folded domains or loops are pulled open. The elongation events occur for forces of a few hundred piconewtons, which are smaller than the forces of over a nanonewton required to break the polymer backbone in the threads. This general mechanism could well prove to be a relatively common means of conveying toughness and elasticity to natural fibres and adhesives.

Number of Fibrillin Molecules in Cross-Section

The automated electron tomography and STEM data both demonstrate that there are about eight fibrillin molecules in a microfibril cross-sectional diameter. The reconstructions of untensioned microfibrils contain volumetric information indicating how many linearly aligned fibrillin molecules are packed within the interbead. The arms emerging from the bead each measure ∼6 × 5 nm (30 nm^2) in cross-section. Since the cross-sectional diameter of calcium-bound cbEGF-like domains, as determined by nuclear magnetic resonance (Downing et al., 1996), is ∼3.6 nm^2, this corresponds to about eight domain arrays per arm. In our model, since each molecule in the arms is folded back, there would be four molecules per arm, possibly arranged as dimer pairs or tetramers. The cross-sectional diameter of the narrow interbead region measures a minimum of ∼10 × 6 nm (60 nm^2), so at least about 16 cbEGF domain arrays may be aligned within this region, which is consistent with predicted domain array folding. STEM has also established here that the mass per unit length (MUL) of the extended interbead is 14.25 kDa/nm. Determination of the MUL of a fibrillin molecule was based on exons 23–36 (13 cbEGF domains and a TB module). The predicted mass of this peptide is 67.18 kDa. Its length is 37.4 nm, based on domain dimensions (37.8 nm) determined by nuclear magnetic resonance (Downing et al., 1996; Yuan et al., 1997, 1998) and published measurements (35.4 nm and 38.3 nm; average 36.9 nm; Reinhardt et al., 1996b, 2000). Thus, the MUL of each molecule is ∼1.80 kDa/nm, which indicates 7.92 molecules in an extended interbead cross-section. Moreover, the actual mass per repeat is ∼2490–2510 kDa compared with a predicted 2510 kDa for eight aligned molecules.

Microfibril Bundle Extensibility and Organisation

Following this in-depth consideration of the packing and extensibility of individual microfibrils, it is important to bear in mind that the in vivo arrangement of microfibrils in bundles and in elastic fibres, will profoundly influence the ability of individual microfibrils within such higher order arrays to extend and retract. X-ray fibre diffraction has proved to be a useful experimental tool for investigating microfibril arrays (Wess et al., 1997, 1998a, 1998b). The X-ray diffraction of microfibrils in the form of hydrated bovine zonular filaments exhibited meridional diffraction peaks indexing on a fundamental periodicity of ∼56 nm in the relaxed state (Wess et al., 1997). These arrays are capable of reversible extension within a defined extension range (up to ∼50%). However, when zonules were

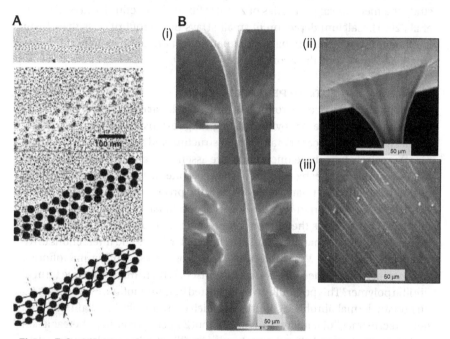

Figure 5.8: Ciliary zonular microfibrillar arrays. (A) An electron micrograph of a native microfibrillar array isolated from nuchal ligament in the presence of Ca^{2+}. Together with published X-ray diffraction studies (Wess et al., 1998a), these observations suggest that 'supramicrofibrillar junctions' may occur in zonules and in microfibril bundles in other tissues. Such juxtaposed regions may influence transmission of force through bundles. Reproduced with permission from 'The Supramolecular Organization of Fibrillin-Rich Microfibrils', by C. Baldock et al., 2001, *J. Cell Biol.* **152,** 1045–56. © 2001 The Rockefeller University Press. (B) Environmental scanned electron microscopy of the anterior (i) and posterior (ii, iii) aspects of human zonules running from the ciliary process to the lens capsule in a 68-year-old male. The length of the individual zonular fibre in (I) is approximately 600 μm. Insertion of the zonule into the lens capsule is shown in (ii).

stretched beyond this range, there was extensive deterioration in the quality of diffraction and the microfibril bundles were no longer reversibly extensible (Wess et al., 1998b). Microfibril extensibility was shown to be independent of the presence or absence of calcium, although the starting periodicities of the microfibrils were different (~56 nm vs. ~39 nm), in line with similar observations made by STEM and rotary shadowing on isolated microfibrils (Wess et al., 1998a). Interestingly, the periodicity of a small proportion of microfibrils always remained unchanged on bundle extension, leading to the prediction that one-third (0.33-D) staggered microfibril 'junction' arrays exist within discrete regions of zonular bundles that may be important in modulating force transmission within the microfibril bundle (Figure 5.8). More recently, a microneedle technique with a resolution power of nN and μm has been used to study the mechanical properties of zonular fibrillin microfibril arrays (Eriksen et al., 2001). Calcium depletion of small arrays of microfibrils less than 1 μm in diameter were shown to have a dramatic effect on reducing microfibril rest length and increasing their compliance.

SUMMARY AND FUTURE PERSPECTIVES

The unique fibrillin-rich microfibrils of the extracellular matrix have been shown to play an essential role in provision of long-range extensibility to many dynamic connective tissues. Recent details of their structure and assembly have begun to reveal precisely how these microfibrils are assembled and the molecular basis of their unique biomechanical properties. Despite this recent progress, many questions still remain unanswered. What is the precise molecular composition of microfibrils in different tissues? How is intracellular assembly regulated and limited? What controls the number of molecules in cross-section? Are fibrillin molecules flexible? What intra- and intermolecular forces and interactions drive extension and recoil? What intermolecular cross-links stabilise microfibrils? Another interesting question has recently arisen: Is there more than one form of fibrillin polymer? This possibility was suggested by immunofluorescence studies on human dermal fibroblast cultures, in which one anti–fibrillin-1 antibody did not detect extracellular microfibrils until about 2 weeks in culture, whereas a second antibody detected abundant extensive fibrillin-1 microfibrils within 3 days (Baldock et al., 2001) [Figure 5.6(g)]. One explanation for this observation could be that conformational changes occur during maturation of beaded microfibrils that expose or unmask cryptic epitopes; such structural changes could correspond to conversion from head-to-tail fibrillin arrays to mature approximately one-third staggered beaded microfibrils. Future studies will undoubtedly focus on defining the molecular motor that drives microfibril extension and recoil, and on establishing how these unique extensible properties can be modified by interactions with other matrix molecules and by incorporation into microfibril bundles.

ACKNOWLEDGMENTS

C.M.K. (with C.B. and M.J.R.) is funded by the Medical Research Council.

REFERENCES

Ashworth, J. L., Murphy, G., Rock, M. J., Sherratt, M. J., Shapiro, S. D., Shuttleworth, C. A., and Kielty, C. M. (1999a) Fibrillin degradation by matrix metalloproteinases: Implications for connective tissue remodelling. *Biochem. J.* **340,** 171–81.

Ashworth, J. L., Kelly, V., Wilson, R., Shuttleworth, C. A., and Kielty, C. M. (1999b) Fibrillin assembly: Dimer formation mediated by amino-terminal sequences. *J. Cell Sci.* **112,** 3549–58.

Ashworth, J. L., Kelly, V., Rock, M. J., Shuttleworth, C. A., and Kielty, C. M. (1999c) Regulation of fibrillin carboxy-terminal furin processing by N-glycosylation, and association of amino- and carboxy-terminal sequences. *J. Cell Sci.* **112,** 4163–71.

Ashworth, J. L., Kielty, C. M., and McLeod, D. (2000) Fibrillin and the eye. *Br. J. Ophthalmol.* **84,** 1312–17.

Baldock, C., Koster, A. J., Ziese, U., Sherratt, M. J., Kadler, K. E., Shuttleworth, C. A., and Kielty, C. M. (2001) The supramolecular organisation of fibrillin-rich microfibrils. *J. Cell Biol.* **152,** 1045–56.

Cardy, C. M., and Handford, P. A. (1998) Metal ion dependency of microfibrils supports a rod-like conformation for fibrillin-1 calcium-binding epidermal growth factor-like domains. *J. Mol. Biol.* **276,** 855–60.

Cleary, E. G., and Gibson, M. A. (1983) Elastin-associated microfibrils and microfibrillar proteins. In: *International Review of Connective Tissue Research,* vol. 10, D. A. Hall and D. S. Jackson (eds.). London, UK: Academic Press, pp. 97–209.

Dallas, S. L., Keene, D. R., Bruder, S. P., Saharinen, J., Sakai, L. Y., Mundy, G. R., and Bonewald, L. F. (2000) Role of the latent transforming growth factor beta binding protein 1 in fibrillin-containing microfibrils in bone cells in vitro and in vivo. *J. Bone Miner. Res.* **15,** 68–81.

Doliana, R., Mongiat, M., Bucciotti F., Giacomello, E., Deutzmann, R., Volpin, D., Bressan, G. M., and Colombatti, A. (1999) EMILIN, a component of the elastic fiber and a new member of the C1q/tumor necrosis factor superfamily of proteins. *J. Biol. Chem.* **374,** 16773–81.

Downing, A. K., Knott, V., Werner, J. M., Cardy, C. M., Campbell, I. D., and Handford, P. A. (1996) Solution structure of a pair of calcium binding epidermal growth factor-like domains: Implications for the Marfan syndrome and other genetic disorders. *Cell* **85,** 597–605.

Eriksen, T. A., Wright, D. M., Purslow, P. P., and Duance, V. C. (2001). Role of Ca^{2+} for the mechanical properties of fibrillin. *Proteins: Struct. Funct. Genet.* **45,** 90–95.

Fleischmajer, R., Perlish, J. S., and Faraggiana, T. (1991) Rotary shadowing of collagen monomers, oligomers and fibrils during tendon fibrillogenesis. *J. Histochem. Cytochem.* **39,** 51–58.

Handford, P. A., Downing, A. K., Reinhardt, D. P., and Sakai, L. Y. (2000) Fibrillin: From domain structure to supramolecular assembly. *Matrix Biol.* **19,** 457–70.

Hermanson, G. T. (1996) Preparation of colloidal-gold-labelled proteins. In: *Bioconjugate Techniques.* San Diego, CA: Academic Press, pp. 593–604.

Keene, D. R., Maddox, B. K., Kuo, H. J., Sakai, L. Y., and Glanville, R. W. (1991) Extraction of beaded structures and their identification as fibrillin-containing matrix microfibrils. *J. Histochem. Cytochem.* **39,** 441–49.

Kielty, C. M., Cummings, C., Whittaker, S. P., Shuttleworth, C. A., and Grant, M. E. (1991) Isolation and ultrastructural analysis of microfibrillar structures from foetal bovine elastic tissues. Relative abundance and supramolecular architecture of type VI collagen assemblies and fibrillin. *J. Cell Sci.* **99,** 797–807.

Kielty, C. M., and Shuttleworth, C. A. (1993) The role of calcium in the organisation of fibrillin microfibrils. *FEBS Lett.* **336,** 323–26.

Kielty, C. M., and Shuttleworth, C. A. (1995) Fibrillin-containing microfibrils – Structure and function in health and disease. *Int. J. Biochem. Cell Biol.* **27,** 747–60.

Kielty, C. M., Whittaker, S. P., and Shuttleworth, C. A. (1996) Fibrillin: Evidence that chondroitin sulphate proteoglycans are components of microfibrils and associate with newly synthesised monomers. *FEBS Lett.* **386,** 169–73.

Lee, B. L., Godfrey, M., Vitale, E., Hori, H., Mattei, M.-G., Sarfarazi, M., Tsipouras, P., Ramirez, F., and Hollister, D. W. (1991) Linkage of Marfan syndrome and a phenotypically related disease to two different genes. *Nature* **352,** 330–34.

Maddox, B. K., Sakai, L. Y., Keene, D. R., and Glanville, R. W. (1989) Connective tissue microfibrils: Isolation and characterisation of three large pepsin-resistant domains of fibrillin. *J. Biol. Chem.* **264,** 21381–85.

Maslen, C. L., Corson, G. M., Maddox, B. K., Glanville, R. W., and Sakai, L. Y. (1991) Partial sequence of a candidate gene for the Marfan-syndrome. *Nature* **352,** 334–37.

McConnell, C. J., Wright, G. M., and DeMont, M. E. (1996) The modulus of elasticity of lobster aorta microfibrils. *Experientia* **52,** 918–21.

Mecham, R. P., and Heuser, J. E. (1991) The elastic fiber. In: *Cell Biology of the Extracellular Matrix,* 2nd ed., E. D. Hay (ed.). New York: Plenum Press, pp. 79–109.

Pereira, L., D'Alessio, M., Ramirez, F., Lynch, J. R., Sykes, B., Pangilinan, T., and Bonadio, J. (1993) Genomic organization of the sequence coding for fibrillin, the defective gene product in Marfan syndrome. *Human Mol. Genet.* **2,** 961–68.

Qian, R. Q., and Glanville, R. W. (1997) Alignment of fibrillin molecules in elastic microfibrils is defined by transglutaminase-derived cross-links. *Biochemistry* **36,** 15841–47.

Raghunath, M., Tschodrich-Rotter, M., Sasaki, T., Meuli, M., Chu, M. L., and Timpl, R. (1999a) Confocal laser scanning analysis of the association of fibulin-2 with fibrillin-1 and fibronectin define different stages of skin regeneration. *J. Invest. Dermatol.* **112,** 97–101.

Raghunath, M., Putnam, E. A., Ritty, T., Hamstra, D., Park, E. S., Tschodrich-Rotter, M., Peters, R., Rehemtulla, A., and Milewicz, D. M. (1999b) Carboxy-terminal conversion of profibrillin to fibrillin at a basic site by PACE/furin-like activity required for incorporation in the matrix. *J. Cell Sci.* **112,** 1093–1100.

Reinhardt, D. P., Sasaki, T., Dzamba, B. J., Keene, D. R., Chu, M. L., Göhring, W., Timpl, R., and Sakai, L. Y. (1996a) Fibrillin-1 and fibulin-2 interact and are colocalized in some tissues. *J. Biol. Chem.* **271,** 19489–96.

Reinhardt, D. P., Keene, D. R., Corson, G. M., Pöschl, E., Bächinger, H. P., Gambee, J. M., and Sakai, L. Y. (1996b) Fibrillin-1: Organisation in microfibrils and structural properties. *J. Mol. Biol.* **258,** 104–16.

Reinhardt, D. P., Mechling, D. E., Boswell, B. A., Keene, D. R., Sakai, L. Y., and Bächinger, H. P. (1997) Calcium determines the shape of fibrillin. *J. Biol. Chem.* **272,** 7368–73.

Reinhardt, D. P., Gambee, J. E., Ono, R. N., Bächinger, H. P., and Sakai, L. Y. (2000) Initial steps in assembly of microfibrils. Formation of disulphide bonded cross-linked multimers containing fibrillin-1. *J. Biol. Chem.* **275,** 2205–10.

Ren, Z. X., Brewton, R. G., and Mayne, R. (1991) An analysis by rotary shadowing of the structure of the mammalian vitreous humor and zonular apparatus. *J. Struct. Biol.* **106,** 57–63.

Ritty, T. M., Broekelmann, T., Tisdale, C., Milewicz, D. M., and Mecham, R. P. (1999) Processing of the fibrillin-1 carboxyl-terminal domain. *J. Biol. Chem.* **274,** 8933–40.

Ritty, T. M., Broekelmann, T. J., and Mecham, R. P. (2000) Fibrillin 1 contains a proteolytically activated heparin binding site important for extracellular matrix deposition. *Mol. Biol. Cell* **11,** S1377.

Robinson, P. N., and Godfrey, M. (2000) The molecular genetics of Marfan syndrome and related microfibrillopathies. *J. Med. Genet.* **37,** 9–25.

Sakai, L. Y., Keene, D. R., and Engvall, E. (1986) Fibrillin, a new 350-kd glycoprotein, is a component of extracellular microfibrils. *J. Cell Biol.* **103,** 2499–2509.

Sherratt, M. J., Holmes, D. F., Shuttleworth, C. A., and Kielty, C. M. (1997) Scanning transmission electron microscopy mass analysis of fibrillin-containing microfibrils from foetal elastic tissues. *Int. J. Biochem. Cell Biol.* **29,** 1063–70.

Sherratt, M. J., Wess, T. J., Baldock, C., Ashworth, J., Purslow, P. P., Shuttleworth, C. A., and Kielty, C. M. (2000) Fibrillin-rich microfibrils of the extracellular matrix: ultrastructure and assembly. *Micron* **32,** 185–200.

Sinha, S., Nevett, C., Shuttleworth, C. A., and Kielty, C. M. (1998) Cellular and extracellular biology of the latent transforming growth factor-beta binding proteins. *Matrix Biol.* **17,** 529–45.

Smith, B. L., Schaffer, T. E., Viani, M., Thompson, J. B., Frederick, N. A., Kindt, J., Belcher, A., Stucky, G. D., Morse, D. E., and Hansma, P. K. (1999) Molecular mechanistic origin of the toughness of natural adhesives, fibres and composites. *Nature* **399,** 761–63.

Thurmond, F. A., and Trotter, J. A. (1996) Morphology and biomechanics of the microfibrillar network of sea cucumber dermis. *J. Exp. Biol.* **199,** 1817–28.

Trask, T. M., Ritty, T. M., Broekelmann, T., Tisdale, C., and Mecham, R. P. (1999) N-terminal domains of fibrillin 1 and fibrillin 2 direct the formation of homodimers: A possible first step in microfibril assembly. *Biochem J.* **340,** 693–701.

Trask, B. C., Trask, T. M., Broekelmann, T., and Mecham, R. P. (2000a) The microfibrillar proteins MAGP-1 and fibrillin-1 form a ternary complex with the chondroitin sulfate proteoglycan decorin. *Mol. Biol. Cell* **11,** 1499–1507.

Trask, T. M., Trask, B. C., Ritty, T. M., Abrams, W. R., Rosenbloom, J., and Mecham, R. P. (2000b) Interaction of tropoelastin with the amino-terminal domains of fibrillin-1 and fibrillin-2 suggests a role for the fibrillins in elastic fiber assembly. *J. Biol. Chem.* **275,** 24400–6.

Unsold, C., Hyytiainen, M., Bruckner-Tuderman, L., and Keski-Oja, J. (2001) Latent TGF-beta binding protein LTBP-1 contains three potential extracellular matrix interacting domains. *J. Cell Sci.* **114,** 187–97.

Wallace, R. N., Streeten, B. W., and Hanna, R. B. (1991) Rotary shadowing of elastic system microfibrils in the ocular zonule, vitreous and ligamentum nuchae. *Curr. Eye Res.* **10,** 99–109.

Wess, T. J., Purslow, P., and Kielty, C. M. (1997) Fibrillin-rich microfibrils: an X-ray diffraction study and elastomeric properties. *FEBS Lett.* **413,** 424–28.

Wess, T. J., Purslow, P. P., Sherratt, M. J., Ashworth, J., Shuttleworth, C. A., and Kielty, C. M. (1998a) Calcium determines the supramolecular organization of fibrillin-rich microfibrils. *J. Cell Biol.* **141,** 829–37.

Wess, T. J., Purslow, P. P., and Kielty, C. M. (1998b) X-ray diffraction studies of fibrillin-rich microfibrils: Effects of tissue extension on axial and lateral packing. *J. Struct. Biol.* **122,** 123–27.

Wright, D. M., Duance, V. C., Wess, T. J., Kielty, C. M., and Purslow, P. P. (1999) The supramolecular organisation of fibrillin-rich microfibrils determines the mechanical properties of bovine zonular filaments. *J. Exp. Biol.* **202,** 3011–20.

Wright, D. W., and Mayne, R. (1988) Vitreous humor of chicken contains two fibrillar systems: An analysis of their structures. *J. Ultrastruct. Mol. Struct. Res.* **100,** 224–34.

Yuan, X., Downing, A. K., Knott, V., and Handford, P. A. (1997) Solution structure of the transforming growth factor beta-binding protein-like module, a domain associated with matrix fibrils. *EMBO J.* **16,** 6659–66.

Yuan, X., Werner, J. M., Knott, V., Handford, P. A., Campbell, I. D., and Downing, K. (1998) Effects of proline *cis-trans* isomerisation on TB domain secondary structure. *Protein Sci.* **7,** 2127–35.

Zhang, H., Apfelroth, S. D., Hu, W., Davis, E. C., Sanguineti, C., Bonadio, J., Mecham, R. P., and Ramirez, F. (1994) Structure and expression of fibrillin-2, a novel microfibrillar component preferentially located in elastic matrices. *J. Cell Biol.* **124,** 855–63.

SIX

Spinning an Elastic Ribbon of Spider Silk

David P. Knight and Fritz Vollrath

INTRODUCTION

Spider silks show a remarkable range of mechanical properties (Madsen et al., 1999; Vollrath et al., 2001), with dragline silks in some species of orb web spider exhibiting elastomeric properties when wet and quite exceptional toughness either wet or dry (Gosline et al., 1986; Vollrath and Edmonds, 1989). The wide range of mechanical behaviour results from adapted design over an evolutionary time-span of circa 400 Mio years (Shear et al., 1989) in both the spinning solution (protein dope) and the spinning conditions (Vollrath and Knight, 2001). Recent studies are beginning to use sequence information of the spidroin dopes to reconstruct phylogenetic trees of silk sequence evolution (e.g., Gatesy et al., 2001). We have now begun a comparative study of silk gland evolution. And, if all goes well, the two trees should overlap, although both sides are still a long way away from this. For such a comparative study of spinneret evolution, the Brown Recluse spider *Loxosceles* sp. is of particular interest because, firstly, it is not closely related (Hormiga et al., 2000) to the orb-weaving spiders most studied for their silk and glands. And, secondly, its major silk is an extremely thin ribbon (Stern and Kullmann, 1975). The extrusion of a broad ribbon instead of a cylindrical thread suggested that the mechanism of formation must be different from that used to form the dragline threads in orb web spiders. *Loxosceles* silk, moreover, is rather 'sticky' and closely 'hugs' any objects it touches (Gardner, 1998; M. Ramirez, 2001; personal communication), as in the comparably thin, but round hackled threads of the cribellate spiders (Opell, 1995), possibly through electrostatic or Van der Waals forces. The Brown Recluse uses its silk to form a small mat of tangled threads which extends from a retreat under stones, scraps of wood, dead cactus, and similar dry sites (Smith, 1982). To date, neither the structure of the silk gland nor the duct has been described in detail for *Loxosceles*, although Kovoor (1987) mentioned the existence of a large pair of tubular silk glands, assigning them to the ampullate type because of their shape. Here, we describe the structure of the silk ribbons. We discuss the form and function of the spinneret and report that the size and shape of the principal silk gland and the location of its spigot strongly suggests homology with the major ampullate gland of orb web spiders. Furthermore, we describe the highly specific structure of the gland cells in the two transverse zones making up the secretory part of the gland. Finally, we discuss the evolutionary significance of the similarities and differences in construction of the glands in *Loxosceles* and orb web spiders.

EXAMINATION OF SILKS AND SPIDERS

Brown Recluse (Fiddle back) spiders, *Loxosceles laeta*, of the Family Sicaridae (Araneomorphae), are found mainly in the American Midwest (Smith, 1982). We were supplied adult female spiders bred in captivity. In the laboratory, the spiders readily spin retreats in the holding containers. To mount silk for scanning electron microscopy (SEM), stubs covered with double-sided adhesive tape were gently pushed through a sheet of silk forming the retreat. The resulting specimens were air-dried and sputter-coated with gold before viewing at 15 kV and a magnification up to 12,000×. A similar technique was used to mount silk ribbons on lightly carboned formvar-coated grids for transmission electron microscopy (TEM). Grids were stained at 20°C with freshly prepared saturated uranyl acetate in 70% v/v ethanol (20 minutes) and undiluted Reynold's lead citrate (12 minutes). They were lightly annealed in a widely spread beam and then imaged at 80 kV using low-dose technique and magnifications up to 40,000×.

Spiders were immediately dissected and fixed on arrival in the laboratory. The abdominal cavity was promptly irrigated with a Karnovsky fixative containing a final concentration of 2% glutaraldehyde, 2% formaldehyde, 0.1 M sodium cacodylate, 0.35 M sodium chloride, and 0.01% calcium chloride, finally adjusted to pH 7.4 with hydrochloric acid. Fixation was continued for 30 minutes, after which the major ampullate glands were dissected out and fixation continued for another hour at 4°C. After washing in buffer, material was postfixed in buffered 1% osmium tetroxide, dehydrated in alcohols, en bloc stained with uranyl acetate and embedded in Spurr's resin using an extended embedding schedule. Ultrathin sections were cut with glass or diamond knives and were examined at 80 kV after staining with alcoholic uranyl acetate and lead citrate as described. For SEM, tissue was fixed in the same way and examined after critical point drying and sputtering with gold. For differential interference microscopy, material fixed in the Karnovsky fixative as described was mounted in Farrant's gum without postfixation in osmic acid.

Structure of the Retreat and Silk Ribbons

The spider *Loxosceles laeta* spins a retreat underneath objects. These retreats usually consist of two sheets of silk threads with a space between them for the spider and a tangle of looser threads outside the sheets. The lower sheet is in

Figure 6.1: SEM micrograph showing the major ampullate silk from a final instar female *Loxosceles laeta* spider. Plaques of adhesion between ribbons (arrows; see text). Scale bar = 10 μm.

contact with the substratum while the upper sheet, attached to the underside of objects, has a small exit hole (5–10 mm diameter) for the spider. The sheets are composed of a meshwork of silk ribbons (see herein) anchored to surfaces by attachment plaques formed from numerous fine threads probably derived, as in orb web spiders (Vollrath, 1992), from the pyriform glands. Similar plaques are also evidently used to stick new ribbons to the surface of an existing sheet. Probing the meshwork with two pairs of watchmaker's forceps showed that the fine silk ribbons often adhered tightly to one another where they crossed and that the whole network was markedly elastic with individual ribbons capable of stretching up to twice their resting length and returning to it quite rapidly when released. Individual fibres were too short to mount on our mechanical testing instrument.

When viewed with the SEM, the silk sheet consisted of broad but extremely thin ribbons (Figure 6.1) with occasional patches of very narrow cylindrical filaments (not shown). Each straight length of ribbon had a remarkably constant width. The mean width of the ribbons was 7.0 ± 0.7 μm ($n = 20$), with some of the variation in width probably resulting from the variable extension produced by mounting on SEM stubs. The two edges of the ribbons scattered more electrons than their middles (Figure 6.1), suggesting that a narrow strip along both edges was appreciably thicker than the main part of the ribbon. At the point at which two ribbons crossed, they often showed flattened patches surrounded at their edges by narrowed necks of ribbon (Figure 6.1), suggesting that the ribbons' surfaces adhered tightly when brought close together. The surfaces of the ribbons appeared almost completely smooth even at the highest

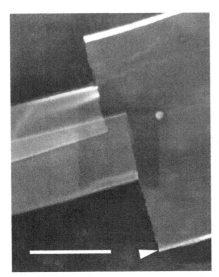

Figure 6.2: As Figure 6.1, but showing a partial fracture of the ribbon (see text). The ribbon is thickened at its edge (arrow). Scale bar = 5 μm.

magnification, showing no evidence of a special adhesive coating. These observations suggest that, as in the fine hairs on the surface of the Gecko's feet (Autumn et al., 2000) or the hackled threads of the cribellate spiders (Opell, 1995; Vollrath, 1994), van der Waal's forces are responsible for the adhesion of the ribbons to one another and to the prey.

Numerous fractured or partly fractured ribbons were seen in SEM preparations. The fracture appeared brittle and always ran perpendicularly across the ribbon with no observable deviation around the nanofibrils (see herein) of the ribbon (Figure 6.2). This fracture behaviour is similar to that of synthetic composites containing elastomeric polymers (Chang et al., 1992).

The ribbons were remarkably stable in the TEM. Their extreme thinness, probably of the order of tens of nanometers, is confirmed by the observation that double-stained preparations show high transparency to electrons, excellent resolution, and are apparently constructed from only a single incomplete layer of nanofibrils (Figures 6.3 and 6.4).

In the TEM, the ribbons show an increase in density at their very edges, confirming an increase in thickness here (Figure 6.3). This would help to prevent crack initiation at the edges of the ribbon. In the TEM, the ribbons are seen to contain numerous fine filaments with an apparent diameter of about 2–4 nm (Figure 6.4). These frequently appeared to run in pairs to give a fibril with a diameter of 12 ± 2 nm ($n = 10$), though this appearance could arise from a fine coating of stained material on the surface of a narrow, poorly-stained nanofibril. The nanofibrils are orientated fairly closely parallel to the long axis of the ribbon

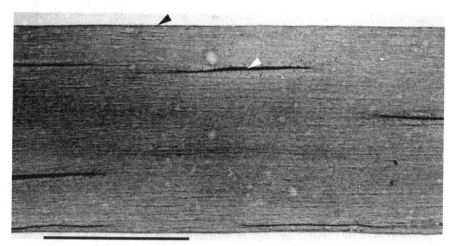

Figure 6.3: TEM micrograph of a stained silk ribbon. The ribbon is denser at its edge (black arrow) and tends to fold longitudinally (white arrow). Scale bar = 5 μm.

forming a markedly anisotropic meshwork with rather large spaces between the nanofibrils. In places, the filaments deflect around convex lens-shaped areas with lower density (Figure 6.4), suggesting that there are at least two components to the ribbon: the nanofibrils and an amorphous less dense component between them. The low packing density of nanofibrils in *Loxosceles* ribbons is in marked contrast to the high density seen in fibres dispersed from the dragline silk of orb web spider *Nephila edulis* (Knight and Vollrath, 2001) and may reflect the fact that the latter, unlike the former, is designed for extreme toughness.

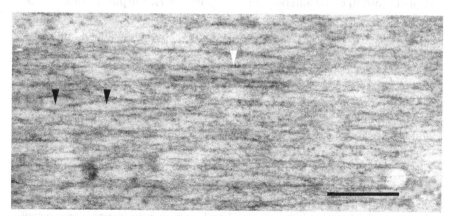

Figure 6.4: As Figure 6.3, but at higher magnification showing nanofibrils deflected around less dense lens-shaped areas (black arrows). The nanofibrils frequently appear double (white arrow). Scale bar = 250 nm.

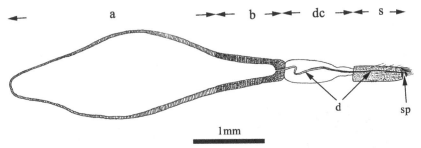

Figure 6.5: Drawing from micrographs of the major ampullate gland of the adult female *Loxosceles laeta*. A zone (a), B zone (b), duct cell region (dc), spinneret (s), duct (d), and spigot (sp).

The marked anisotropy in the arrangement of nanofibrils in *Loxosceles* ribbons probably accounts for its tendency to form longitudinal folds (Figure 6.3).

Gross Morphology of Major Ampullate Gland and Duct

Figure 6.5 summarises the parts of the major ampullate gland and associated duct in *Loxosceles laeta*. Although the glandular portion occupies about two-thirds of the combined length of the gland and duct, it was much shorter in relation to the length of the abdomen, compared with the equivalent highly folded structure in orb web spiders (Knight and Vollrath, 1999a). This probably correlates with a much smaller production of silk, compared with orb web spiders. Figure 6.6 shows a pair of the somewhat sac-shaped glands running for much of the length of the abdomen and located close to the ventral surface. The gland contributes its secretion to a duct that can be followed by dissection to a single large spigot located subterminally and medially on the short terminal segment (turret) of the anterior spinneret (Figure 6.7). The position and size of the spigot relative to that of the 9–11 pyriform spigots found apically on the anterior spinneret (see Figure 6.13) closely resemble that of the major ampullate gland of the araneomorph spiders (Coddington, 1989). The major ampullate duct in *Loxosceles* has a small S-bend (Figure 6.8) much less pronounced than that in orb web spiders (Knight and Vollrath, 1999a) and tapers only slightly as it runs towards the spigot. Thus, the morphology of the gland and duct portions and the position and size of the spigot indicate a close homology to the major ampullate gland of araneomorphs, although proportions of the gland and duct are markedly different from those of orb web spiders.

The duct in *Loxosceles* simply emerges from between the columnar cells at the distal end of the glandular portion (Figure 6.8) without the marked cuticular thickening ('funnel') (Vollrath and Knight, 2001) seen in orb web spiders. This gives the junction between gland and duct an appearance more closely resembling that of the presumed ancestral pyriform-type gland (Glatz, 1973;

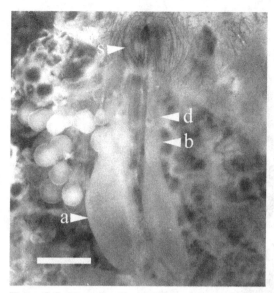

Figure 6.6: Dissection showing ventral view of the abdominal contents of a female spider fixed in Karnovsky solution. The major ampullate gland and associated duct run for much of the length of the abdomen and are connected to the anterior spinneret (s). The glandular portion has separate A and B zones (a and b, respectively). The duct region (d) is sharply divided from the glandular portion. Scale bar = 1 mm.

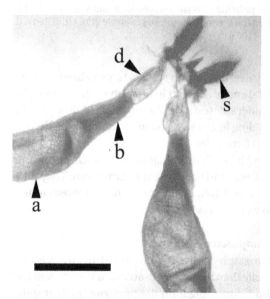

Figure 6.7: Cleared unstained whole mount of the major ampullate gland (bright field illumination). Labels as in Figure 6.6. The duct region is attached to the anterior spinneret. Scale bar = 1 mm.

121

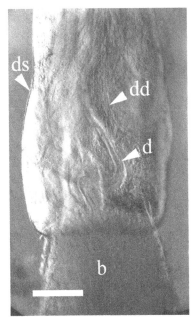

Figure 6.8: Differential interference micrograph of preparation as Figure 6.7. The duct (d) is S-shaped and remarkably short in this species. The cuticle lining of the duct appears to be secreted by a cylinder of duct epithelial cells surrounded by a thin sheath (ds). The duct cells are clearly demarcated from the B-zone cells (b). Degenerating duct (dd) derived from the previous instar. Scale bar = 100 μm.

Haupt and Kovoor, 1993). The 'valve' of orb web spiders is also absent in *Loxosceles*. This structure at the distal end of the duct in orb web spiders is thought to act as a clamp to control spider's descent on the dragline (Wilson, 1962). This function is vital to web building in orb web spiders, but not in the far less acrobatic and more earth-bound *Loxosceles*.

The flasked-shaped cells found beneath thinnings of duct cuticle just before the valve (Vollrath and Knight, 1998, 2001) are absent in *Loxosceles*. They may secrete the pheromones (Schultz and Toft, 1993) known to be present in the frame and radial threads of orb web spiders.

Morphology of the Major Ampullate Spigot
Figures 6.9 and 6.10 show micrographs of the spigot. This structure is about 45 μm long and is much less gracile than the pyriform, asciform spigots, and the single minor ampullate or cylindriform spigot on the posterior median spinnerets in this species. The lumen of the major ampullate duct is practically circular in cross-section, where it enters the spigot and continues without appreciable change in diameter through the annulated basal part. Thereafter, it quite

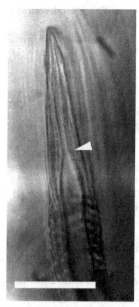

Figure 6.9: As Figure 6.8, but showing the major ampullate spigot sideways on (compare with Figure 6.10). The lumen constricts markedly in the plane of the optical section (arrow) to give the narrow but highly elongated opening slit seen in Figure 6.10. Scale bar = 10 μm.

Figure 6.10: SEM showing the major ampullate spigot. The outer opening (the top of the micrograph) is a markedly elongated slit. Corrugations at the base of the spigot may prevent kinking when the spigot flexes during spinning while longitudinal ridges on the distal part may stiffen the crucial part of the extrusion die containing the constriction (see Figure 6.5). The flexible lips (arrow) lack ridges. Scale bar = 15 μm.

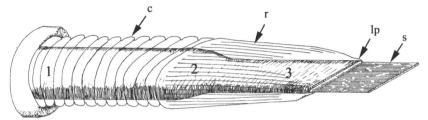

Figure 6.11: Isometric drawing of the major ampullate spigot (dorsal side up) illustrating the proposed method of action (see text). The spigot is about 45 μm long, and the exit slit is about 10×0.275 μm. Parallel-sided region of duct with circular cross-section (1), dorso-ventral constriction (2), highly flattened part (3), silk ribbon (s), corrugations (c), ridges (r), and flexible lips (lp).

suddenly constricts dorso-ventrally to about one-third of its diameter. After this constriction, the lumen continues without a marked change in shape to discharge at a narrow exit slit measuring approximately 10 μm wide by about 0.275 μm thick (Figure 6.10), although the internal functional opening of the exit is probably much thinner. The outermost ends of the exit slit are appreciably wider than the rest, and this is thought to form the thicker edges of the ribbon (as discussed previously). The widest axis of the slit is orientated transverse to the animal's antero-posterior axis. This probably helps the spider to apply an even pull across the width of the slit as it draws silk from the spigot. The spigot's duct is thought to function in the following way (Figure 6.11): Liquid dope is fed to the spigot and reaches the dorso-ventral constriction. Here the convergence of the lumen induces elongational flow that would help to orient the molecules. Thereafter, wall shear in the elongated and highly flattened lumen in the last part of the spigot may further orient the molecules and initiate a strain-induced refolding and aggregation of the silk protein molecules (Knight et al., 2000; Chen et al., 2001) to produce a partially plastic solid. Further draw down of the extremely thin silk ribbon in the air gap may account for the observation that the silk thread is narrower and thinner than the opening of the spigot. This final air draw down is likely to contribute to molecular alignment and further solidification of the ribbon as in orb web major ampullate silk (Vollrath and Knight, 2001).

Fixed whole mounts of the duct and spinneret show no evidence of an internal draw down similar to that seen in orb web spiders (Work, 1977; Knight and Vollrath, 1999a; Vollrath and Knight, 2001).

The extreme flatness and smoothness of the silk ribbon extruded in this way may enable it to stick efficiently to itself and to prey using van der Waals force's rather than a water-based adhesive (as aforementioned). An adhesive would tend to dry out in the relatively dry environment in which the spider lives. Alternatively, the large surface area of the ribbon may enable it to carry a high electrostatic charge. This in turn may help the silk ribbons to fly apart to

Figure 6.12: SEM showing highly modified comb-like hairs (c) found only on the apex of the posterior lateral spinneret. These may help the spider to position and press together the silk ribbons spun from the major ampullate spigot. Scale bar = 20 μm.

deploy effectively in the space used for the spider's retreat. The greatly broadened comb-like hairs found only on the tip of the posterior spigot (Figures 6.12 and 6.13) may help to press the ribbons together or direct the silk as it merges from the major ampullate spigot, with the pointed teeth of the comb acting as a low friction surface or possibly an electrostatic brush.

The annular constrictions in the outer surface of the basal part of the spigot (Figures 6.9 and 6.10) may serve to provide flexibility yet prevent kinking of the lumen when the spigot bends during spinning while longitudinal ridges in the cuticle of the outer part may stiffen the crucial constricted region of the die (Figures 6.9 and 6.11).

Ultrastructure of the Major Ampullate Gland

The glandular sac, like that of most arachnid silk glands (Glatz, 1973; Work, 1984; Kovoor, 1987) contains two completely distinct transverse zones. The A zone (located furthest from the spigot) is considerably larger than the more distal B zone. Both zones are constructed from tall columnar cells whose height increases distally in the gland. A single incomplete layer of small basal cells is

Figure 6.13: SEM showing the size and position of the comb-like spines (arrows) on the posterior lateral spinneret in relation to the major ampullate spigot (ma) and a group of pyriform spigots (p) mounted on the anterior spinneret. Scale bar $= 10$ μm.

found between the columnar cells and the basement membrane of the gland (Figure 6.14). These cells are closely similar in appearance to the 'regenerative cells' described by Plazaola and Candelas (1991) in the sac of the major ampullate gland in *Nephila clavipes*. The nuclei of the columnar cells (Figure 6.15) are located basally and are large (up to 50 μm in diameter), approximately five times the diameter of the basal cell nuclei. They contain at least six nucleoli and a very large number of islands of heterochromatin. These features (Luszczek et al., 2000) suggest that they may be polyploid, unlike the equivalent nuclei in *Nephila*. A secretory apparatus is present at the apical pole in both A- and B-zone columnar cells (Figure 6.16) and is similar to that described by Plazaola and Candelas (1991), but contains a much deeper apical pit.

A-zone columnar epithelial cells are packed with numerous relatively large secretory vesicles (Figure 6.17), with a mean diameter of 9 ± 2 μm ($n = 20$) (measured in 5-μm survey sections with a light microscope). In TEM sections, the vesicles frequently contain numerous fine nanofibrils (Figure 6.18), often arranged in a peripheral band surrounding a central clear zone in which myelin figures can often be distinguished (Figure 6.17). The nanofibrils have a diameter of 2–5 nm and are similar to those seen in the secretory vesicles in the

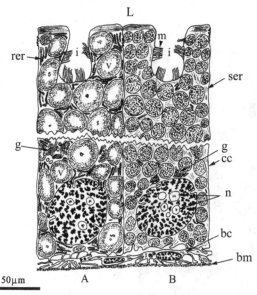

Figure 6.14: Drawing summarising the ultrastructure of the junction between the A and B zones. A zone (A), B zone (B), lumen (L), columnar epithelial cell (cc), basal cell (bc), basement membrane (bm), Golgi apparatus (g), secretory vesicle (V), invagination (i), microvilli (m), nucleoli (n), rough endoplasmic reticulum (rer), and smooth endoplasmic reticulum (ser). The main differences between the A- and B-zone columnar epithelial are different secretory vesicles and the predominance of rough endoplasmic reticulum in the A-zone cells and smooth in the B zone. Drawing shows about half the height of the epithelial cells.

proximal tail of the A zone in *Nephila edulis* (D. P. Knight and K. Davies, unpublished data). Regions of sufficiently low fibril density enable the length of the nanofibrils to be estimated at about 150 nm, in *Loxosceles*. These are thinner and much shorter than the nanofibrils seen in the final silk (see above). The presence of nanofibrils in the secretory vesicles of the A zone suggests that, as in *Nephila* (Vollrath and Knight, 2001), this zone elaborates the nanofibrillar component of the finished silk (see aforementioned). The A-zone secretory vesicles in *Loxosceles* as in *Nephila* (Vollrath and Knight, 2001) appear to be formed in a large Golgi apparatus (Figure 6.19). This can be found rapidly by looking in a supranuclear region for a concentration of numerous small vesicles (Figure 6.19), some of which have contents similar to those of the mature secretory vesicles. The Golgi apparatus is closely similar to that described by Plazoala and Candelas (1991). The cytoplasm in between the A-zone secretory vesicles contains abundant rough endoplasmic reticulum (RER).

The B-zone columnar epithelial cells are also packed with secretory vesicles (Figures 6.16 and 6.20), but these are smaller ($4 \pm 0.5\ \mu$m) and are somewhat

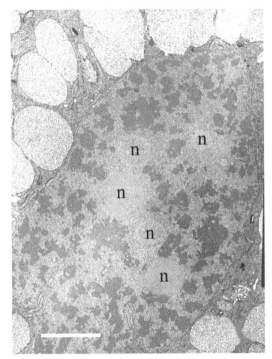

Figure 6.15: TEM micrograph showing the basal part of a B-zone columnar cell. The large nucleus contains many nucleoli (n) and islands of heterochromatin (see text). Scale bar = 5 μm.

more electron dense than those of the A zone. They contain dense granular material (Figure 6.20) in some places arranged like beads on as string. This material shows no resemblance to that of the B-zone secretory granules in *Nephila* (Knight and Vollrath, 1999b). The cytoplasm contains much smooth endoplasmic reticulum, some RER with a low concentration of ribosomes, and only a little RER with abundant ribosomes. This, together with the dense granular appearance of the material in the secretory vesicles, could be explained if the cells secreted a phospholipid and a small quantity of protein, but we have no histochemical evidence for this. The Golgi apparatus of the B zone (Figure 6.20) contains numerous flattened and dilated cisternae, giving it an appearance somewhat different from that of the A zone. The B-zone secretion may provide a smooth surface to the ribbon necessary for sticking two ribbons together by van der Waal's forces.

CONCLUSIONS

The spider *Loxosceles laeta* draws remarkably broad yet thin ribbons of elastomeric silk from its spigot that clearly functions as a sophisticated extrusion die.

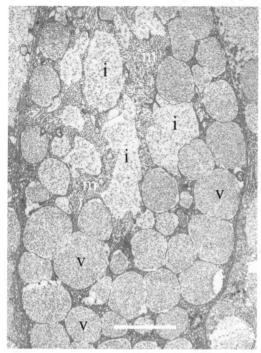

Figure 6.16: TEM micrograph showing the apical part of a B-zone columnar cell. An apical secretory apparatus contains clusters of microvilli (m) surrounding an invagination containing dense granules similar to those found in the numerous mature secretory vesicles (v) packing the cytoplasm. Scale bar = 5 μm.

The large surface area of the ribbon and smoothness of its surface are thought to enable the ribbons to stick together by van der Waal's forces to help the web function as an elastic net. This method of microfabrication may have biomimetic potential.

The observations reported here demonstrate that, although the major ampullate gland of *Loxosceles* possess the same basic units (separate transverse zones in the gland, a duct, and spigot) seen in other spider silk glands, the detailed construction is remarkably different from that of orb web spiders. One is immediately struck by the relative shortness of the duct and the fact that the opening of the spigot is a highly elongated slit and not a circular opening. We suggest that these design features depend on a fundamental difference between the spinning mechanisms in *Loxosceles* and the orb web spiders. The major ampullate gland of orb web spiders, typified by that of *Nephila* spp., produces a cylindrical thread by a mechanism involving an internal draw down starting well inside the distal part of the duct followed by an external draw down in the air

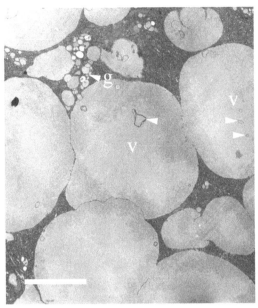

Figure 6.17: TEM of an ultrathin section showing part of the cytoplasm of an A-zone epithelial cell. The cytoplasm is packed with numerous large secretory vesicles (v) often containing a tangle of fine filaments (see Figure 6.10) and a central clear zone sometimes containing myelin figures (arrows). Golgi apparatus (g, arrow). Scale bar = 5 μm.

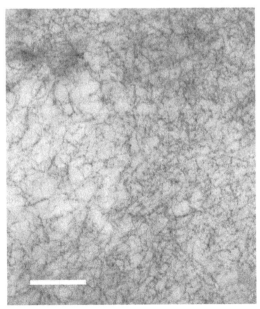

Figure 6.18: TEM micrograph showing nanofibrils in an A-zone secretory vesicle at high magnification. Scale bar = 250 nm.

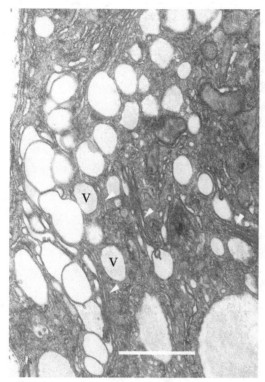

Figure 6.19: TEM micrograph showing part of the Golgi apparatus of an A-zone colum-nar epithelial cell. Putative developing secretory vesicles (v) contain filaments closely similar to those seen in the mature secretory vesicles. Stacks of flattened Golgi cisternae (arrowed). Scale bar = 1 μm.

after the spigot (Work, 1977; Vollrath and Knight, 2001). In contrast, *Loxosceles* draws a highly flattened ribbon of silk directly from the spigot without an internal draw down. To do this, the dope fed to the spigot must be highly fluid in contrast to orb web spiders where the dope is thought to be largely solid by the time it reaches the spigot (Knight et al., 2000). Thus, *Loxosceles* can manage without the extreme length of the spinning duct required in orb web spiders for forming a solid thread within the spinning duct by secreting K^+ and H^+ ions and recovering water and Na^+ ions (Vollrath et al., 1998; Chen et al., 2001; Knight and Vollrath, 2001; Vollrath and Knight, 2001). The long tapering duct in orb web spiders is also thought to produce the extreme elongation of filler particles seen in their dragline silk (Knight and Vollrath, 1999) and to spin silk over a huge range of draw rates (Vollrath and Knight, 2001). There is no evidence that *Loxosceles* does either of these. The fluidity of the dope in the duct before the spigot in *Loxosceles* probably also correlates with the absence of the clamp-like

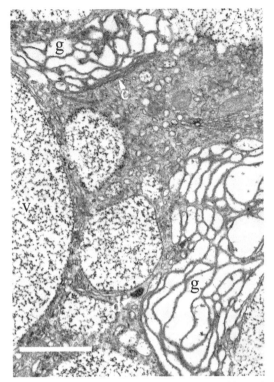

Figure 6.20: TEM micrograph showing part of the Golgi apparatus (g) of a B-zone columnar epithelial cell. Mature secretory vesicles (v) contain dense granules that frequently
appear to have clumped into chains. Occasional small vesicles (d, arrow) contain what
appears to be dense aggregations of these granules. Golgi cisternae (c). Scale bar $= 1 \mu$m.

'valve' which orb web spiders use like the rock climber's 'descendeur' to control their descent on the dragline (Wilson, 1962). Such a clamp would be of no
use on the still-fluid silk in *Loxosceles*. The absence the 'funnel' in *Loxosceles* is
intriguing. This structure is a marked thickening of the cuticle between the gland
and duct in silk glands with highly elongated ducts (major and minor ampullate
and cylindriform) in orb web spiders, but is absent from glands with a shorter
duct (aggregate, pyriform, and aciniform). The funnel may serve as a mechanical
reinforcement between the elongated and extensible duct and the less mobile
glandular portion to prevent premature shear-induced coagulation of the dope
when the spider wiggles its abdomen (Vollrath and Knight, 2001) or draws silk
from the duct. The shortness of the duct and much smaller cross-sectional area
of the silk in *Loxosceles* may result in less deformation of the junction between
gland and duct, allowing the spider to do without this structure.

Thus, differences between the structure of the major ampullate gland in *Loxosceles* and orb web spiders may reflect differences in the spinning mechanism. This in turn probably reflects the different uses to which the two spiders put their major ampullate silks: forming a retreat in *Loxosceles* and extremely tough threads for major engineering works in orb web spiders. Given the antiquity of the Sicaridae to which *Loxosceles* belongs (Hormiga et al., 2000), the simplicity of the major ampullate gland and its spinning technology, compared with that of orb web spiders, may represent primitive features (Glatz, 1973; Haupt and Kovoor, 1993) derived from an ancestral spider and subsequently modified in orb web spinners.

ACKNOWLEDGEMENTS
We thank the Danish Science Research Council, the Science Faculty of Aarhus University, the Carlsberg Foundation, the Danish Academy of Science, the British Biological and Engineering Research Councils (BBSRC, EPSRC), and the European Science Foundation for funding; Thomas Vinnman for his generous gift of spiders; David Kaplan, Beat Meier, and John Gosline for helpful discussions; and the Biological Imaging Centre, Southampton University, for technical assistance.

REFERENCES
Autumn, K., Liang, Y. A., Hsieh, S. T., Zesch, W., Chan, W. P., Kenny, T. W., Fearing, R., and Full, R. J. (2000). Adhesive force of a single gecko foot-hair. *Nature* (Lond.) **405,** 681–85.

Chang, F. C., Wu, J. S., and Chu, L. H. (1992) Fracture and impact properties of polycarbonate and MBS elastomer-modified polycarbonates. *J. Appl. Polym. Sci.* **44,** 491–504.

Chen, X., Knight, D. P., and Vollrath, F. (2001) Rheological characterization of *Nephila* spidroin solution. *Biomacromolecules* **3,** 644–48.

Coddington, J. A. (1989) Spinneret silk spigot morphology: Evidence for the monophyly of orb weaving spiders, Cyrtophorinae (Aranaeidae) and the group Theriidae plus Nesticidae. *J. Arachnol.* **17,** 71–95.

Gardner, K. (1998) Talk given at the Second International Symposium on Silk, June 1998, University of Virginia, Charlottesville.

Gatesy, J., Hayashi, C., Montriuk, D., Woods, J., and Lewis, R. (2001) Extreme diversity, conservation and convergence of spider silk fibroin sequences. *Science* **291,** 2603–5.

Glatz, L. (1973) Der spinnapparat der orthognatha (Arachnida, Araneae). *Z. Morph. Tiere* **75,** 1–50.

Gosline, J. M., DeMont, M. E., and Denny, M. W. (1986) The structure and properties of spider silk, *Endeavour, N.S.* **10,** 31–43.

Haupt, J., and Kovoor, J. (1993) Silk-gland system and silk production in mesothelae (araneae). *Ann. Sci. Nat.-Zool. Biol. Anim.* **14,** 35–48.

Hormiga, G., Scharff, N., and Coddington, J. A. (2000) The phylogenetic basis of sexual size dimorphism in orb-weaving spiders (Araeae, Orbiculariae). *Syst. Biol.* **49**, 435–62.

Knight, D. P., and Vollrath, F. (1999a) Liquid crystals and flow elongation in a spider's silk production line. *Proc. R. Soc. B* **266**, 519–23.

Knight, D., and Vollrath, F. (1999b) Hexagonal columnar liquid crystal in the cells secreting spider silk. *Tiss. Cell* **31**, 617–20.

Knight, D. P., Knight, M. M., and Vollrath, F. (2000) Beta transition and stress-induced phase separation in the spinning of spider dragline silk. *Int. J. Biol. Macromol.* **27**, 205–10.

Knight, D. P., and Vollrath, F. (2001) Changes in element composition in the secretory pathway for dragline silk in a *Nephila* spider. *Naturwissenschaften* **88**, 179–82.

Kovoor, J. (1987) Comparative structure and histochemistry of silk-producing organs in Arachnids. In: *Ecophysiology of Spiders*, chap. IV, W. Nentwig (ed.). Berlin, Germany: Springer-Verlag, p. 176.

Luszczek, D., Swierczynska, J., and Bohdanowicz, J. (2000) Polyploidization of suspensor basal cell in *Triglochin maritimum* L. (Juncaginaceae). *Acta Biol. Crac. Ser. Bot.* **42**, 131–37.

Madsen, B., Shao, Z., and Vollrath, F. (1999) Variability in the mechanical properties of spider silks three levels: Interspecific, intraspecific and intraindividual. *Int. J. Biol. Macromol.* **24**, 301–6.

Opell, B. D. (1995). Do static electric forces contribute to the stickiness of a spider's cribellar prey capture threads? *J. Exp. Zool.* **273**, 186–89.

Plazaola, A., and Candelas, G. C. (1991) Stimulation of fibroin synthesis elicits ultrastructural modifications in spider silk secretory cells. *Tiss. Cell* **23**, 277–84.

Shear, W. A., Palmer, J. M., Coddington, J. A., and Bonamo, P. M. (1989) A Devonian spinneret: Early evidence of spiders and silk use. *Science* **246**, 479–81.

Schultz, S., and Toft, S. (1993) Branched long chain alkyl methyl ethers: A new class of lipids from spider silk. *Tetrahedron* **49**, 6805–20.

Smith, R. L. (1982) *Venomous Animals of Arizona*. Tucson, AZ: The University of Arizona Press.

Stern, H., and Kullmann, E. (1975). *Leben am seidenen Faden: die rätselvolle Welt der Spinnen*. Munich, Germany: Bertelsmann Verlag.

Vollrath, F. (1992) Spider webs and silk. *Sci. Am.* **266**, 70–76.

Vollrath, F. (1994) General properties of spider silk. In: *Silks Polymers: Materials Science and Biotechnology*, D. L. Kaplan, W. W. Adams, C. Viney, and B. L. Farmer (eds.). Washington, USA: ACS Books, pp. 17–28.

Vollrath, F., and Edmonds, D. (1989) Modulation of normal spider silk by coating with water. *Nature* **340**, 305–7.

Vollrath, F., Hu, X. W., and Knight, D. P. (1998) Silk production in a spider involves acid bath treatment. *Proc. R. Soc. Lond. B* **263**, 817–20.

Vollrath, F., and Knight, D. P. (1998) Structure and function of the silk production pathway in the spider *Nephila edulis*. *Int. J. Biol. Macromol.* **24**, 243–49.

Vollrath, F., and Knight, D. P. (2001) Liquid crystalline spinning in Nature. *Nature* (Lond.) **410**, 541–48.

Vollrath, F., Madsen, B., and Shao, Z. (2001) The effect of spinning conditions on silk mechanics. *Proc. Roy. Soc. Lond. B* (in press).

Wilson, R. S. (1962) The structure of the dragline control valves in the garden spider. *Q. J. Microsc. Sci.* **103,** 549–55.

Work, R. W. (1977) Mechanisms of major ampullate silk fiber formation by orb-web-spinning spiders. *Trans. Am. Microsc. Soc.* **96,** 170–98.

Work, R. W. (1984) Duality in major ampullate silk and precursive material from orb-web-building spiders (Araneae). *Trans. Am. Microsc. Soc.* **103,** 113–21.

SEVEN

Sequences, Structures, and Properties of Spider Silks

R. Lewis

INTRODUCTION

Spiders are unique in the animal world due to the use of silk throughout their life span and a nearly total dependence on silk for their evolutionary success (Lucas, 1964; Vollrath, 1992). There were periods of fairly intense study of spider silk prior to World War II and in the late 1950s. However, progress was relatively meager, especially when compared with research on silkworm silk. Beginning in the 1970s, studies carried out in the laboratories of Work, Gosline, and Tillinghast reinvigorated interest in spider silk, with several papers describing their physical, mechanical, and chemical properties. These papers set the stage for the current advances in our understanding of these fibers.

BIOLOGICAL ASPECTS OF SPIDER SILK PRODUCTION

Typical spider webs are constructed from several different silks, each of which is produced in a separate gland. Non-orb web-weaving spiders produce fewer silks and use them differently in many cases. The non-orb weavers constitute the majority of spiders and include those species that do not make the usual orb-shaped web. The various silks produced by the orb web-weaving spiders, the glands that produce them, and the uses of the spiders for each silk are listed in Table 7.1.

Although each of the silk glands has its own distinctive shape and size, they are all functionally organized in a similar pattern. The majority of the gland serves as a reservoir of soluble silk protein that is synthesized in specialized cells at the distal end of the gland. The soluble silk is then pulled down a narrow duct during which the physical and chemical changes which produce the solid silk fiber occur. A muscular valve is present at the exit to the spinneret that can control the flow rate of the fiber and may control the fiber diameter. The silk exits through the spinnerets, of which there are three pairs: anterior, median, and posterior.

Table 7.1. Spider Silks and Their Uses

Silk	Use
Major ampullate	dragline, web frame, radii
Minor ampullate	reinforcing, web initial spiral
Flagelliform	core fibers of capture spiral
Aggregate	adhesive silk of capture spiral
Tubuliform	covering of the egg case
Aciniform	swathing prey and inner egg sack lining
Perform	attachment disk and joining fibers

Due to their large size and ease of study, the major ampullate glands have received the majority of attention, and most of what is known about the synthesis of silk proteins is based on this gland. However, morphological and histochemical studies of the other glands support the ideas developed from the major ampullate gland research. Synthesis of the silk protein(s) takes place in specialized columnar epithelial cells (Bell and Peakall, 1969). There appear to be at least two different types of cells producing protein (Kovoor, 1972), which is consistent with data which show two proteins in the silk from these glands. The newly synthesized protein appears as droplets within the cell which are secreted into the lumen of the gland.

The protein in the lumen of the gland is believed to be in a liquid crystal state (Willcox et al., 1996) that prevents fiber formation until passage down the duct. This state is probably determined by a combination of protein structure and concentration that prevents aggregation into large protein arrays. While fluid silk in the lumen is not birefringent, it becomes increasingly birefringent as it passes down the duct (Work, 1977). Thus, the ordered array of protein seen in the final fiber is assembled in the duct. Protein alignment appears to be due to the mechanical and frictional forces altering the secondary structure to the form found in the fiber. Iazuka (1983) has proposed a similar mechanism for silkworm silk formation. The ability to draw silk fibers directly from the lumen of the major, minor, and cylindrical glands (M. Hinman, 1999, personal communication) provides experimental evidence in support of the hypothesis that the physical force of drawing the solution is sufficient for fiber formation. A recent review by Vollrath and Knight (2001) summarizes much of the current state of understanding of this process (see also Chapter 6 in this volume).

MECHANICAL PROPERTIES

One of the features attracting the attention of scientists to spider's silks is their unusual mechanical properties. Orb-web weaving spiders use the minimum amount of silk in their webs to catch prey. The web has to stop a rapidly flying

Table 7.2. Comparisons of Mechanical Properties of Spider Silk

Material	Strength (N m^{-2})	Elasticity (%)	Energy to Break (J kg^{-1})
Major amp. silk	4×10^9	35	4×10^5
Minor amp. silk	1×10^9	5	3×10^4
Flagelliform silk	1×10^9	>200	4×10^5
KEVLAR	4×10^9	5	3×10^4
Rubber	1×10^6	600	8×10^4
Tendon	1×10^9	5	5×10^3

Data are derived from Gosline et al. (1986) and Stauffer et al. (1994)

insect in such a manner that the prey becomes entangled and trapped. To do this, the web must absorb the energy of the insect without breaking and yet not act as a trampoline to bounce the insect away from the web. Gosline et al. (1986) have reviewed several aspects of this property and concluded that spider silk and the web are nearly optimally designed for each other.

Major and minor ampullate and egg case silks from both *Nephila clavipes* and *Araneus gemmoides* were tested using standard mechanical testing methods (Stauffer et al., 1994). These silks were found to exceed the published data for tensile strength by a substantial margin. This was due to the use of an average minimum diameter measured at ten points along the tested fiber instead of the average diameter calculated from the density, length, and weight. This minimum diameter is about 50% of the average diameter. Since silks are likely to break at the narrowest point, these values may be more typical of the actual properties of these silk fibers. Since that publication, these fibers have been examined using scanning electron microscopy, confirming the presence of large variation in their diameter (Theil et al., 1994).

As with any polymer, especially those composed of protein, there are numerous factors that can affect the tensile strength and elasticity. These can include temperature, hydration state, and rate of extension. Even with these caveats, it is clear that dragline silk is a unique biomaterial. As shown in Table 7.2, dragline silk will absorb more energy prior to breaking than almost any other commonly used material. It is nearly as strong as several of the current synthetic fibers, but can outperform them in many applications in which kinetic energy absorption is required.

PROTEIN SEQUENCES

Spider silks has been known to be composed predominantly of protein since the studies by Fischer (1907). In fact, except for the sticky material deposited on the catching spiral thread, no significant amount of any other compound has been show to be covalently linked to the silk proteins, including sugars, minerals, and

(A)

MaSp 1 and 2 Consensus Repeats

MaSp1 GGAGQGGYGGLGGQGAGRGGLGGQGAG<u>AAAAAAA</u>
MaSp2 (GPGGYGPGQQ)₂GPSGPGS<u>AAAAAAAAAA</u>

(B)

MiSp 1 and 2 Consensus Amino Acid Repeats

MiSp 1

AGGAGGYGRGAGAGAGAAAGAGA---GYGGQGGYGAGAGAGAAAAAGAG

MiSp 2

AGGYGRGAGAG<u>S</u>GAAAGAGA<u>GSA</u>GGYGGQGGYGAGAGAAAAGAG<u>S</u>

(C)

MiSp 1 and 2 Spacer Regions

GSSAGNAFAQSLSSNLLSSGDFVQMISSTTSTDHAVSVATSVAQNVGSQLGLDANAMNNLLGAV
SGYVSTLGNAISDASAYANALSSAIGNVLANSGSISESTASSAASSAASSVTTTLTSYGPAVFY
APSASSGG

Figure 7.1: Amino acid sequences of major (MaSp1/2) and minor (MiSp1/2) silk proteins of *N. clavipes*. The repeat sequences shown for MaSp1 and 2 in (A) are the consensus of the repeat sequences in each of the proteins. Most individual repeats show variation in the length of the poly-ala segment or the number of GGX or GPGGX motifs. The repeat sequences shown for MiSp1 and 2 in (B) are the consensus of the repeat sequences in each of the proteins. Most individual repeats show variation in the length of the poly-GA segment or the number of GGX motifs. The underlined residues in MiSp 2 are the key consistent differences between the repeats in the two proteins. The sequence of the non–silk-like spacer region in the MiSp proteins shown in (C) is highly conserved in both proteins, where it is repeated as part of the overall repeat with the sequences in (B).

lipids. The various silks have significantly different amino acid compositions, as do the same silks from different spiders. In major ampullate silks from different species, the combination of E, P, G, and A comprise 80% of the silk. However, the proportion of proline is significantly different in each. As will be discussed later, these differences can be accounted for by differing ratios of two proteins.

When detailed studies of spider silk protein were initiated, all that was known was the amino acid compositions for a number of spider silks and mechanical test data. In 1990, the first cDNA for a spider silk protein was cloned and sequenced from the major ampullate silk of *N. clavipes*, termed MaSp 1 (Figure 7.1; Xu and Lewis, 1990). The partial sequence contains a number of similar but non-identical repeats with unique elements, including stretches of poly-A of up to seven residues alternating with (GGX)_n sequences, where the X residues are Y, L, and Q in that sequence order. There are differences in the number of

GGX units in each repeat, but there are virtually no substitutions for the three X amino acids. No matches greater than five amino acids were found when these sequences were compared with the current protein sequence databases, indicating that they represent novel structures.

The subsequent discovery of the second major ampullate silk protein was based on the identification of a proline-containing peptide in the silk, which was absent from protein encoded by the cDNA for MaSp 1. This led to the cloning and sequencing of the cDNA for the second major ampullate silk protein, MaSp 2 (Hinman and Lewis, 1992). The partial sequence and predicted structure of this protein prompted a rethinking of the mechanism for elasticity and tensile strength and led to the models described later. The novel sequence features in the consensus repeats of this protein (Figure 7.1) include the poly-A regions as in MaSp 1 that alternate with sequences of GPGQQGPGGY. As with MaSp 1, virtually no amino acid substitutions occur in the repetitive regions. The repeats shown in Figure 7.1 are the consensus of the hundreds of repeats in the protein, and most repeats have minor differences from the consensus either in the number of three or five amino acid motifs or the number of alanine residues in the poly-A motif.

Using northern blotting, restriction digestion, and Southern blots of genomic DNA, the sizes of the mRNA and genes for all the aforementioned proteins were determined. The mRNA sizes for MaSp 1 and 2 were shown to be about 12.5 and 10.5 kb, respectively, while analyses of genomic DNA indicated the absence of large introns in the coding regions and lack of any detectable introns in the main repetitive portions of the genes.

Shortly after the previous data were published, cDNAs representing minor ampullate silk protein transcripts from *N. clavipes* were sequenced (Colgin and Lewis, 1998). The sequence obtained (Figure 7.1) shows similarity to that of the major ampullate silk proteins, but shows striking differences as well. GGX and short poly-A sequences are present, but the long poly-A stretches in the MaSps are replaced by $(GA)_n$ repeats. The consensus repeats show very similar organizations, but the number of GGX and GA repeats varies greatly.

A second cDNA from the same library shows substantial similarity to the first minor ampullate protein, but clearly encodes a different protein (MiSp 2). Multiple copies of a unique conserved 130 amino acid non-repetitive sequence were found in both proteins which were termed "spacer" regions, as they break up the repetitive regions (Figure 7.1). These serine-rich spacer regions are reminiscent of the 30 amino acid non-repetitive amorphous domains of silkworm silk that also interrupt repetitive regions. Interestingly, the MiSp spacer regions are highly conserved in sequence and thus differ from the repetitive regions which show variation as seen in the MaSp proteins. The function of this spacer region is currently unknown, but it may serve to separate crystalline regions as well as participate in interchain associations through charged residues.

Northern blotting confirmed that minor ampullate silk is composed of at least these two highly repetitive proteins. The MiSp 1 and MiSp 2 transcripts are about 9.5 kb and 7.5 kb, respectively. Genomic DNA analysis by Southern blotting using both the 3′ end of the clone and a probe to the repetitive region demonstrated the presence of only two separate MiSp genes. Restriction digestion and Southern blotting of the genomic DNA demonstrated that the entire fragment, which corresponds closely in size to the transcript, is composed of the same sequences as found in the partial cDNA. This indicates the lack of any substantial introns or different coding regions within this gene (Colgin and Lewis, 1998).

Next to be cloned were cDNAs from the *N. clavipes* flagelliform gland mRNA encoding a protein for the catching spiral silk (Hayashi and Lewis, 1998). Flagelliform silk is highly elastic (>200%), but lower in tensile strength than major ampullate silk (Table 7.2). The cDNA clones contain sequences encoding a 5′ untranslated region and a secretory signal peptide, numerous iterations of a five amino acid motif, and the C-terminal end. Northern blot analysis indicates an mRNA transcript of ~15 kb. The amino acid sequence predicted from the gene sequence suggests a model of protein structure that explains the physical basis for the elasticity of spider silk and similar to that described for MaSp 2 (described later).

In comparison with the other known silk genes, the flagelliform protein has four very distinctive features: (1) It has the simplest repeat unit, a pentapeptide (GPGGX) that also appears as a motif within the MaSp 2 repeat unit. (2) The pentapeptide repeat units have more sequence variation than seen in any other silk gene. When the tandem repeats are aligned with each other, the fifth codon often results in an amino acid substitution. While this variability suggests that the identity of the fifth amino acid is not critical to the protein structure, only a limited set of amino acids (A, V, S, Y) are common in this position. Curiously, there are three predominant patterns to the strings of repeats: non-alternating As, alternating (V,S), and alternating (Y,S). (3) A highly conserved 34 amino acid spacer unit (sequence TITEDLDITIDGADGPITISEELTISGA) occurs among the basic repeat unit which is reminiscent of MiSp 1 and 2. This region is also non-silk-like, with charged and hydrophilic amino acids predominating. This region has no nucleotide similarity with any known gene, and its significance to the structure of the protein is unknown. (4) In contrast, the C-terminal non-repetitive region of flagelliform silk shows no homology to that of the MaSp and MiSp proteins.

The first genomic sequence data for spider silk were also from the flagelliform silk (Hayashi and Lewis, 2000). This gene was of particular interest as it showed a pattern of repeating introns and exons with the introns more highly conserved than the exons. When two closely related species (*N. clavipes* and *N. madagascariensis*) were compared, the introns within the genes of each were more similar than the corresponding introns compared between species. Thus, a significant homogenization must be occurring in these genes.

Two other papers have been published describing spider silk protein and DNA sequences. The first used polymerase chain reaction (PCR) to identify sequences in an *Araneus* species based on sequences published for major ampullate silk proteins from *Nephila* (Beckwitt and Arcidiacono, 1994) while Gosline's group (Guerette et al., 1996) sequenced clones from gland-specific cDNA libraries from an *Araneus* species. Both publications demonstrated that the sequences of the silk proteins from these species were very similar to those in *Nephila*.

Figure 7.2 shows recently published sequences for these same three silks from other orb weaver or derived orb weavers (Gatesy et al., 2001). The data clearly

MaSp1

```
Nep.c.         GGA--GQGGYGGLGSQGA-------GRGGLGGQ---GA--GAAAAAA----
Nep.m.         GGA--GQGGYGGLGSQGA-------GRGGYGGQ---GA--GAAAAAA----
Nep.s.         GGA--GQGGYGGLGGQGA--------------------GAAAAAA----
Tet.k.         GGLGGGQ-GAGQGGQQGAGQGGYGSGLGGAGQ--------GASAAAAAAAA
Tet.v.         GGLGGGQGGY------------GSGLGGAGQGGQGGQGAGQGAAAAAASAAA
Lat.g.         GGA--GQGGY--------GQ-------GGQGGA-------GAAAAAAAAA-
Arg.a.         GGQ-GGXGGYGGLGSQGAGQ-GYGSGLGGQGGAGQG----GAAAAAAAAAA
Arg.t.         GGQ-GGQGGYGGLGXQGAGQ-GYGAGSGGQGGXQG----GAAAAAAAA--
Ara.d.(ADF-2)  GGQ-GGQGGQGGLGSQGAG----GAGQGGY-GAGQG----GAAAAAAAA--
```

MaSp2

```
Nep.c.         ---GPG--QQGPGGYGPG---QQGPGGYGPGQQGPSGPGSAAAAAAAA
Nep.m.1        ---GPG--QQGPGGYGPG---QQGPGGYGPGQQGPSGPGSAAAAAAAA-
Nep.s.         ---GPG--QQGPGXY----------------GPSGPGSAAAAA---
Lat.g.         ----------GPGGYGPGPGXQQGY--------GPGGSGAAAAAAAA-
Arg.a.         GGYGPGAGQQGPGSQGPGSGGQQGPGGX-----GPYGPSAAAAAAAA-
Arg.t.1        GGYGPGAGQQGPGSQGPGSGGQQGPGGQ-----GPYGPSAAAAAAAA-
Gas.m.         GGYGPGSGQQGPGQQGPGSGGQQGPGGQ-----GPYGPGAAAAAAAA-
Ara.b.         GGYGPGSGQQGPGQQ--------GPGQQ-----GPYGPGASAAAAAA-
Ara.d.1(ADF-3) GGYGPGSGQQGPGQQ--------GPGGQ-----GPYGPGASAAAAAA-
Nep.m.2        -GRGPGGY--GPGQQ----------------GPGGPGAAAAAA---
Arg.t.2        ---GPGGQ--GPGQQ--------GPGGYGPS--GPGGASAAAAAAAA-
Ara.d.2(ADF-4) ---GPGGY--GPGSQGPS-----GPGAYGPG--GP-GSSAAAAAAAAS
```

MiSp

```
Nep.c.1        [GAGGAGGYGR--GAGAGAGAAAGAGAGAGGYGGQGGYGAGAGAGAAAAAGA-]₁₀ [spacer]₁
Nep.c.2        [-----GGYGRGVGAGAGAGAAAGXGAGAGGYGGQGGYGAGXGA---AAAGAG]₁₀ [spacer]₁
Ara.d.(ADF-1)  [GAGAAGGYGG--GAGAGAG------GAGGY-GQ-GYGAGAGAGAAAAAGA-]₅ [spacer]₁
```

Flag

```
Nep.c.         [GPGGX]₄₁ [GGX]₇ TIIEDLDITIDGADGPITISEELTIS--GAGGS   [GPGGXₙ]₂₆
Nep.m.         [GPGGX]₃₆ [GGX]₇ TVIEDLDITIDGADGPITISEELTIGGAGAGGS   [GPGGXₙ]₁₉
Arg.t.§†       [GPGGXₙ]₆      GPVTVDVDVSVGGAPGG   [GPGGXₙ]₅ [GGX]₆ [GPGGXₙ]₇
```

Figure 7.2: Consensus amino acid sequence repeats from various orb-weaving spiders. Consensus repeat sequences for major and minor ampullate and flagelliform silks from various orb-weaving spiders are shown. From Guerette et al. (1996) and Gatesy et al. (2001).

Dolomedes cDNA1
GGAGSGQGGYGNQGGLGGYGQGAGAGAAAAAAA

Dolomedes cDNA2
GGAGSGQGGYGGQGGLGGYGQGAGAGAAAAAAA

Zorocrates cDNA
AAAAASAAAAGGRGSQGGYGDDGGAAAAAAAAAAAAAAAGSGGTGGGQGGRGDGGAAAAAAAAAAEAAAGGKGRQGSYGDDGGAAVAA
AAAAAAAAGRGGSGRGQGLRRDKGSYGVDGG

Phidippus cDNA
GAGAGSGAGYGAGSGAGAGAGAGAGYGAGAGAGAGAGSGAGAGYGAGAGAGYGVGAGAGYGAGAGA

Plectreurys cDNA1
GAGAGAGAGAGAGAGAGSGASTSVSTSSSSGSGAGAGAGSGAGSGAGAGSGAGAGAGAGGAGAGFGSGLGLGYGVGLSSAQAQAQA
QAAAQAQAQAQAQAYAAAQAQAQAQAQAQAAAAAAAAAAA

Plectreurys cDNA2
TIAGLGYGRQGQGTDSSASSVSTSTSVSSSATGPDTGYPVGYYGAGQAEAAASAAAAAAASAAEAA

Plectreurys cDNA3
repeat type 1:
AISSSLYAFNYQASAASSAAAQSSAQTASTSAKQTAASTSASTAATSTTQTAATTSASTAASSQTVQKASTSSAASTAASKSQSSS
VGSSTTSTAAASASSSYAFAQSLSQYLLSSQQFTTAFASSTAVASSQQYAEAMAQSVATSLGLGYTYTSALSVAMAQAISGVGGGA
SAYSYATAISQAISRVLTSSGVSLSSSQATSVAS
repeat type 2:
SSQQSSYDTSSDLSSASSAAAAAASASSYESQFSDASSSSNAAAAA

Plectreurys cDNA4
SQQGPIGGVGGSNAFSSSFASALSLNRGFTEVISSASATAVASAFQKGLAPYGTAFALSAASAAADAYNSIGSGANAFAYAQAFAR
VLYPLVRQYGLSSSAKASAFASAIASSFSSGTSGQGPSIGQQQPPVTISAASASAGASAAAVGGGQVGQGPYGGQQQSTAASASAA
AATATSGGAQKQPSGESSVATASAAATSVTSAGAPVGKPGVPAPIFYPQGPLQQGPAPGPSNVQPGT

Euagrus cDNA
ASQIAASVASAVASSASAAAAAASSSAAAAGASSAAGAASSSSTTTTTSTSSSAAAAAAAAAAAASASGASSASAAASASAAASAFS
SALISDLLGIGVFGNTFGSIGSASAASSIASAAAQAALSGLGLSYLASAGASAVAGVGVGAGAYAYAYAIANAFASILANTG
LLSVSSAASVASSVASAIATSVSSSSAAAAASASAAAAASASASAAASSASASSSASAAAAAGASAAAGAASSASASAAASAFSSAF
ISDLLGFSQFNSVFGSITSSSLGLGIAANAVQSGLASLGLRAAASAAASAVANAGLNGSGAYAYATAIASAIGNALLGAGFLTAGN

Figure 7.3: Amino acid sequence repeats for major ampullate silks from non-orb weavers. Sequences are for presumed major ampullate silk proteins from a variety of non–orb-weaving spiders. From Gatesy et al. (2001) and Van Kirk, C. (personal communication, 2000).

show that high sequence conservation has been maintained over the 125 million years since these various species diverged from each other.

When the major ampullate silk cDNAs from more primitive spiders were examined, the results were very different (Figure 7.3). In fact, except for *Dolomedes*, there is little similarity between the various species or to the orb weavers. Interestingly, the only sequence that appears to be highly conserved is the non-repetitive sequence located at the C-terminus of the MaSp proteins. There is at least 45% amino acid identity between the most divergent pair in this group.

This conservation in the C-terminal region of major and minor ampullate silks was noted previously (Hayashi and Lewis, 1998) and was used by others as part of their cloning strategy (Guerette et al., 1996). It remains surprising in view of the major sequence changes occurring in these species that this sequence was conserved, but it confirms the identity of these proteins as silk proteins. In addition, the protein sequences of the repetitive regions correlate very well with the amino acid compositions of the silks determined from proteins isolated from the same glands as the cDNA.

A key question that remains to be answered about these sequences is whether they are truly major ampullate silk proteins or whether the glands producing them are unrelated to the major ampullate glands of the orb weavers and are from a different lineage. The glands and species were all chosen because the morphological data were consistent with the glands being major ampullate, but the incredible diversity of sequences indicates this may not be the case. If they are indeed major ampullate glands, then a tremendous diversification in sequences of non-orb weavers must have occurred during the same time frame as high conservation was occurring during the species diversification of the orb weavers and derived orb weavers. Either way, it is clear that the constraints of web capture have substantially suppressed protein divergence in the orb weavers.

BIOPHYSICAL STUDIES

The majority of data on spider silk has been obtained from major ampullate (dragline) silk. There are two reasons for this: it is easily obtained and exhibits the combination of elasticity and high tensile strength that makes it an extraordinary material.

Several early studies of silk fibers using X-ray diffraction provided some information, much of which was interpreted based on the structure of silkworm silk. These studies led to the classification of dragline silk as β-sheet proteins. However, it was also demonstrated using this methodology that much of the structure was not β-sheet, but appeared to be unordered.

The structure of the dragline silk fiber was probed in the relaxed and extended states using fourier transform infra-red (FT-IR) spectroscopy (Dong et al., 1991). Data confirm the presence of significant amounts of β–sheet-like structure that appear to be the same in both relaxed and extended forms. However, in the extended state, a helical structure appears to form that returns to the original form when tension is released. The parallel-polarized spectra showed that the orientation of protein in the sheets and helices is parallel to the fiber axis for both *N. clavipes* and *A. gemmoides*. The helical regions appear to arise from random or non-oriented regions as no decrease in intensities was observed in the β-sheet regions.

Two papers have demonstrated the presence of β-sheet structure using solid-state nuclear magnetic resonance (NMR) (Simmons et al., 1996; Parkhe et al., 1997). This structure was assigned to the alanine-rich regions based on the resonances of alanine being consistent with a β-sheet structure. Solid-state NMR using ^{13}C shows chemical shifts of α, β, and carbonyl carbons for amino acid residues in polypeptides are sensitive to secondary structure and therefore can be used to predict the peptide backbone conformations of solid-state biomaterials in their native environments. The alanine Cα (48.9 ppm) and Cβ (20.7 ppm) chemical shifts for the relaxed major ampullate silk indicate that the alanine residues reside in β-sheet regions, in agreement with previous results (Simmons et al., 1994). In addition, no change in the alanine chemical shifts occurs upon stretching or supercontraction in major ampullate or in minor ampullate silk. These results indicate that the poly-A regions of both major and minor ampullate silks adopt a β-sheet structure that is retained upon stretching and supercontraction. However, a change in the orientation of the β-sheets as demonstrated by the aforementioned diffraction experiments would not be detected by magic angle spinning NMR experiments.

The GGX-rich regions of MaSp 1 were originally thought to be responsible for elasticity, but later studies with molecular modeling suggested that the structures in MaSp 2 were more likely candidates. The GGX structure was proposed by us to be poly-glycine II helix based on molecular modeling (Parkhe et al., 1997). NMR data supported a helical structure with a 3_1 helix best fitting the data (Kummerlen et al., 1996).

A β-turn structure for the proline-rich regions of MaSp 2 has been proposed to describe silk elasticity based on computer modeling studies (Hayashi et al., 1999). In this model, the pentapeptide segments of MaSp 2 form β-turns which are organised to form a β-spiral. The properties of the flagelliform silk protein are consistent with this model as the silk is highly elastic. In addition, other elastic proteins – such as elastin, the mussel byssus thread, and wheat gluten – are believed to form similar structures (see Chapters 4, 9, and 13).

STRUCTURE–FUNCTION RELATIONSHIPS

To understand the basis for the mechanical properties of spider silks, especially elasticity, it is necessary to recognize that proteins form distinct three-dimensional structures. This is in contrast to many other polymers that form networks of molecules with differing individual structures. It therefore follows that understanding the structures of these protein molecules will lead to understanding their role(s) in the various mechanical properties of spider silks.

All of the spider silk proteins sequenced thus far show the presence of multiple clearly defined sequence motifs that are discussed previously and summarized

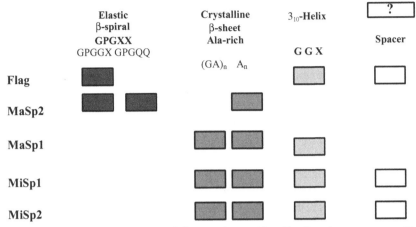

Figure 7.4: Protein sequence motifs from various spider silks. Protein sequence motifs for the spider silk proteins that have been sequenced from multiple species are shown.

in Figure 7.4. As several different groups concluded, the poly-A segments predominantly form β-sheets. These β-sheets then form the crystallites seen in X-ray diffraction studies. They are also responsible for the interactions between the individual protein molecules in major ampullate silk that lead to insolubility. These regions are likely to be responsible, in large part, for the high tensile strength by intermolecular hydrogen bonding and hydrophobic interactions that hold the polymer molecules together. Because these motifs are spread throughout MaSp 1 and 2, which are very large proteins, it is likely that the molecules arrange themselves in a random alignment pattern. This further enhances the strength as the formation of gaps is minimised. This is especially important for a fiber that can be converted from soluble to insoluble in less than 0.1 s.

Minor ampullate silks have similar β-sheet regions, but these are composed of poly-GA repeats. This type of crystallite will be inherently weaker due to the lack of interactions on both sides of the sheet. The lack of the alanine methyl groups will leave 'holes' on one side of each motif in the crystallite (Figure 7.5). We predict this is likely to be the cause of the four-fold lower tensile strength observed for minor ampullate silk.

The GGX regions are also present in all three of these silk types, but the flexibility of glycine makes it unlikely that these regions can form a β-sheet. A more likely possibility is that these regions exist as some form of poly-glycine helix. Evidence supporting a similar helical structure was recently published as described previously. Interestingly, these regions of the protein molecules could also form protein:protein interactions regions since the X residues are all

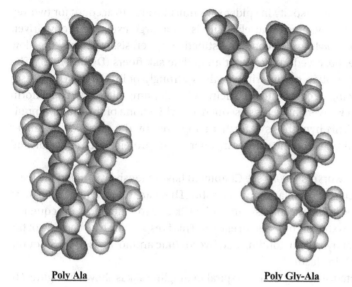

Poly Ala   Poly Gly-Ala

Figure 7.5: Computer models of the poly-A and poly-GA segments. Computer-predicted models for four antiparallel strands of the poly-A or poly-GA motifs are shown. Note the large vacant area in the poly-GA when compared with the poly-A strands.

located along one side of the helix. This could lead to structures that form almost crystallite-like regions much like the β-sheets.

The basis for the interactions of the flagelliform silk protein is less well defined due to its apparent lack of β-sheet motifs. At this point, two plausible mechanisms can be proposed. The first is that the GGX regions, as described, could form intermolecular crystallites. The conservation of sequence of the X amino acids lends support to this mechanism. The second is that the amino acids in the X position of the GPGGX repeats can also serve as the source of interactions between protein molecules. The distinct pattern of the X residues in the repeat of consecutive A or alternating S and Y support this possibility. Either or both of these proposed mechanisms may be correct. As stated previously (Vollrath and Knight, 2001), better structural studies will be needed to more fully understand the structure–function relationships of spider silk proteins.

The combination of these β-sheet and interhelix interactions between protein chains in all the silks sequenced to date would produce high tensile strength. Ideally, these interactions required would exceed the energy of a typical peptide bond (about 80 kcal), making the fiber as strong as its monomers allow. That amount of energy corresponds to approximately 16 regions of 6–8 amino acids if average hydrogen bond energies are used. The protein sequences deduced to this point all exceed this number of possible interacting motifs.

Explanations of elasticity in spider silks must be able to account for two key facts. The first is evidence that elasticity is, to a large extent, entropy driven (Gosline et al., 1984) and second, a distinctive hysteresis, especially the slow recovery time, observed in the major ampullate silk fibers (Denny, 1976). The correlation of protein motifs with elasticity strongly points to the GPGXX sequences as being necessary for elasticity. All major ampullate and flagelliform silks from orb weavers have these sequences, while none of the minor ampullate silks do. This hypothesis is further supported by the presence of similar proline-containing sequence motifs in other elastic proteins such as elastin and gluten.

Urry and coworkers (2002) (see Chapter 3) have extensively studied the basis for the elasticity of a similar motif in elastin. There are reasons to believe there are differences in the mechanism in spider silk due to differences in sequence, as well as the hysteresis not seen in elastin. This is especially true because of the substitution of hydrogen bonding and hydrophilic amino acids (Q, Y, S) for the hydrophobic V in elastin.

The expected structure of the β-spiral in dragline silk is shown in Figure 7.6. The structure is very much a 'nano-spring'. It can be imagined that the longer the spiral, the higher the extensibility of the protein and the fiber will be. This

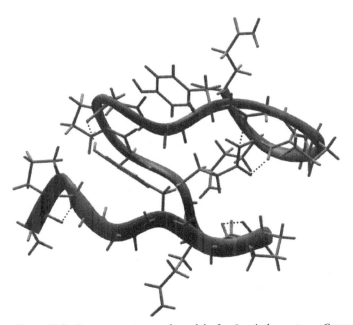

Figure 7.6: Computer-generated model of a β-spiral structure. Computer-generated model for a pair of GPGQQGPGGY repeats is shown. Hydrogen bonds (shown as dashed lines) occur both between residues in the same β-turn and between different β-turns.

correlates with the presence of 3–6 pentamer repeats in major ampullate silks, with elasticities in the 25–35% range and 43–65 pentamer repeats in flagelliform silks, with elasticities exceeding 200%.

Entropy-based elasticity is a process in which elongation of the fibre leads to an increased order of the molecules in the fiber. Some interpretations imply the conversion from a random structure to a defined one, but this seems less likely in a protein where the molecules are already in a single well-defined structure. In this case, the entropy force probably arises from another mechanism.

The consistent presence of proline in the β-turns of spider silk suggests a key role for this amino acid since a significant fraction of β-turns in proteins do not contain proline. If the β-spiral is extended, then a significant torque will be applied to the proline causing the protein main chain to attempt to straighten out. This torque would lead to a large decrease in entropy of the proline due to the loss of various vibration and rotational states. Release of tension on the fiber will lead to relaxation of the proline with a concomitant loss of energy. Data from Denny (1976) show a loss of nearly 65% of the absorbed energy, with only a 20% decrease in length for flagelliform silk and a 25% decrease in length for major ampullate silk. Thus, the vast majority of energy is released rapidly as heat without a corresponding decrease in the fiber length. The loss of energy with little retraction in length is consistent with a major role for proline in the spider silk elasticity mechanism.

The second part of the hysteresis is the slow retraction after initial loss of energy. This phenomenon is probably due to the reformation of hydrogen bonds that were broken in the initial extension. These hydrogen bonds are not only inside the β-turns, but also between layers of the spiral and between protein molecules (dotted lines in Figure 7.6) which may contribute even more to the mechanism. This slow retraction prevents these silks and the web from acting like a trampoline to catapult the intercepted insect back off without trapping it.

CONCLUSIONS

The past 15 years have seen an explosion of information about spider silk sequences, structures, and mechanical properties. Despite this, there is still much to be learned from the more than 450 million years of arachnid evolution. Structures of the proteins from the non–orb-weaving spiders have yet to be determined; and considering the diversity of sequences in this group, it is highly likely that new sequence motifs are yet to be discovered. Only the proteins from three of the six silks produced by orb weavers have been sequenced, and the amino acid compositions of the remaining three indicate they may not be composed exclusively of the currently known motifs.

The detailed relationships between structure and function are not fully understood. This holds true for tensile strength, elasticity, fiber formation, and the liquid crystalline state. The evolutionary relationships between different silk types and the mechanism for generating the diversity of major ampullate gland silk proteins are also important unsolved questions. The tools to answer these questions are now becoming available and will almost certainly uncover more intruiging questions as current ones are answered.

REFERENCES

Beckwitt, R., and Arcidiacono, S. (1994) Sequence conservation in the C-terminal region of spider silk proteins (Spidroin) from *Nephila clavipes* (Tetragnathidae) and *Araneus bicentenarius* (Araneidaie). *J. Biol. Chem.* **269,** 6661–63.

Bell, A. L., and Peakall, D. B. (1969) Changes in the fine structure during silk protein productions in the ampullate gland of the spider *Araneus sericatus. J. Cell. Biol.* **42,** 284–95.

Colgin, M., and Lewis, R. V. (1998) Spider minor ampullate silk proteins contain new repetitive sequences and highly conserved non-silk-like "spacer regions." *Protein Sci.* **7,** 667–72.

Denny, M. (1976) The physical properties of spider's silk and their role in the design of orb-webs. *J. Exp. Biol.* **65,** 483–506.

Dong, Z., Lewis, R. V., and Middaugh, C. R. (1991) The molecular basis of spider silk elasticity. *Arch Biochem. Biophys.* **284,** 53–58.

Fischer, E. (1907) About spider silk. *Hoppe-Seyler's Z Physiol. Chem.* **53,** 440–50.

Gatesy, J., Hayashi, C. Y., Motriuk, D., and Lewis, R. V. (2001) Extreme diversity, conservation, and convergence of spider silk fibroin sequences. *Science* **291,** 2603–5.

Gosline, J. M., Denny, M. W., and DeMont, M. E. (1984) Spider silk as rubber. *Nature* **309,** 551–52.

Gosline, J. M., DeMont M. E., and Denny, M. W. (1986) The structure and properties of spider silk. *Endeavor* **10,** 37–43.

Guerette, P. A., Ginzinga, D. G., Weber, B. H. F., and Gosline, J. M. (1996) Silk properties determined by gland specific expression of a spider silk fibroin gene family. *Science* **272,** 112–15.

Hayashi, C., and Lewis, R. V. (1998) Spider flagelliform silk proteins. *J. Mol. Biol.* **275,** 773–84.

Hayashi, C. Y., Shipley, N. H., and Lewis, R. V. (1999) Hypotheses that correlate the sequence, structure, and mechanical properties of spider silk proteins. *Int. J. Biol. Macromol.* **24,** 271–75.

Hayashi, C. Y., and Lewis, R. V. (2000) Molecular architecture and evolution of a modular spider silk protein gene. *Science* **287,** 1477–79.

Hinman, M. B., and Lewis, R. V. (1992) Isolation of a clone encoding a second dragline silk fibroin. *J. Biol. Chem.* **267,** 19320–24.

Iazuka, E. III. (1983) The physico-chemical properties of silk fibers and the fiber spinning process. *Experientia* **39,** 449–54.

Kovoor, J. (1972) Etude histochimique et cytologique des glandes sericigenes de quelques Argiopidae. *Ann. Sci. Nat. Zool. Biol. Anim.* **14,** 1–40.

Kummerlen, J., van Beek, J. D., Vollrath, F., and Meier, B. H. (1996) Local structure in spider dragline silk investigated by two-dimensional spin-diffusion NMR. *Macromolecules* **29,** 2920–28.

Lucas, F. (1964) Spiders and their silk. *Discovery* **25,** 20–26.

Parkhe, A. J., Seeley, S. K., Gardner, K., Thompson L., and Lewis, R. V. (1997) Structural studies of spider silk proteins in the fiber. *J. Mol. Recogn.* **10,** 1–6.

Simmons, A., Ray, E., and Jelinski, L. W. (1994). Solid-state ^{13}C NMR of *Nephila clavipes* dragline silk establishes structure and identity of crystalline regions. *Macromolecules* **27,** 5235–37.

Simmons, A., Michal, C. A., and Jelinski, L. W. (1996) Molecular orientation and two-component nature of the crystalline fraction of spider dragline silk. *Science* **271,** 84–87.

Stauffer, S. L., Coguill, S. L., and Lewis, R. V. (1994) Comparison of physical properties of three silks from *Nephila clavipes* and *Araneus gemmoides. J. Arachnol.* **22,** 5–11.

Theil, B. L., Kunkel, D. D., and Viney, C. (1994). Physical and chemical microstructure of spider dragline: A study by analytical transmission electron microscopy. *Biopolymers* **34,** 1089–97.

Urry, D. W., Hugel, T., Seitz, M., Gaub, H. E., Sheiba, L., Dea, J., Xu, J., and Parker, T. (2002) Elastin: a representative ideal protein elastomer. *Phil. Trans. R. Soc. London B* **357,** 169–84.

Vollrath, F. (1992) Spider webs and silks. *Sci. Am.* **266,** 70–76.

Vollrath, F., and Knight, D. P. (2001) Liquid crystalline spinning of spider silk. *Nature* **410,** 541–48.

Willcox, P. J., Gido S. P., Muller, W., and Kaplan, D. L. (1996) Evidence for a cholesteric liquid crystalline phase in natural spinning processes. *Macromolecules* **29,** 5109–10.

Work, R. W. (1977) Dimensions, birefringence and force-elongation behavior. *Textile Res. J.* **47,** 650–62.

Xu, M., and Lewis, R. V. (1990) Structure of a protein superfiber: Spider dragline silk. *Proc. Natl. Acad. Sci. U.S.A.* **87,** 7120–24.

EIGHT

The Nature of Some Spiders' Silks

Fritz Vollrath and David P. Knight

INTRODUCTION

Lepidopteran silk has a long tradition of commercial exploitation (Anquetil and Walter, 1996). The traditional standing of spider silk, on the other hand, has until now been mainly that of a mystical material made by a rather dubious creature (Hilliard, 1994). Recently, however, spider silks have attracted considerable attention as potential blueprints for designing modern 'techno-fibres' (Vollrath and Knight, 2001). Web silks have unusual characteristics for structural proteins. Instead of growing inside the body, they are extruded and body contact is not important for their function. Moreover, most silks function best in the dry and not in the aqueous state. A single fibre typically is of considerable purity and is many hundreds of meters long. Indeed, a single fibre is also thick enough to be handled, allowing its mechanical properties to be measured and analysed with traditional methods (Madsen et al., 1999). Finally, partly because of their proven commercial track record, silks have attracted considerable attention and funding. Hence, novel insights are rapidly accumulating (Kaplan et al., 1994); especially in the areas of silk genomics (e.g., Prince et al., 1995; Guerette et al., 1996; Lewis, 1999; Hayashi and Lewis, 2000), silk chemistry (Lewis, 1992; Winkler and Kaplan, 2000), silk structure (Grubb and Jelinski, 1997; Beek et al., 1999; Sirichaisit et al., 1999; Riekel and Vollrath, 2001), silk spinning (Knight and Vollrath, 1999b, 2001a, 2001b), silk mechanics (Thiel and Kunkel, 1994a; Vollrath et al., 2001), silk modelling (Fossey and Kaplan, 1994; Gosline et al., 1994; Termonia, 1994; Thiel and Viney, 1995), and silk copying (O'Brien et al., 1994, 1998; Lazaris et al., 2002). Thus silks, probably more than any other elastomer, might lead the way to understanding how biological elastomers are formed and how they function (Knight and Vollrath, 2002a).

There are two reasons for the rapid ascent of spider silks over silkworm silks as objects for study and prototypes for modern materials: firstly, spider silks generally have more remarkable mechanical properties and, secondly, they can be drawn from the spider under highly controlled conditions. The first of these statements 'superb properties' is by now well accepted and will be discussed later. The second statement 'controlled spinning' is less obvious and requires immediate attention.

Commercial silkworm silk is obtained by unraveling a pupal cocoon spun virtually without restriction by the lepidopteran larva; this means that production conditions during thread spinning are rather variable, leading to considerable variability in the mechanical properties along a single length of thread (Perez-Rigueiro et al., 2001; Shao and Vollrath, 2002). Spider silk (especially dragline silk), however, can be drawn artificially (and under highly controlled conditions) from the immobilised spider (Vollrath et al., 2001). This allows us to reel from one individual spider long lengths of thread that are continuous yet spun under highly diverging conditions (e.g., by heating or cooling the spider or by changing the speed of reeling out the silk). Moreover, such experiments can be rather 'portable', with the silk being made to pass through the imaging windows of an X-ray beam (Riekel and Vollrath, 2001) or a Raman spectrometer (Shao et al., 1999a) for on-line bandshift analysis. In either case, if the experimental conditions are carefully recorded, then the mechano-chemical properties of each length of thread can later be correlated to specific spinning conditions (Vollrath et al., 2001). Such experiments (only very recently achieved for silkworm silks; see Shao and Vollrath, 2002) allow us to ultimately correlate the mechanical properties of a spider's silk with its molecular composition, orientation, and interaction. Hence, the attraction of spider silk for the study and design of advanced biopolymer fibres. In addition, such 'silking' experiments, as mentioned previously, are beginning to shed light on another aspect of biomimetic fibre design, namely the interaction of spinning conditions and 'raw' silk feedstock in shaping the fibre (Chen et al., 2001).

THE NATURAL ROLE OF SPIDER SILK
To understand the properties of spider silk, we must attempt to understand the function of this material in its natural state. In the following, we shall concentrate on presenting the requirements for spider threads in nature and, in a purely exemplary way, discuss the biology and structure (the function and form) of two specific silks that evolved in response to these requirements. Of course, not all spiders make webs, and many web and dragline silks will have rather different chemico-physical properties from the ones presented here (Zemlin, 1968; Denny, 1980). Moreover, virtually nothing is known about the cocoon silks of spiders, their cement and wrapping silks, or indeed any other unusual silks such as, for example, the underwater silks of *Argyroneta aquatica* or the nanometer thick ribbons of the brown recluse *Loxosceles* sp. (Knight and Vollrath, 2002b).

Nevertheless, the web silks of orb-weaving spiders (Figure 8.1) give a good introduction to the interaction of two very diverse silks with the environment

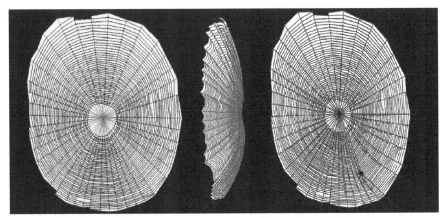

Figure 8.1: Typical orb web of the garden cross spider *A. diadematus*, at rest (frontal view, left), bowing out during a wind of 5 metres per second (in side view middle), and in maximum deflection during the impact of a 50 mg prey (oblique view, right). Different behaviour of the stiff radial threads and the soft capture spiral threads is apparent.

and one another. For the purpose of prey capture, in the spider's web the builder (the spider), the architecture (the web structure), and the building materials (which are all silks) form a unit. Only if all parts interact well can the unit intercept and hold firm the prey that, after all, has evolved in a constant life-dinner arms race with the spider and is highly adapted to escape.

A fly is food, but typically this food tends to travel at speed and thus has considerable kinetic energy. Hence, it is important that the web strands do not easily break on impact (i.e., are tough), which entails a combination of strength (force required to break) and stretchability (extension required to break). At the same time, each thread has a function in the web as an engineered structure, with a lifetime that typically (although not in all spiders) is designed to 'survive' the elements (encountered in a normal day), as well as capture (ideally) several prey items before web-rebuilding or any major repairs are required. Thus, in addition to prey impact, there are other design requirements for a web thread (Vollrath et al., 1997). Wind, for example (Figure 8.1), directly exerts a strong tensile load on a thread (when the thin strands bow out under laminar flow), and wind indirectly affects web strands (when the web's anchor threads are stretched because the attachment branches flutter in the wind). Clearly, web threads should be rather elastic (i.e., resume their original length quickly rather than have slow hysteresis) and thus, by 'hanging through', turn the whole structure from a tight filter net into a bag when the wind gusts. On the other hand, the strands should not be so elastic as to become a trampoline, flinging the fly back out to from whence it came. Finally, there are the combined design requirements of stretchability and softness for the holding strands to prevent a fly that has been stopped from

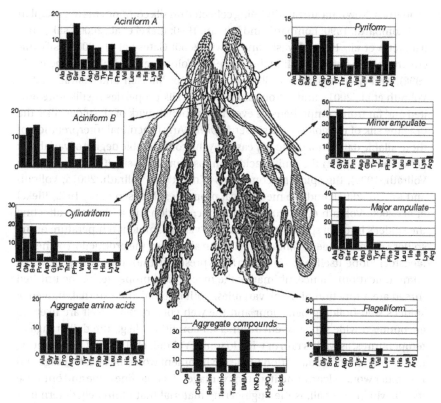

Figure 8.2: Seven specialized glands of a typical Araneid orbweaver showing the marked difference in amino acid compositions. The aciniform glands make the silk for wrapping prey, the cylindriform glands make the cover silk for the egg sac, and the pyriform glands make the silk for attachments and for joining threads. The major and minor ampullate glands are of similar shape, with the smaller minor glands providing the radial silk and the larger major ones the dragline and frame threads. The flagelliform glands provide the core threads of the capture spiral while their coating is supplied by the aggregate glands. Further, note that all silk glands are bilaterally paired, having evolved from limb glands (after Vollrath and Knight, 2001).

getting a hold and walking straight out of the web. At the same, time, however, the web must allow the spider to get a grip so that it can attack and drag the fly to its lair. Clearly, web strands have to fulfil a multitude of functions, and we could do worse than to look for web silks to explore the functional diversity of spider silks.

All spiders produce more than one silk, but web spiders have evolved an amazing battery of silks, each adapted for a particular task (Figure 8.2). These spiders use their complement of different silk types for many different tasks and accordingly have evolved a wide range of different chemico-mechanical properties. New data are rapidly being collected on silk components (e.g., Kaplan and Lombardi, 1990; Lombardi and Kaplan, 1990; Riekel et al., 2000), silk genes (Guerette et al., 1996; Hayashi and Lewis, 2000; Gatesy et al., 2001), silk microstructure (Li et al., 1994; Mahoney, 1994; Vollrath et al., 1996; Frische et al., 1998), and fibre mechanics (Shao et al., 1999b; Madsen and Vollrath, 2000; Vollrath et al., 2001), and yet our knowledge of silk properties is still quite limited and far from comprehensive even for the benchmark of all spider silks: the dragline silk of the golden orb spider *Nephila*. Any functional interpretation of the available data is not made any easier by the considerable degree of variability apparently found in the amino acid composition of the silk (Craig et al., 1999; Vollrath, 1999), the spinning conditions (Knight and Vollrath, 2001a; Vollrath and Knight, 2001), and the mechanical properties (Madsen et al., 1999; Riekel et al., 2000; Vollrath et al., 2001). External conditions, in addition to affecting the spider's internal spinning process [e.g., via the ambient temperature or the spider's running speed (Vollrath, 2000)] can also affect the production of the silk feedstock 'dope' itself (e.g., via amino acids present in the food; C. Craig, personal communication). Hence, ultimately the mechanical properties of the finished thread are a function of many variables, both internal and external. Indeed, a spider's web-building behaviour and the web construction itself are likely to encounter a wide range of environmental conditions [e.g., the daily range of temperatures, as well as rain or wind loading (Vollrath et al., 1997)]. Considering this, rapid temporal adaptation to the environment would be an advantage and the spider would clearly find it to its advantage if it could fine-tune and optimise its silk which, after all, is the engineering material that ultimately determines web function (Lin et al., 1995).

The structural and the capture silks of the advanced araneid orb weavers illustrate well the amazing extent of nature's ability to tinker with spider silks through modifying keystone traits. Sometimes, such modifications in material properties are surprisingly 'clever', for example, when a serious phylogenetic design constraint (e.g., plastisation by water, see below) is turned to a crucial advantage that allows adaptive radiation into a new niche. A prime example is the evolutionary 'invention' of the wet windlass system, which allowed orb webs to be manufactured more cheaply than the dry hackled-spring system (Vollrath, 1994). This great savings in production costs (at least 50%) may very well be the main reason for the enormous numerical superiority of the ecribellate (windlass) over the cribellate (hackled silk) orb weavers (Opell, 1997, 1998; Vollrath, 2000).

THE ORB WEB AND ITS TWO MAJOR SILKS
Spider webs provide protective housing, signal lines to monitor walking/flying/swimming prey and single trap lines, as well as extensive nets to ensnare walking

or flying prey (Bristowe, 1958; Foelix, 1996). In many ways, the most interesting of all webs are the orb webs which are two-dimensional aerial filters highly adapted to the task of capturing high-velocity insect 'missiles', and often specialised for very specific types of insects (Stowe, 1986). Indeed, one of the most stunning examples of animal architectural engineering is probably the araneid orb web in its various guises (Witt et al., 1968; Hansell, 1984; Dawkins, 1996). Such webs are constructed by a highly adaptable behaviour (Eberhard, 1988; Vollrath, 1992) that consists of surprisingly simple orientation and decision rules (Krink and Vollrath, 1997; Krink, 2000). The structural complexity of these webs is to some extent the result of emergent properties in the building behaviour (Vollrath, 1992), as well as the complex interactions of the various building materials (Lin et al., 1995).

The typical orb web (Figure 8.1) consists of a frame of extremely tough silk supporting a wheel of tough silken spokes radiating from a hub that balances the tensile forces (Wirth and Barth, 1992); these tightly strung radii in turn support an extremely elastic sticky spiral (Vollrath and Edmonds, 1989). Soft spiral and hard radii work together to absorb the insect's impact and to hold it firm ideally until the spider has made its move (Lin et al., 1995). Two-dimensional orb webs evolved to negotiate substantial impact loads by out-of-plane deflection of the whole lightweight structure. Their astonishing performance is made possible because an orb web incorporates two very different types of principal silks: the supporting radius threads and the sticky holdfast threads.

Although the overall shapes of all typical orb webs are very similar indeed; their fine structure is not. In fact, orb web-builders have evolved two very different ways for their capture threads to hold prey fast (Coddington, 1986; Shear, 1986; Peters, 1987; Eberhard, 1990). Thus, they have been classified into two groups defined by the type of capture spiral stickiness (Foelix, 1996). Cribellate spiders, such as *Uloborus walckenarius*, use very fine silk filaments of nanometer diameter which are combed onto supporting axial threads in hackled strands of many hundred filaments. These hackled strands probably provide elasticity, as well as adhesion by a combination of silk-inherent spring action and combing-transferred electrostatic charges (Eberhard, 1980; Peters, 1984; Opell, 1995). Ecribellate spiders, such the temperate garden cross spider *Araneus diadematus* or the tropical golden silk spider *Nephila clavipes*, lack the cribellum silk. Instead, they coat the axial threads with an aqueous solution that forms the sticky droplets, which are so crucial for the function of these capture threads (Figure 8.3). It has been argued that the ecribellate orb weavers have evolved from cribellate spiders, which lost the cribellum (for a discussion, see Shear, 1986). Irrespective of this, the ecribellate way of orb web construction is significantly more economical and is presumably a key factor in the adaptive radiation of the ecribellate orb weavers (Vollrath, 2000).

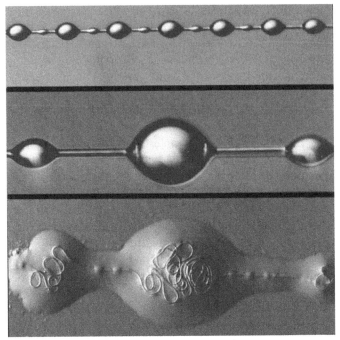

Figure 8.3: Capture thread of *A. diadematus* under increasing magnification. The wind-lass mechanism is seen in the lower panel, where the core fibres after a large extension-contraction cycle have been reeled into a droplet (Vollrath and Edmonds, 1989).

THREADS OF THE GARDEN SPIDER'S ORB

Let us take a closer look at the well-known orb webs of our garden spider *A. diadematus* (Figure 8.1). Essential for the effectiveness of this web are the rather divergent mechanical properties of the two main silks used in construction: stiffness in the dry radii and elasticity in the wet capture spiral. The strong and very tough radius threads of *Araneus* are dry, show good extensibility (ca. 40%), high tensile strength (ca. 1.2 GPa), and considerable elastic recoil (around 80% after loading to 50% breaking strain), which indicates that they function as shock-absorbing structural elements (Köhler and Vollrath, 1995) (Figure 8.4). The soft sticky spiral, however, is wet and has little structural function. It absorbs impact energy by its astonishing extendibility (ca. 500%) of the wetted thread, which develops substantial force only after around 100% extension (Vollrath and Edmonds, 1989; Köhler and Vollrath, 1995) (Figure 8.4). The engineering strength of these capture threads is ca. 1400 MN/m^2 with a breaking energy of 163 J/cm^3, which thus compares well with the radial threads at ca. 1200 MN/m^2 engineering strength with a breaking energy of 194 J/cm^3 (Vollrath, 2000). Note

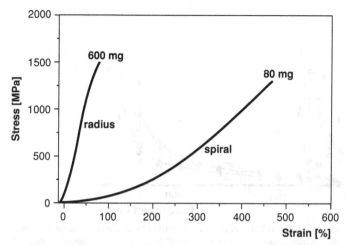

Figure 8.4: Typical stress strain curves of the major web silks of *A. diadematus*. The dry silk of the major ampullate glands, which is used for the radii (web spokes) as well as the safety dragline, is much stiffer than the wetted compound silk of the capture spiral made by the flagelliform and aggregate glands (after Köhler and Vollrath, 1995).

that the double core fibres of a capture thread are much thinner than those of a radius thread; hence, the force required to snap a capture thread is much lower. However, most attempts at diameter normalisation give impossibly inaccurate values because inside their aqueous coating, the core fibres are waterlogged and thus supercontracted (Vollrath and Edmonds, 1989).

Both amino acid compositions (Andersen, 1970; Work and Young, 1987; Kaplan and Lombardi, 1990; Lombardi and Kaplan, 1990) and gene sequences (Gosline et al., 1994; Guerette et al., 1996; Hayashi and Lewis, 2000; Gatesy et al., 2001) of the two fibre types differ significantly. Nevertheless, it seems that this is not responsible for the enormous difference in the material properties of the radius-dragline silk and the capture spiral silk (Vollrath and Edmonds, 1989). Indeed, this is to a large extent due to a rather simple trick during fibre production. It is an 'elegant' trick, a simple adaptation that illuminates nature's way of creative tinkering and is based on the principle of turning an inherent functional weakness into a derived functional advantage. The adaptive modification of a stiff to an elastic thread is based simply on wetting it (Figure 8.5). This is achieved by the spider coating the naturally dry capture threads with a layer of hygroscopic compounds (Vollrath et al., 1990; Tillinghast et al., 1991). The coating is done by adding the aqueous dope of the aggregate gland spigots to the dry core fibres as they emerge from the flagelliform gland spigots (Peters, 1955). The liquid dope surrounds the core fibre like a tight coat (Edmonds and Vollrath, 1992). However, the dope compounds immediately draw water from the atmosphere and thus

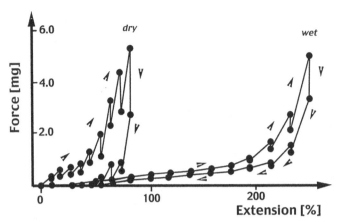

Figure 8.5: Capture spiral thread of *A. diadematus* in the dry and wet state. Such thread loses its extreme stretchiness (right hysteresis curve) if its aqueous coating is washed off (left curve) (after Vollrath and Edmonds, 1989).

the thin coat becomes a rapidly swelling aqueous cylinder. Finally, this water cylinder becomes unstable and 'disintegrates' into interconnected droplets. The rate of swelling and the subsequent droplet size are dominated by a dynamic equilibrium between the water in the sticky coat and the water vapour in the air (Edmonds and Vollrath, 1992). Some of the water enters the core fibres and plasticises them which makes them extremely stretchable (Vollrath and Edmonds, 1989) as water is a strong plasticiser for many (if not most) native silks (Shao et al., 1999a, 1999b). The flagelliform capture thread fibroin proteins assist the plastisation by making ample space for water molecules (Vollrath and Edmonds, 1989; Bonthrone et al., 1992) possibly in the β-turns of the glycine-rich regions (Hayashi and Lewis, 2000). However this may be, without the water trick, the capture spiral would be just another stiff silk (Figure 8.5) and lack the supreme stretchability so necessary for the toughness required to absorb the high-kinetic energy of the prey without breaking.

But this is not the end of the story. The inherent instability of the swollen aqueous coat leads to the formation of droplets. These droplets, in turn, empower the capture threads to 'recoil' instantaneously after even large amounts of stretch (Gosline et al., 1984, 1986; Vollrath and Edmonds, 1989). This is done by an intriguing mini-mechanism, which relies on the use of water not only as a plasticiser to toughen the core fibres, but also to power a mechanical (rather than molecular) elasticity. This is the second part of nature's clever 'trick of the supercontractile capture threads'. After attracting the necessary water from the atmosphere, the hygroscopic compounds hang on the spiral thread in droplets like pearls on a string, but are interlinked by a thin water film (Edmonds and

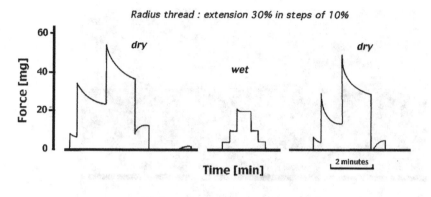

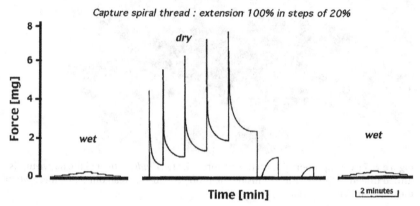

Figure 8.6: Effect on stress-strain characteristics of wetting and drying *A. diadematus* silk. The radius/dragline silk from the major ampullate glands is normally dry, rather stiff, and shows considerable stress relaxation, while the composite capture spiral with its flagelliform gland core fibres and aggregate gland coating is normally wet, soft, and shows little stress relaxation. Wetting a radius fibre induces it to lose stiffness, together with stress-relaxation, while drying a spiral thread induces it to become stiff and show tremendous stress relaxation. In both cases, the change is instantly and totally reversible (after Vollrath and Edmonds, 1989).

Vollrath, 1992). Each swollen droplet now can act like a tiny windlass. Powered by the force of the water's surface tension, each droplet is able to reel in any slack in the core fibre that might result from hysteresis brought about by a rapid stretch-relax cycle (Vollrath and Edmonds, 1989). Thus, the windlasses ensure that, with a strong plastic component of the core fibre, the threads do not sag even after large deformations. The windlass action requires the core fibres to be soft and pliable, which is achieved by water plasticisation. Indeed, without the water, the surprising extensibility of the core fibres would not be possible (Figure 8.6). Consequently, the water of the aqueous coat is at the same time responsible for

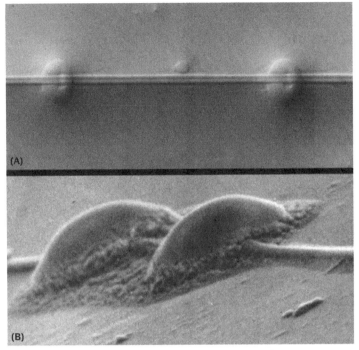

Figure 8.7: Glycoprotein glue on the capture silk of *A. diadematus*. On the native undis-
turbed thread, it forms little tori inside the droplets which spread out and adhere when
the thread touches a surface.

the huge stretchability of the capture spiral and its astonishing elasticity. We say,
"Astonishing," because a 10 mm length of capture spiral taken from the *Araneus*
web can be stretched to 40 mm and then, immediately, without any sag, contract
into a 1-mm droplet upon relaxation.

But this is not all, there is a third part to the 'trick of the droplets'. During
the formation of a droplet, the aqueous solution undergoes a phase transition,
which leads to the separation of a microscopic torus of *N*-acetylgalactosamine
glycoprotein hanging suspended in the middle of each droplet and encircling
the core thread (Vollrath et al., 1990). Due to its sugar component, each torus
is highly adhesive (Figure 8.7) and, together with its neighbours, these minute
sticky nodules provide the famous stickiness of the garden spider's orb web.
Adhesiveness is as important as the silk's mechanical properties, and the fine
threads of the capture spiral must stick firmly to hold the prey and ensnare it
even when it struggles to get free. The discreet distribution of the glue, combined
with the ability of each parcel of glue to slide along on the soft core fibres, allows

for an extremely close fit of the capture thread to the prey surface. Moreover, even as the prey struggles, the pulling-away forces are never localised on one part of the capture thread, but continue to change position as the animal moves and the glue parcels slide on the thread.

As demonstrated, the aqueous coat and droplets are crucial for the function of these ecribellate capture threads. Chemical analysis of the coat shows that, in addition to the 80% or so water, it contains high concentrations of five water-soluble organic compounds and low concentrations of inorganic salts, as well as at least one glycoprotein. Some of the organic compounds are close relatives of neurotransmitters, being either derivatives or hydrolysates (i.e., γ-amino-n-butyricamide, N-acetyltaurine, choline, betaine, and isethionic acid) (Vollrath et al., 1990; Tillinghast et al., 1991). In addition, the coating contains cysteic acid (2%), lysine (2%), serine (2%), potassium nitrate (7%), potassium dihydrogenphosphate (3%), glycine, and highly saturated fatty acids (Fischer and Brander, 1960; Schildknecht and Munzelmann, 1972). Betaine and γ-amino-butyric-acid (GABA) are osmoprotectives and osmolytes in a wide range of organisms, whereas taurine is a protein-stabilising compound (Tillinghast et al., 1991). It may be argued that, by acting osmotically without interference with normal protein function, the organic compounds not only maintain the aqueous coating, but also keep the silk proteins in favourable conformations (Tillinghast et al., 1991). Furthermore, together with the salts, they may act as fungicides and/or bacteriocides (Schildknecht et al., 1972; Vollrath et al., 1990), which is important to keep the coat functional and the core fibres intact.

WATER PLASTICISATION

Water is important not only for the capture threads, but also for many other types of thread as well, and the role of water, and other solvents, in determining (and modifying) the mechanical properties of spider silk cannot be understated. This is especially true if we aim to design and produce bioengineered silks with predictable properties.

Water, like many other solvents, plasticises most silks (Shao and Vollrath, 1999; Shao et al., 1999a, 1999b) and in doing so typically also contracts the fibre. Many but not all silks contract considerably when submersed in water (e.g., Denny, 1980; Work, 1985; Gosline et al., 1984); the dragline thread of *Nephila* is a good example for one that does (Shao and Vollrath, 1999). Although we do not know exactly what the water does to the thread, it is a good guess that it primarily insinuates itself between the microfibrils. Water molecules then enter the fibrils themselves and begin to break hydrogen bonds. Urea is well known for being a strong breaker of hydrogen bonds, and a 8 M urea solution is able to practically dissolve *Araneus* dragline (Shao et al., 1999b). The same solution applied to sections of *Nephila* dragline silk 'merely' supercontracts it to one-tenth of its original length while at the same time swelling it by 500% (Vollrath et al., 1996). Moreover, for a minute or so before contraction, the silk section undergoes violent rotations, always clockwise (Vollrath et al., 1996). Once swollen, such silk shows structures that are not visible in the native state, and there is strong evidence that most of these structures are already present in the native state (Vollrath et al., 1996). *Nephila* dragline threads show surface and marginally subsurface structures indicative of fine fibrils (Li et al., 1994; Thiel et al., 1994a, 1994b; Vollrath et al., 1996) or even a fibril wall surrounding a fibrilless core (Mahoney, 1994). Atomic force microscopy of untreated fibre cross-sections suggests a trinity of coat, tube, and core (Li et al., 1994), confirming light microscopy studies of dragline silk still in the glandular duct (Work, 1984) and electron microscopy studies silk in the gland (Knight and Vollrath, 1999a, 1999b, 2001a; Vollrath and Knight, 1999; Knight et al., 2000) and in the fibre (Frische et al., 1998; Shao et al., 1999c). Indeed, major ampullate silk fibres from the orb-weaving spider *Nephila madagascariensis* under transmission electron microscopy show the complexity of a dragline thread (Figure 8.8). There is a thin outer layer probably lipids (Schultz and Toft, 1993), surrounding a column of apparently homogeneous material, which contains elongated cavities oriented parallel to the silk fibre axis. The cavities appear similar to 'elongate vacuolar droplets' observed in the silk of some moth larva (Akai et al., 1987, 1988; Robson, 1999). The overall skin-core structure is the result of a rheological pattern originating in the two secreting regions recognised in *Nephila* silk glands; the cavities indicate material inhomogeneities (Vollrath and Knight, 1999). These structures (the canaliculi) resemble small filled rifts, which are extremely fine and are elongated in line with the fibre axis. Stressing a fibre until fracture can lead to cracks forming between these canaliculi (Shao et al., 1999c), and it may be argued that by diverting 'crack energy', the canaliculi may contribute further to the exceptional tensile strength and toughness of the threads that have them.

Whatever the details of their fine structure, the material properties of most (if not all) dragline silks are consistently modified by a wide range of solvents of which water and urea are but the most prominent (Shao and Vollrath, 1999; Shao et al., 1999b). Raman spectroscopy of single silk fibres shows that solvents with different polarities affect different regions of the silk's composite microstructure by modifying the conformations of the different molecular chains. The differences in the mechanical properties between native and supercontracted silks can be attributed to variation in β-sheet content (Shao et al., 1999a; Sirichaisit et al., 1999), suggesting that these crystalline regions are somehow unravelled by the action of the solvent. Clearly, the wet capture spirals of the ecribellate orbweavers take full advantage of this action (Figure 8.6).

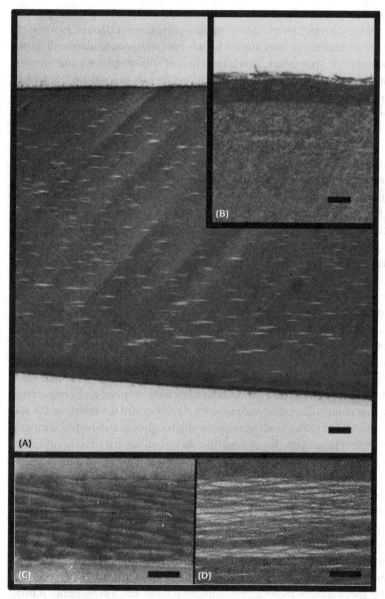

Figure 8.8: Electron micrographs of *Nephila* dragline silk. (A) Longitudinal section of a fully urea swollen fibre. Note the numerous electron-light canaliculi and the electron-dark outer coat (bar = 1 μm). (B) Magnification of the coat showing the various layers (bar = 100 nm). (C) Cryosection of an untreated fibre (bar = 1 μm). (D) Enzyme K-treated fibre (bar = 1 μm). (A/B: modified from Frische et al., 1996. C/D: W. Hu helped with this preparation.)

SPINNING

The mechanical properties of major ampullate dragline silk depend to some extent on the conditions of spinning. At higher temperatures, spiders walk faster and consequently spin faster. This interaction of spinning speed and temperature presents an interesting set of conditions for silk protein folding which, apparently, reveal basic production constraints and trade-offs. *Nephila* spiders that are silked faster (the natural equivalent to a faster moving animal) or are heated up during silking, produce threads with different diameters and mechanical properties than 'slower moving' (or colder) animals (Vollrath et al., 2001). Both the speed of drawing and the temperature at spinning affect the silk with strong non-linear components. Under natural conditions, of course, the effects of temperature and reeling speed (which are production constraints) are closely linked. Thus, they are traded-off against each other and, as it happens, produce silk filaments with similar average mechanical properties. It seems that the metabolic constraints resulting from the typical invertebrate impact of temperature on metabolism and walking speed leads to a surprisingly balanced end product.

How might this be done? The feedstock is prepared in advance, stored in the gland, and spun as required. The speed of drawing this feedstock through the duct affects the time allowed to the silk molecules to align in the different sections while at the same time increasing shear and elongational flow. The temperature of spinning affects both the viscosity of the silk feedstock and the action of the various 'pumping' metabolic processes in the extruder. Thus, changes in silk diameter as a function of both reeling speed and temperature suggest that elongational flow and viscosity are important for 'molding' the silk, with faster drawing rates leading to lower silk toughness and higher temperatures to greater toughness. Toughness (the ability to absorb much energy before breaking) would be a main selective criterion on both dragline and web silk. Thus, toughness (integration of a fibre's strength and its stretchability) would be measured for the optimisation (trade-off of constraints) in a silk's molecular composition and orientation. This line of research, studying the various constraint parameters and the way they are traded-off against each other during feedstock production, storage, and spinning will illuminate the 'design criteria' of natural as well as artificial spider silk. Indeed, first insights are emerging from such experiments: thermal analysis of the glass transition suggests that spinning speed affects molecular alignment (Guess and Viney, 1998), Raman spectroscopy suggests that both spinning speed and temperature affect molecular structure (Sirichasit et al., 1999), and X-ray diffraction demonstrates that spinning speed affects the molecular orientation of a fibre (Riekel et al., 2000; Riekel and Vollrath, 2001).

The exact mechanisms by which the proteins first unfold and then refold, assemble and cross-link in the spider's extrusion duct are still unknown. In principle, the aqueous protein feedstock in the gland is converted into the solid fibre at the spigot by first passing through a short, firm-walled funnel and then a long, soft-walled tapering tube. The core and coat composite structure of the *Nephila* dragline thread, for example, is formed by the codrawing of at least two feedstocks through a single hyperbolic extrusion die accompanied and followed by a number of coatings (Vollrath and Knight, 2001). The dialysis membrane-like cuticle that lines the gland's duct facilitates the ionic exchange and removal of water (Vollrath et al., 1998; Knight and Vollrath, 2001a), which escort the flow of the highly oriented liquid crystalline feedstock through the duct (Vollrath and Knight, 1999; Knight and Vollrath, 2001a). Staining experiments with Congo red suggest that a combination of wall shear and elongational flow induces the β-sheet transitions in the proteins as they pass through the taper (Knight et al., 2000).

THE ROLE OF MANUFACTURE FOR MECHANICS

The stress-strain characteristics of dragline silks show large differences between spiders of different families (Madsen et al., 1999) and are probably on the whole also fairly species specific in their fine details. Indeed, a silk's mechanical properties are significantly affected by the conditions of manufacture. For example, *Nephila* dragline silk produced under conditions of highly controlled for body temperature and speed of extrusion varied consistently in thread diameter, as well as nearly all mechanical properties measured (Vollrath et al., 2001) (Figure 8.9).

The degree of a spider's control over its silk production system can be shown elegantly by on-line X-ray diffraction patterns of single threads as they are being spun (Riekel et al., 2000). In this way, using state-of-the-art micro-X-ray diffraction, molecular structure can be recorded for a wide range of silking speeds and/or spider body temperatures (Riekel et al., 2000). All patterns agree with the model of a small crystalline block in a matrix containing both oriented and unoriented amorphous material (Kümmerlen et al., 1996; Simmons et al., 1996; Grubb and Jelinski, 1997; Beek et al., 1999) with the crystalline fraction showing the pattern of the β-poly(L-alanine) structure (Riekel et al., 2000).

It seems that the molecular chains of the spider silk are aligned parallel to the axis of the fibre (Knight and Vollrath, 2002a). The molecular deformation during stretching of a single *Nephila* dragline can be examined under a Raman microspectroscope (Sirichaisit et al., 1999), and the band shifts under tension can reveal interesting information on structure (Grubb and Jelinski, 1997). In the spider's silk, two prominent bands shift with a strong, positive, and linear

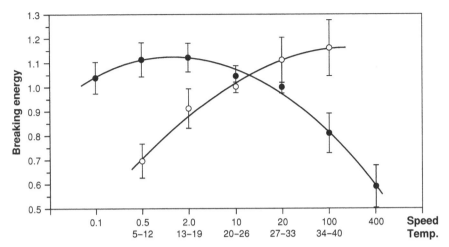

Figure 8.9: Effect of spinning conditions on the breaking energy of *Nephila edulis* dragline silk. Average breaking energy for these samples at control conditions was 165 ± 28 kJ kg^{-1}. Breaking energy is normalised for diameter since silk diameter is affected by silking speed (black dots shown in mm s^{-1}) and body temperature (white dots shown in °C). Control temperature was 25°C when reeling speed was varied, and the control speed was 20 mm s^{-1} when body temperature was varied (after Vollrath et al., 2001).

correlation when the silk is stressed and strained, while in *Bombyx* cocoon silk the correlation is not only weaker, but also non-linear (Sirichaisit et al., 1999). The linear dependence during stretching can be interpreted as the silk's microstructure being subjected to an overall and uniform stress during deformation (Sirichaisit et al., 1999). As yet, this information cannot be fully translated into the mechanical action of the two principal molecular components of typical silks (Riekel and Vollrath, 2001). Nuclear magnetic resonance spectra suggest structural models for spider dragline where the poly-Ala segments adopt highly ordered β-sheet structures and the glycine-rich segments form 3(1) or β-helical structures (McNamee et al., 1994; Simmons et al., 1994, 1996; Hijirida et al., 1996; Kümmerlen et al., 1996; Beek et al., 1999). How these structures are integrated into the microfibrils of the thread, and how they interact with one another, however, are also still open to discussion. Indeed, even the mechanical role of the different levels of the hierarchical microstructure in so many silks, such as the skin-core and the canaliculi, is still far from solved.

CONCLUSIONS

Studying spider silk fibres with the aim to integrate structural and mechanical data with their role in nature have begun to open a window to the evolution of a most remarkable elastomer. Any understanding of the macroscopic properties of a particular spider silk demands not only insights into the environmental conditions to which it is adapted, but also good working hypotheses on molecular as well production constraints and trade-offs.

Spider silk has recently become a 'sexy topic', and we seem daily to be learning more about the molecular composition of the various silks and the structural hierarchy of their molecular interactions. Since thread properties are highly variable, even for the same thread of the same spider (Madsen et al., 1999), unravelling the cause of this diversity could open an interesting and important window into the folding of those proteins and microstructures that are responsible for silk strength.

ACKNOWLEDGMENTS

We gratefully acknowledge previous and present research funding from Aarhus University, the Danish Statens Naturvidenskabelige Fonds, the Carlsberg Foundation, the British Biology & Biotechnology and Engineering & Physical Science Research Councils, the European Synchroton Research Facilities, the UK/China Science Transfer Fund, and the European Science Foundation 'Silk Network'.

REFERENCES

Akai, H., Imai, T., and Tsubochi, K. (1987) Fine-structural changes of liquid silk in the silk gland during spinning stage of Bombyx larvae. *J. Seric. Sci. Jpn.* **56,** 131–37.

Akai, H., Kiuchi, M., and Tamura, T. (1988) Ultrastructure of silk glands and cocoon filaments of wild silkmoths, *Antheraea yamamai* and *Antheraea pernyi. Wild Silkmoths* **88,** 9–23.

Andersen, S. O. (1970) Amino acid composition of spider silks. *Comp. Biochem. Physiol.* **35,** 705–11.

Anquetil, J., and Walter, M. (1996) *Silk.* New York: Abbeville Press.

Beek, J. V., Kümmerlen, D., Vollrath, F., and Meier, B. H. (1999) Solid state NMR on supercontracted spider dragline silk. *Int. J. Biol. Macromol.* **24,** 173–78.

Bonthrone, K. M., Vollrath, F., Hunter, B. K., and Sanders, J. K. M. (1992) The elasticity of spider's webs is due to water-induced mobility at a molecular level. *Proc. R. Soc. Lond.* **248,** 141–44.

Bristowe, W. S. (1958) *The World of Spiders.* London, UK: Collins.

Chen, X., Knight, D. P., Shao, Z. Z., and Vollrath, F. (2001) Regenerated Bombyx silk solutions studied with rheometry and FTIR. *Polymer* **42,** 9969–74.

Coddington, J. (1986) The monophyletic origin of the orb web. In: *Spiders, Webs, Behaviour and Evolution,* W. A. Shear (ed.). Stanford, CA: Stanford University Press, pp. 319–63.

Craig, C. L., Hsu, M., Kaplan, D., and Pierce, N. E. (1999) A comparison of the composition of silk proteins produced by spiders and insects. *Int. J. Biol. Macromol.* **24,** 109–18.

Dawkins, R. (1996) *Climbing Mount Improbable.* Penguin, Ltd. Hasmbudsworth England.

Denny, M. W. (1980) Silks—Their properties and functions. In: *The Mechanical Properties of Biological Materials,* vol. 34, J. D. Currey (ed). Cambridge, UK: Cambridge University Press, pp. 245–71.

Eberhard, W. G. (1980) Persistent stickiness of cribellum silk. *J. Arachnol.* **8**, 283.

Eberhard, W. G. (1988) Behavioural flexibility in orb web construction: Effects of supplies in different silk glands and spider size and weight. *J. Arachnol.* **16**, 295–302.

Eberhard, W. G. (1990) Function and phylogeny of spider webs. *Ann. Rev. Ecol. Syst.* **21**, 341–72.

Edmonds, D., and Vollrath, F. (1992) The contribution of atmospheric water vapour to the formation and efficiency of a spider's web. *Proc. R. Soc. Lond.* **248**, 145–48.

Fischer, F., and Brander, J. (1960) Eine Analyse der Gespinste der Kreuzspinne. *Hoppe-Seyler's Z. Physiol. Chem.* **320**, 92–102.

Foelix, R. (1996) *Biology of Spiders.* Oxford, UK: Oxford University Press.

Fossey, S. A., and Kaplan, D. (1994) Molecular modeling studies on silk peptides. In: *Silk Polymers. Materials Science and Biotechnology,* D. Kaplan (ed.). Washington, DC: American Chemical Society, Symposium Series, pp. 270–82.

Frische, S., Maunsbach, A., and Vollrath, F. (1998) Elongate cavities and skin-core structure in *Nephila* spider silk observed by electron microscopy. *J. Microscopy* **189**, 64–70.

Gatesy, J., Hayashi, C., Motriuk, D., Woods, J., and Lewis, R. (2001) Extreme diversity, conservation and convergence of spider silk fibroin sequences. *Science* **291**, 2603–5.

Gosline, J., Denny, M., and DeMont, M. (1984) Spider silk as rubber. *Nature* (Lond.) **309**, 551–52.

Gosline, J. M., DeMont, M. E., and Denny, M. W. (1986) The structure and properties of spider silk. *Endeavour* **10**, 31–43.

Gosline, J. M., Pollak, C. C., Guerette, P., Cheng, A., DeMont, M. E., and Denny, M. W. (1994) Elastomeric network models for the frame and viscid silks from the orb web of the spider *Araneus diadematus.* In: *Silk Polymers. Materials Science and Biotechnology,* D. Kaplan, W. W. Adams, B. Farmer, and C. Viney (eds.). Washington, DC: American Chemical Society, Symposium Series, pp. 328–41.

Grubb, D. T., and Jelinski, L. W. (1997) Fibre morphology of spider silk: The effects of tensile deformation. *Macromolecules* **30**, 2860–67.

Guerette, P., Ginzinger, D., Weber, B., and Gosline, J. (1996) Silk properties determined by gland-specific expression of a spider fibroin gene family. *Science* **272**, 112–15.

Guess, K. B., and Viney, C. (1998) Thermal analysis of major ampullate (dragline) spider silk: The effect of spinning rate on tensile modulus. *Thermochim. Acta.* **315**, 61–66.

Hansell, M. H. (1984) *Animal Architecture & Building Behaviour.* London, UK: Longmann.

Hayashi, C. Y., and Lewis, R. V. (2000) Molecular architecture and evolution of a modular spider silk gene. *Science* **287**, 1477–79.

Hijirida, D. H., Do, K. G., Michal, C., Wong, S., Zax, D., and Jelinski. L.W. (1996) C-13 NMR of *Nephila clavipes* major ampullate silk gland. *Biophys. J.* **71**, 3442–47.

Hilliard, P. (1994). *The Book of Spiders.* London, UK: Hutchinson.

Kaplan, D. L., and Lombardi, S. J. (1990) The *Nephila clavipes* major ampullate gland silk protein: Amino acid composition and detection of silk gene-related nucleic acids in the genome. *Acta Zool. Fennica.* **190**, 243–48.

Kaplan, D. L., Adams, W. W., Viney, C., and Farmer, B. L. (1994) *Silk Polymers: Materials Science and Biotechnology.* Washington, DC: ACS Books.

Knight, D. P., and Vollrath, F. (1999a) Liquid crystals in the cells secreting spider silk feedstock. *Tissue Cell* **31**, 617–20.

Knight, D. P., and Vollrath, F. (1999b) Liquid crystals and flow elongation in a spider's silk production line. *Proc. R. Soc. Lond.* **266,** 519–23.

Knight, D. P., Knight, M., and Vollrath, F. (2000) Beta sheet formation and stress-induced phase separation in the spinning of spider dragline silk. *Int. J. Biol. Macromol.* **27,** 205–10.

Knight, D. P., and Vollrath, F. (2001a) Changes in element composition in the secretory pathway for dragline silk in the golden orb web spider, *Nephila edulis. Naturwissenschaften* **88,** 179–82.

Knight, D. P., and Vollrath, F. (2001b) Comparison of liquid crystal spinning of Selachian egg case ply sheets and orb web spider dragline filaments. *Biomacromolecules* **2,** 323–34.

Knight, D. P., and Vollrath, F. (2002a) Biological liquid crystal elastomers. *Phil. Trans. R. Soc. Lond.* **357,** 155–64.

Knight, D. P., and Vollrath, F. (2002b) Spinning an elastic ribbon of spider silk. *Phil. Trans. R. Soc. Lond.* **357,** 219–28.

Köhler, T., and Vollrath, F. (1995) Thread biomechanics in the two orb weaving spiders *Araneus diadematus* (Araneae, Araneidae) and *Uloborus walckenaerius* (Araneae, Uloboridae). *J. Exp. Zool.* **271,** 1–17.

Krink, T., and Vollrath, F. (1997) Analysing spider web-building behaviour with rule-based simulations and genetic algorithms. *J. Theor. Biol.* **185,** 321–31.

Krink, T. F. V. (2000) Optimal area use in orb webs of the garden cross spider. *Naturwissenschaften* **87,** 90–93.

Kümmerlen, J., Vanbeek, J., Vollrath, F., and Meier, B. (1996) Local-structure in spider dragline silk investigated by 2-dimensional spin-diffusion nuclear-magnetic-resonance. *Macromolecules* **29,** 2920–28.

Lazaris, A., Arcidiacono, S., Huang, Y., Zhou, J. F., Duguay, F., Chretien, N., Welsh, E. A., Soares, J. W., and Karatzas, C. N. (2002) Spider silk fibers spun from soluble recombinant silk produced in mammalian cells. *Science* **295,** 472–76.

Lewis, R. V. (1992) Spider silk: The unraveling of a mystery. *Acc. Chem. Res.* **25,** 392–98.

Lewis, R. (1999). Hypotheses that correlate the sequence, structure and mechanical properties of spider silk proteins. *Int. J. Biol. Macromol.* **24,** 271–75.

Li, S. F. Y., McGhie, A. J., and Tang, S. L. (1994) New internal structure of spider dragline silk revealed by atomic force microscopy. *Biophys. J.* **66,** 1209–12.

Lin, L., Edmonds, D., and Vollrath, F. (1995) Structural engineering of a spider's web. *Nature* (Lond.) **373,** 146–48.

Lombardi, S. J., and Kaplan, D. L. (1990) The amino acid composition of major ampullate gland silk (dragline) of *Nephila clavipes* (Araneae, Tetragnathidae). *J. Arachnol.* **18,** 297–306.

Mahoney, D. V., Vezie, D. L., Eby, R. K., Adams, W. W., and Kaplan, D. (1994) Aspects of the morphology of dragline silk of *Nephila clavipes.* In: *Silk Polymers. Materials Science and Biotechnology,* D. Kaplan, W. W. Adams, B. Farmer, and C. Viney (eds.). Washington, DC: American Chemical Society, Symposium Series, 196–210.

Madsen, B., Shao, Z., and Vollrath, F. (1999) Variability in the mechanical properties of spider silks on three levels: Interspecific, intraspecific and intraindividual. *Int. J. Biol. Macromol.* **24,** 301–6.

Madsen, B., and Vollrath, F. (2000) Mechanics and morphology of silk reeled from anaesthetised spiders. *Naturwissenschaften* **87,** 148.

McNamee, S. G., Ober, C. K., Jelinski, L. W., Ray, E., Xia, Y., and Grubb, D. T. (1994) Toward single-fiber diffraction of spider dragline silk from *Nephila clavipes*. In: *Silk Polymers. Materials Science and Biotechnotogy*, D. Kaplan, W. W. Adams, B. Farmer, and C. Viney (eds.). Washington, DC: American Chemical Society, Symposium Series, pp. 176–84.

O'Brien, J. P., Hoess, R. H., Gardner, K. C. H., Lock, R. L., Wasserman, Z. R., Weber, P. C., and Salemme, F. R. (1994) Design, synthesis and fabrication of a novel self-assembling fibrillar protein. In: *Silk Polymers. Materials Science and Biotechnology*, D. Kaplan, W. W. Adams, B. Farmer, and C. Viney (eds.). Washington, DC: American Chemical Society, Symposium Series, pp. 104–17.

O'Brien, J. P., Fahnestock, S. R., Termonia, Y., and Gardner, K. C. H. (1998) Nylons from nature: Synthetic analogs to spider silk. *Adv. Mater.* **10,** 1185–95.

Opell, B. D. (1995) Do static electric forces contribute to the stickiness of a spider's cribellar prey capture threads? *J. Exp. Zool.* **273,** 186–89.

Opell, B. D. (1997) The material cost and stickiness of capture threads and the evolution of orb weaving spiders. *Biol. J. Linn. Soc.* **62,** 443–58.

Opell, B. D. (1998) Economics of spider orb-webs: The benefits of producing adhesive capture thread and of recycling silk. *Funct. Ecol.* **12,** 613–24.

Perez-Rigueiro, J., Elices, M. J., Llorca, J., and Viney, C. J. (2001) Tensile properties of silkworm silk obtained by forced silking. *J. Appl. Polymer Sci.* **82,** 1928–35.

Peters, H. M. (1955) Über den Spinnapparat von *Nephila madagascariensis* (Radnetzspinnen, Fam. Argiopidae). *Z. Naturforsch.* **10b,** 395–404.

Peters, H. M. (1984) The spinning apparatus of Uloboridae in relation to the structure and construction of capture threads (Arachnida, Araneida). *Zoomorphology* **104,** 96–104.

Peters, H. M. (1987) Fine structure and function of capture threads. In: *Ecophysiology of Spiders,* W. Nentwig (ed.). Berlin, Germany: Springer, pp. 187–202.

Prince, J. T., McGrath, K. P., DiGirolamo, C. M., and Kaplan, D. L. (1995) Construction, cloning and expression of synthetic genes encoding spider dragline silk. *Biochemistry* **34,** 10879–85.

Riekel, C., Madsen, B., Knight, D., and Volrath, F. (2000). *In-situ* x-ray diffraction during biopolymer extrusion: Forced silking of *Nephila edulis* spider silk. *Biol. Macromol.* **24,** 179–86.

Riekel, C., and Vollrath, F. (2001) Spider silk fibre extrusion: Combined wide- and small-angle X-ray microdiffraction experiments. *Int. J. Biol. Macromol.* **29,** 203–210.

Robson, R. M. (1999) Microvoids in Bombyx mori silk. An electron microscope study. *Int. J. Biol. Macromol.* **24,** 145–50.

Schildknecht, H., and Munzelmann, P. (1972) Über die Chemie der Spinnwebe, I. *Naturwissenschaften* **59,** 98–99.

Schultz, S., and Toft, S. (1993) Branched long chain alkyl methyl ethers: A new class of lipids from spider silk. *Tetrahedron* **49,** 6805–20.

Shao, Z., and Vollrath, F. (1999) The effect of solvents on the contraction and mechanical properties of spider silk. *Polymer* **40,** 1799–1806.

Shao, Z., Vollrath, F., Sirichaisit, J., and Young R. J. (1999a) Analysis of spider silk in native and supercontracted states using Raman spectroscopy. *Polymer* **40,** 2493–500.

Shao, Z., Young, R. J., and Vollrath, F. (1999b) The effects of solvents on spider silk studied by mechanical testing and single-fibre Raman spectroscopy. *Int. J. Biol. Macromol.* **24,** 295–300.

Shao, Z., Wen Hu, X., Frische, S., and Vollrath, F. (1999c) Heterogeneous morphology in spider silk and its function for mechanical properties. *Polymers* **40,** 4709–11.

Shao, Z., and Vollrath, F. (2002). The surprising strength of silkworm silk (submitted for publication). *Nature* **418,** 741.

Shear, W. A. (1986) The evolution of web-building behavior. In: *Spiders: Webs, Behavior and Evolution,* W. A. Shear (ed.). Stanford, CA: Stanford University Press, pp. 364–400.

Simmons, A., Ray, E., and Jelinski, L. (1994) Solid-state c-13 nmr of nephila-clavipes dragline silk establishes structure and identity of crystalline regions. *Macromolecules* **27,** 5235–37.

Simmons, A., Michal, C., and Jelinski, L. (1996) Molecular-orientation and 2-component nature of the crystalline fraction of spider dragline silk. *Science* **271,** 84–87.

Sirichaisit, S., Young, R. J., and Vollrath, F. (1999) Molecular deformation in spider dragline silk subjected to stress. *Polymers* **41,** 1223–27.

Stowe, M. K. (1986) Prey specialisation in the Araneidae. In: *Spiders: Webs, Behavior and Evolution,* W. A. Shear (ed.). Stanford, CA: Stanford University Press, pp. 101–31.

Termonia, Y. (1994) Molecular modeling of spider silk elasticity. *Macromolecules* **27,** 7378–81.

Thiel, B., and Viney, C. (1995) A nonperiodic lattice model for crystals in *Nephila-clavipes* major ampullate silk. *MRS Bull.* **20,** 52–56.

Thiel, B. L., and Kunkel, D. (1994a). Composite microstructure of spider (*Nephila clavipes*) dragline. In: *Biomolecular Materials by Design,* MRS Symposium, H. B. M. Alper, D. Kaplan, and M. Navia (eds.). Pittsburgh, PA: Materials Research Society, pp. 21–30.

Thiel, B. L., Kunkel, D., and Viney, C. (1994b) Physical and chemical microstructure of spider dragline: A study by analytical transmission electron microscopy. *Biopolymers* **34,** 1089–97.

Tillinghast, E. K., Townley, M. A., Bernstein, D. T., and Gallagher, K. S. (1991) Comparative study of orb web hygroscopicity and adhesive spiral composition in three araneid spiders. *J. Exp. Zool.* **259,** 154–65.

Vollrath, F. (1992) Analysis and interpretation of orb spider exploration and web-building behavior. *Adv. Study Anim. Behav.* **21,** 147–99.

Vollrath, F. (1994) General properties of some spider silks. In: *Silk Polymers Materials Science and Biotechnology,* C. Viney (ed.). Washington, DC: American Chemical Society, pp. 17–28.

Vollrath, F. (1999) Biology of spider silk. *Int. J. Biol. Macromol.* **24,** 81–88.

Vollrath, F. (2000) Coevolution of behaviour and material in the spider's web. In: *Biomechanics in Animal Behaviour,* P. Domenici and R. W. Blake. Oxford, UK: Bios, pp. 315–34.

Vollrath, F., and Edmonds, D. (1989) Modulation of the mechanical properties of spider silk by coating with water. *Nature* **340,** 305–7.

Vollrath, F., Fairbrother, W. J., Williams, R. J. P., Tillinghast, E. K., Bernstein, D. T., Gallagher, K. S., and Townley, M. A. (1990) Compounds in the droplets of the orb spider's viscid spiral. *Nature* **345,** 526–28.

Vollrath, F., Holtet, T., Thogersen, H., and Frische, S. (1996) Structural organization of spider silk. *Proc. R. Soc. Lond. B* **263,** 147–51.

Vollrath, F., Downes, M., and Krackow, S. (1997) Design variables in web geometry of an orb weaving spider. *Physiol. Behav.* **62,** 735–43.

Vollrath, F., Wen Hu, X., and Knight, D. P. (1998) Silk production in a spider involves acid bath bath treatment. *Proc. R. Soc. Lond.* **263,** 817–20.

Vollrath, F., and Knight, D. (1999) The silk press of the spider *Nephila edulis. Int. J. Biol. Macromol.* **24,** 243–9.

Vollrath, F., and Knight, D. P. (2001) Liquid crystal silk spinning in nature. *Nature* **410,** 541–48.

Vollrath, F., Madsen B., and Shao, Z. (2001) The effect of spinning conditions on silk mechanics. *Proc. R. Soc. Lond.* **268,** 2339–46.

Winkler, S., and Kaplan, D. (2000) Molecular biology of spider silk. *Rev. Mol. Biotech.* **74,** 85–93.

Wirth, E., and Barth, F. G. (1992) Forces in the spider orb web. *J. Comp. Physiol. A* **171,** 359–71.

Witt, P. N., Reed, C. F., and Peakall, D. B. (1968) *A Spider's Web: Problems in Regulatory Biology.* Heidelberg, Germany: Springer.

Work, R. W. (1984) Duality in major ampullate silk and percursive material from orb-web-building spiders (Araneae). *Trans. Am. Microsc. Soc.* **103,** 113–21.

Work, R. W. (1985) Viscoelastic behaviour and wet super contraction of major ampullate silk fibres of certain orb web building spiders (Araneae). *J. Exp. Biol.* **118,** 379–404.

Work, R. W., and Young, C. T. (1987) The amino acid compositions of major and minor ampullate silks of certain orb-web-building spiders (Araneae, Araneidae). *J. Arachnol.* **15,** 65–80.

Zemlin, J. C. (1968) *A Study of the Mechanical Behavior of Spider Silks.* Clothing and Organic Materials Laboratory, U.S. Army, Natick, MA.

NINE

Collagen: Hierarchical Structure and Viscoelastic Properties of Tendon

Peter Fratzl

INTRODUCTION

Tendon is a hierarchically structured collagenous tissue which has outstanding mechanical properties. A sketch of the stress-strain curve is shown in Figure 9.1. Most remarkably, the stiffness increases with strain up to an elastic modulus in the order of 1 to 2 GPa. Moreover, tendons are viscoelastic, and their deformation behaviour depends on the strain rate, as well as on the strain itself. In vivo, it is very likely that tendons are always somewhat prestrained (even if the muscles are at rest); hence, they are normally working in the intermediate ('heel'; see Figure 9.1) and high modulus regions (Vincent, 1990). In this context, it is also interesting to compare the maximum stress generated in muscle [in the order of 300 kPa (Abe et al., 1996)] to the strength of tendon which is about 300 times larger. This explains why tendons and ligaments can be much thinner than muscle. Obviously, the remarkable mechanical properties of the tendons are linked to their complex hierarchical structure (Figure 9.2).

This chapter reviews some of the well-known (Diamant et al., 1972; Kastelic and Baer, 1980; Mosler et al., 1985; Sasaki et al., 1996, 1999; Misof et al., 1997; Fratzl et al., 1997) and more recently discovered (Puxkandl et al., 2001) structural principles giving rise to the mechanical behaviour of tendons.

DEFORMATION MECHANISMS OF COLLAGEN FIBRILS

The stress-strain curve of tendons usually shows three distinct regions (Vincent, 1990) which can be correlated to deformations at different structural levels (Figure 9.3). In the 'toe' region, at small strains, a very small stress is sufficient to elongate the tendon. This corresponds to the removal of a macroscopic crimp of the fibrils (Diamant et al., 1972) visible in polarised light (Figure 9.3, left). In the second region, at higher strains (Figure 9.3, centre), the stiffness of the tendon increases considerably with the extension. An entropic mechanism, where disordered molecular kinks in the gap region of collagen fibrils are straightened out, has been proposed to explain the increasing stiffness with increasing strain

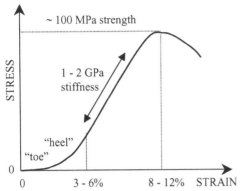

Figure 9.1: After 'toe' and a 'heel' regions (Vincent, 1990), where the stiffness increases considerably, there is a linear dependence between strain and stress with a slope corresponding to a stiffness in the order of 1–2 GPa. The strength of tendons is typically around 100 MPa. The extensibility depends on the rate of deformation. For slow stretching, the maximum strain is in the order of 8–10%.

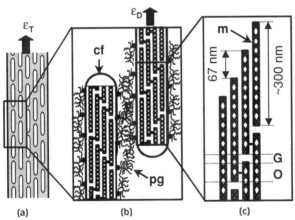

Figure 9.2: Simplified tendon structure (see, e.g., Vincent, 1990). A tendon is made of a number of parallel fascicles (a) containing collagen fibrils (b). The fibrils (**cf**) are typically coated with proteoglycans (**pg**). They have a thickness of several hundred nanometres and a length in the order of 10 micrometres. Triple-helical collagen molecules (**m**) are packed within fibrils (c) in a staggered way with an axial spacing of $D = 67$ nm, when there is no load on the tendon (Hodge and Petruska, 1963). Since the length of the molecules (300 nm) is not an integer multiple of the staggering period, there is a succession of gap (**G**) and overlap (**O**) zones. The lateral spacing of the molecules is around 1.5 nm, but the full three-dimensional arrangement is not yet fully clarified (Hulmes et al., 1995; Wess et al., 1998). ε_T is the total strain of the tendon and ε_D the strain in the fibrils, which can be measured via the change in the axial D-period due to tensile loading.

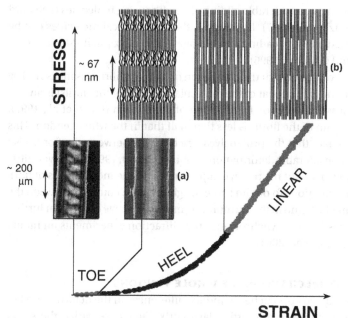

Figure 9.3: Schematic behavior of the normal collagen fibril structure from rat-tail tendon (marked cf in Figure 9.5) during tensile deformation. From 'Fibrillar Structure and Mechanical Properties of Collagen', by P. Fratzl et al., 1997, *J. Struct. Biol.* **122,** 119–22. The experiment was performed at a strain rate where the actual strain of the fibril (ε_D) was about 40% of the total stain of the tendon (ε_T) in the linear region. Plotted on the horizontal axis is the total strain of the tendon (ε_T). Previous synchrotron X-ray scattering experiments (Misof et al., 1997) have shown that the tendon structure goes through a sequence of changes upon stretching. First, a macroscopic crimp in the tendon (Diamant et al., 1972) is straightened out, as visible in the polarised light (a). Then, microscopic kinks in the collagen molecules (located mostly in the gap region of the fibril structure) are removed, leading to an entropic contribution to elasticity (b, left). Finally, the molecules start to glide past each other in the linear region of the stress-strain curve (b, right).

(Misof et al., 1997). When all the kinks are straightened, another mechanism of deformation must prevail and explain the linear dependence of stress and strain in this region of the force-elongation curve (Figure 9.3, right). The most likely processes are thought to be the stretching of the collagen triple-helices and the cross-links between the helices, implying a side-by-side gliding of neighbouring molecules, leading to structural changes at the level of the collagen fibrils. This has previously been investigated by use of synchrotron radiation diffraction experiments (Mosler et al., 1985; Folkhard et al., 1986; Sasaki and Odajima, 1996; Fratzl et al., 1997; Sasaki et al., 1999). By monitoring the structure factors of the second and third order maxima, it could be shown that the length ratio of the gap to the overlap region may increase during stretching by as much as ~10%, implying a considerable gliding of neighbouring molecules (Folkhard et al., 1986; Fratzl et al., 1997). In addition, the triple-helical molecules can be slightly stretched as well, leading to a change of the helix pitch (Mosler et al., 1985; Sasaki and Odajima, 1996).

Moreover, it was found that the elongation of collagen fibrils (as measured by the increase in the axial repeat of the molecular packing of the fibrils) is always considerably less than the total elongation of the tendon (Fratzl et al., 1997). Typically, the strain of the fibrils is less than half that in the whole tendon. This emphasises the fact that the proteoglycan-rich matrix between the fibrils also experiences a considerable deformation (Cribb and Scott, 1995). Relatively little is known, however, about the relative importance and the mechanisms of fibril and matrix deformation. To obtain some insight into the coupling between the deformation in fibrils and in the surrounding matrix, we have carried out further in situ tensile testing and synchrotron X-ray diffraction experiments on rat tail tendons (Puxkandl et al., 2001).

DEFORMATION MECHANISMS OF WHOLE TENDONS

We have pursued two approaches to induce differences in the deformation behaviour between fibrils and interfibrillar matrix: Firstly, we varied the strain rate systematically, assuming that the viscosity of fibrils and matrix could be different. Secondly, we compared normal collagen with collagen deficient in cross-links, the latter being from rats fed with β-aminopropionitrile (β-APN). It is well-known that the formation of covalent cross-links between adjacent molecules may be inhibited by β-APN (Eyre et al., 1984), which acts by the enzymatic process of aldehyde formation (Tang et al., 1983; Lees et al., 1990). It was also shown previously that cross-links have a large influence on the mechanical behaviour of collagen; an increased number of cross-links increases its stiffness (Bailey et al., 1998; Davison, 1989; Eyre et al., 1984) [and also brittleness (Light and Bailey, 1979)], while a reduction due to β-APN reduces its stiffness dose-dependently (Lees et al., 1990).

In Situ Tensile Testing and X-ray Diffraction with Synchrotron Radiation

Native fibres from rat tail tendons were measured. The tail tendons were obtained from a 7-week-old rat fed with 0.17 g/day β-APN for the last 3 weeks and from a 6-week-old rat fed with 0.4 g/day β-APN for the last 3 weeks. For comparison, tails from two 6-week-old rats not fed with β-APN were used as a control group. The tails were frozen and stored at approximately $-18°C$ until dissection. After dissection, the tail tendons were immediately stored in phosphate-buffered saline (PBS) solution at pH 7.4 to keep them from deterioration. There is no observable difference in the behaviour of a specimen stored in PBS at $4°C$

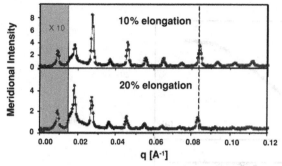

Figure 9.4: Evolution of the axial (meridional) diffraction peaks with the macroscopic elongation. Peaks arise from the periodic arrangement of collagen molecules within fibrils (Hodge and Petruska, 1963). The period, D, was evaluated by fitting the peak positions using Gaussian functions for each peak. Peak positions are clearly seen to lie at smaller q values (longer real spacings) at the higher extensions.

for several months and a specimen measured directly after dissection (Misof et al., 1997).

The mechanical testing was performed with a special device developed for this purpose (Fratzl et al., 1997; Misof et al., 1997), while time-resolved X-ray diffraction patterns were taken simultaneously at intervals which depended on the strain rate and lasted between 30 and 50 s. The X-ray diffraction experiments were carried out at the SAXS Beamline of the synchrotron radiation source ELETTRA in Trieste (Amenitsch et al., 1995), because a large X-ray flux was needed to obtain X-ray diffraction patterns of sufficient counting statistics in such a short measuring time. Data were collected using a two-dimensional position-sensitive X-ray CCD camera (AXS, Karlsruhe, Germany). The first twelve reflections of the axial period of the collagen fibrils were recorded. A simultaneous fit of all peak positions (weighted with the intensity of the peaks to improve statistical accuracy) yielded the axial repeat, D (Figure 9.4).

At the same time, the macroscopic strain in the whole tendon was determined by the movement of the clamps holding the tendon and the applied load measured by a force sensor. The tendons were kept partially immersed in PBS during the whole experiment to avoid drying. Each experiment was carried out at constant velocity of elongation, which was chosen between 0.0001 mm/s and 0.01 mm/s for the control group and between 0.0001 mm/s and 0.005 mm/s for the two β-APN treated rats. Each tendon was stretched to failure, and a load/extension curve was recorded. The initial length of the fibre was determined by stretching it until the smallest force (0.005 N) was detected. The fibre length at this load was then taken as the initial length. By correlating the time from the start of each tensile test, the structural information contained in the diffraction patterns (i.e., the D-period of the fibrils) could be matched to the corresponding point on the load/extension curve.

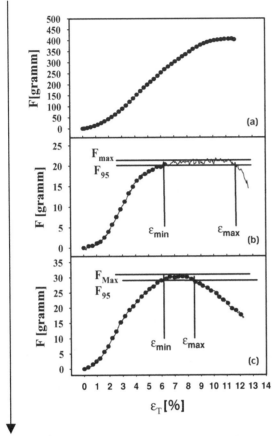

Figure 9.5: Typical load/extension curves from the rat-tail tendons used in this experiment. Dots represent the mechanical values at the times where an X-ray diffraction pattern was taken. The time needed between two points varied between 30 s and 50 s. There were three types of mechanical behaviour: (a) the tendons from the normal control group, (b) tendons from β-APN treated rats which showed a marked plateau and (c) some β-APN treated which did not show a plateau. A plateau was defined to occur when the strain at 95% of the maximum load, e_{max} and e_{min}, followed the condition $(e_{max} - e_{min})/e_{min} > 0.5$.

Results from In Situ Experiments

Typical load/elongation curves for three different types of samples are shown in Figure 9.5(a–c). It is readily seen that all three curves show a similar behaviour in the first part of the plot at low strains. All samples developed a toe region, a heel region, and a linear region, although the corresponding macroscopic strains were slightly lower for tendons taken from the β-APN treated rats.

Figure 9.5(a) shows the typical curve for normal collagen. Twelve of the twenty tendons from β-APN treated rats showed a distinctive mechanical behaviour, such as in Figure 9.5(b); a plateau is clearly visible, where the strain increases at constant load. This is very different from the behaviour of normal collagen, which fractured very rapidly after reaching the maximum load [Figure 9.5(a)]. The other 8 of 20 tendons from the β-APN treated rats did not show a pronounced plateau, but a load/extension curve typified by Figure 9.5(c). We divided the tendons from β-APN treated rats in two groups according to their mechanical behaviour [as in Figure 9.5(b) and Figure 9.2(c), respectively] by estimating the difference between the two strain values ε_{max} and ε_{min} measured at 95% of the maximum load [Figure 9.5(b) and (c)]. Where $(\varepsilon_{max} - \varepsilon_{min})/\varepsilon_{min}$ was larger than 0.5, we considered the tendon to show a plateau-like behaviour. No significant effect due to the higher dose of β-APN in one of the animals could be detected. Both types of load/elongation curves [Figure 9.5(b) and (c)] were observed in animals treated with both doses (but not in controls). This is most probably because the β-APN treated rats have a mixture of normal and cross-link deficient collagen, so that the defective collagen content may vary greatly between individual tendons in these animals.

The applied stress σ is simply the applied force F divided by the cross-sectional area of the tendon. As it was extremely difficult to determine the exact diameter of each tendon, we did not attempt to convert the force/elongation into stress-strain curves, thus avoiding the introduction of additional uncertainty into the data. However, it is possible to make a reasonable estimation of the average fracture stress values within the groups because, even though each individual measurement of the tendon diameter is affected with a large uncertainty, the average value for each group is more reliable. The average maximum force from the control group was about 390 g, whereas the average failure force of the β-APN treated samples was about 28 g. The average diameter of the tendons was about 200 μm for the controls and about 150 μm for the β-APN treated animals. With these values, the average maximum stress of the normal collagen was 120 MPa, whereas the average maximum stress for the β-APN treated rats was about 15 MPa, that is, roughly one order of magnitude smaller. The strain at maximum stress was on average 6.7% for the β-APN treated rats and 9.5% for the control group, with a large variation between individual tendons. This means that cross-link–deficient collagen has a dramatically reduced fracture load without a large decrease in extensibility. Axial D-periods determined by X-ray diffraction increased between 0.2 nm and 2 nm before fracture. There was no significant difference in the average values for the β-APN treated rats (0.81 nm) and the control group (0.87 nm).

Total elongation of the tendons was used to compute the macroscopic strain ε_T, and the change in D-period was taken to compute the strain in the fibrils (ε_D)

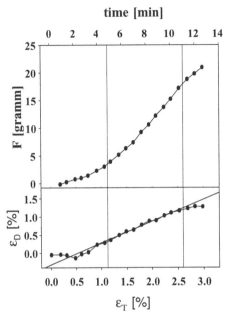

Figure 9.6: A typical force/elongation curve (top) and the corresponding change of the collagen D-period, ε_D, as a function of the macroscopic strain of the tendon, ε_T (bottom). In the region where force and elongation were related linearly (upper curve), the slope of the lower graph was use to determine $d\varepsilon_D/d\varepsilon_T$.

as a function of load within each tendon. The relation between the two strains was analysed in graphs such as Figure 9.6. Using the slope of the plot ε_D vs. ε_T in the region where the load/elongation curve was roughly linear, the ratio of fibril to tendon strain, $d\varepsilon_D/d\varepsilon_T$, was determined for each specimen. Figure 9.7 shows the summary of all values obtained for $d\varepsilon_D/d\varepsilon_T$ as a function of the applied strain rate $d\varepsilon_T/dt$ (t being the time). It is clear from Figure 9.7(a) that $d\varepsilon_D/d\varepsilon_T$ increases with strain rate in normal collagen, while it decreases in the cross-link–deficient collagen showing a plateau in the force/elongation curve [such as in Figure 9.7(b)]. In normal collagen, the lowest value for $d\varepsilon_D/\varepsilon_T$ was about 10% for strain rate of 0.001%/s, and the highest value was about 40% at a strain rate of 0.08%/s [Figure 9.7(a)]. In contrast, the values for 0.0012%/s were as high as 60% in cross-link deficient collagen [Figure 9.7(b)]. Data from cross-link deficient collagen without a plateau scatter considerably [Figure 9.7(c)], and no clear trend can be seen.

VISCOELASTIC MODEL FOR TENDON ELONGATION
The strain-rate dependence of the mechanical properties of rat tail tendons was investigated by simultaneous tensile testing and synchrotron X-ray diffraction characterisation. The main results can be summarised as follows:

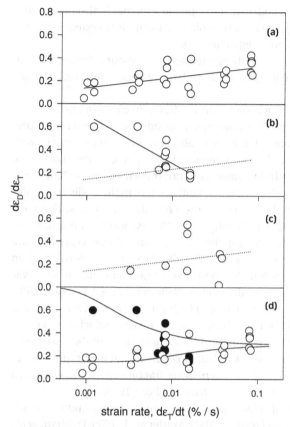

Figure 9.7: Strain-rate dependency of the ratio of fibril elongation to total tendon elongation $d\varepsilon_D/d\varepsilon_T$, for (a) normal collagen taken from the control group, (b) collagen from β-APN treated rats which showed a plateau in their force-elongation curves and (c) collagen from β-APN treated rats which did not. The full straight lines in (a) and (b) are the result of linear regression to show the tendency to increase and to decrease with macroscopic strain rate $d\varepsilon_T/dt$, respectively. The dotted line in (b) and (c) reproduces the regression line from (a) for easier comparison of the data, (d) compares the data taken from (a) (open symbols), and data from (b) (solid symbols) to the model given by Eq. (A4). Values chosen for the free parameters in this model are: $\alpha = 0.15$, $\beta = 0.3$, and $\gamma <\varepsilon_T> = 0.01$ for the lower curve (normal collagen) and $\alpha^* = 0.8$, $\beta^* = \beta$, $\gamma^* = \gamma\alpha/\alpha^*$ for the upper one (cross-link deficient collagen).

1. The extension of collagen fibrils inside the tendon is always considerably less than the total extension of the tendon. At slow deformations, this effect is much more pronounced in normal than in cross-link deficient collagen.

2. In normal collagen, the ratio between the extension of the fibrils and the tendon increases with the strain rate.

3. In cross-link deficient collagen, the opposite tendency is observed in connection with the appearance of a plateau in the load/extension curve indicating pronounced creep behaviour.
4. The fracture stress is dramatically reduced (by about a factor of 10) in cross-link deficient collagen, without a large change in maximum strain.

Clearly, some qualitative conclusions can be drawn from these observations. Firstly, considerable deformation must occur outside the collagen fibrils, presumably in the proteoglycan-rich matrix (observation 1), in agreement with previous work (Fratzl et al., 1997). In cross-link deficient collagen, additional slippage of molecules or subfibril structures (not prevented by covalent cross-linking) may be responsible for the larger relative strain in the fibrils at slow deformation rates. Secondly, the proteoglycan-rich matrix becomes stiffer when the applied strain rate is larger (observation 2), which is most likely due to a substantial viscous component in its response. Thirdly, the collagen cross-links are crucial in determining the stiffness of the fibrils, since deformation at a given load is much larger in cross-link deficient than in normal collagen (observation 3), also in agreement with earlier studies (Bailey et al., 1998; Davison, 1989; Eyre et al., 1984; Lees et al., 1990; Light and Bailey, 1979). Finally, the cross-link deficiency also enhances the creep behaviour of collagen, which further emphasises the importance of the cross-links for the stability of the tissue. The mechanical behaviour of cross-link deficient collagen is in this respect somewhat similar to immature collagen, where lower fracture stress and enhanced creep behaviour were also observed (Kastelic and Bear, 1980).

Combining the wealth of structural and mechanical information available for collagen fibril (Hodge and Petruska, 1963; Mosler et al., 1985; Folkhard et al., 1986; Fratzl et al., 1993, 1997; Hulmes et al., 1995; Sasaki and Odajima, 1996; Misof et al., 1997; Sasaki et al., 1999), the proteoglycan-rich matrix (Scott, 1991; Cribb and Scott, 1995), as well as the covalent cross-links (Light and Bailey, 1979; Eyre et al., 1984; Davison, 1989; Lees et al., 1990; Bailey et al., 1998), we propose a simple model for the strain-rate dependent effects reported in this chapter. This model is based on the simplified tendon structure shown in Figure 9.2 and depicted in Figure 9.8. The tendon is considered as a composite material with collagen fibrils embedded in a proteoglycan-rich matrix. This matrix is mostly loaded under shear. Since the spacing between fibrils is much smaller than their length, the shear stress τ effectively applied to the matrix is much smaller than the tensile stress σ on the tendon. In fact (Hull, 1981), $\tau \approx \sigma H/L$, where L is the length of the fibrils and H their spacing. The aspect ratio L/H is typically in the order of 100 to 1000. One may suppose that the elastic response of the matrix is mostly due to the entanglement of molecules attached to the collagen fibrils, such as proteoglycans. In addition, there will be a considerable viscosity in the matrix due to the many hydrogen bonds that can form in this glassy structure.

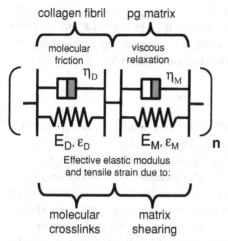

collagen fibril   pg matrix

molecular
friction   viscous
relaxation

η_D η_M

E_D, ε_D E_M, ε_M **n**

Effective elastic modulus
and tensile strain due to:

molecular   matrix
crosslinks   shearing

Figure 9.8: Depiction of a mechanical model, where fibrils and matrix are considered as viscoelastic systems arranged in series. E_D is the elastic modulus of the fibrils which depends critically on the covalent cross-links. η_D is the viscosity of the fibrils, possibly due to friction between molecules. E_M is the effective elastic modulus of the matrix which relates to its shear modulus G by $E_M = G(L/H)^2$, where L is the length and H the lateral spacing between collagen fibrils (Hull, 1981; Vincent, 1990). The order of magnitude for the length and diameter of collagen fibrils are 10 μm and 0.1 μm, respectively. If one supposes that the lateral distance between fibrils, less than their diameter, then L/H is certainly larger than 100. Finally, η_M is the viscosity of the matrix.

Although actually deforming in shear, this matrix can be mathematically characterised by an *effective* elastic modulus E_M (linked to the shear modulus G roughly by $E_M = G(L/H)^2$ (Hull, 1981; Vincent, 1990) and by the viscosity η_M (Figure 9.8). In terms of the observed mechanical response when the tendon is stretched, this matrix is effectively in series with the collagen fibrils. The elastic modulus of the fibrils, E_D, is mostly determined by the covalent cross-links and is therefore likely to be considerably decreased in cross-link deficient collagen. The fibrils will generally also have some viscosity η_D, provided by friction between molecules due to interactions via charges and hydrogen bonds. In total, this provides the very simple model shown in Figure 9.8.

This viscoelastic model has been treated mathematically in (Puxkandl et al., 2001) and a simple equation is derived for the dependence of $d\varepsilon_D/d\varepsilon_T$ on the strain rate

$$d\varepsilon_D/d\varepsilon_T = \alpha + (\beta - \alpha)e^{-\gamma \varepsilon_T/\dot{\varepsilon}_T}$$

where we have used:

$$\alpha = \frac{E_M}{E_D + E_M}; \qquad \beta = \frac{\eta_M}{\eta_D + \eta_M}; \qquad \gamma = \frac{E_D + E_M}{\eta_D + \eta_M}.$$

The parameters are defined in Figure 9.8. Assuming further that the β-APN treatment changes only the elastic modulus of the fibrils from E_D to a smaller value E_D^* (due to fewer covalent cross-links), then the corresponding constants for the β-APN treated animals relate to the normal values by:

$$\alpha^* = \frac{E_M}{E_D^* + E_M}; \qquad \beta^* = \beta; \qquad \gamma^* = \gamma\alpha/\alpha^*.$$

This model predicts that, at very small strain rates, tendon behaviour is dominated by the elastic moduli; that is, $d\varepsilon_D/d\varepsilon_T = \alpha = E_M/(E_D + E_M)$, while at large strain rates it is dominated by the viscosity, that is, $d\varepsilon_D/d\varepsilon_T = \beta = \eta_M/(\eta_D + \eta_M)$. The consequence is that $d\varepsilon_D/d\varepsilon_T$ increases with increasing $d\varepsilon_T/dt$ when α is smaller than β and decreases in the opposite case. Hence, if in normal collagen $\alpha < \beta$, the behaviour seen in Figure 9.7(a) is qualitatively reproduced. Furthermore, the cross-link deficiency would decrease E_D and, therefore, increase α. If the increase of α is sufficient, the behaviour of $d\varepsilon_D/d\varepsilon_T$ would be reversed, such as seen in Figure 9.7(b).

In Figure 9.7(d), the prediction for $d\varepsilon_D/d\varepsilon_T$ was plotted with suitably chosen constants (given in the figure legend), together with data from Figure 9.7(a) and 9.7(b). There is a reasonable agreement between the model and the experiments, but this should not be overinterpreted, since we have no independent estimate of the correct values for the constants α, β, and γ. Nevertheless, the model shows that an increase of $d\varepsilon_D/d\varepsilon_T$ with increasing strain rate in normal collagen and a decrease of $d\varepsilon_D/d\varepsilon_T$ with increasing strain rate in cross-link deficient collagen are not unlikely, due to the fact that the mechanical behaviour becomes more and more dependent on the viscosities when the strain rate increases.

This model relates to the mechanical characteristics of the intact tendon before yielding and failure take place, and so does not relate to the different fracture behaviours clearly evident in the right-hand portions of Figure 9.5(a–c). Normal rat-tail tendon [Figure 9.5(a)] normally fails by fibrils or groups of fibrils becoming debonded from the proteoglycan matrix at the maximum load and then pulling out past each other as the load rapidly drops to zero. In normal tendon, the tensile strength of the fibrils greatly exceeds the shear strength of the interface between fibril and matrix. In β-APN treated animals, a proportion of the fibrils have reduced tensile strength due to reduced cross-link density. This makes it more likely that slippage within fibrils or groups of fibrils can occur while the interface between fibril/fibril group and matrix is still intact. The ultrastructure of fractured ends of tendons may well reveal clues to such differences in fracture mechanisms and will be investigated further. However, the different mechanical behaviour of the two subgroups identified within the cross-link deficient tendons [Figure 9.5(b) and 9.7(b), on the one hand, and Figure 9.5(c) and 9.7(c), on the other] is more difficult to explain. This might possibly be due to the fact that the rats had time to form normal collagen before they were fed with β-APN. Tendons will, therefore, contain a mixture of normal and cross-link deficient fibrils. A variation of the relative amounts in this mixture could possibly be at the origin of the variations seen in mechanical behaviour. Generally speaking, the properties are more variable in β-APN fed rats than in normal controls, which is clearly visible from the scattering of data in Figure 9.7(a) to 9.7(c). For a more reliable correlation between cross-links and mechanical properties, one could envisage in future work to measure the cross-link density within each of the tested fibrils. This was beyond the scope of the present study.

The model in Figure 9.8 treats the tendon as a hierarchical level where collagen fibrils and interfibrillar matrix can be considered as interacting viscoelastic systems. It treats the collagen fibrils as homogeneous entities and neglects the fact that they are assemblies of regularly packed collagen molecules (Hulmes et al., 1995). In conclusion, the complex viscoelastic behaviour of collagen cannot be understood without considering the hierarchical nature of the structure. Indeed, a number of different deformation mechanisms operate at the various levels, providing collagen with its remarkable strength and tensile stiffness.

REFERENCES

Abé, H., Hayashi, K., and Sato, M. (eds.). (1996) *Data Book on Mechanical Properties of Living Cells, Tissues, and Organs*. Tokyo, Japan: Springer.

Amenitsch, H., Bernstorff, S., and Laggner, P. (1995) High flux beamline for small-angle x-ray scattering at ELETTRA. *Rev. Sci. Instr.* **66**, 1624–26.

Bailey, A. J., Paul, R. G., and Knott, L. (1998) Mechanisms of maturation and ageing of collagen. *Mech. Ageing Dev.* **106**, 1–56.

Cribb, A. M., and Scott, J. (1995) Tendon response to tensile stress: An ultrastructural investigation of collagen: Proteoglycan interactions in stressed tendon. *J. Anat.* **187**, 423–28.

Davison, P. F. (1989) The contribution of labile crosslinks to the tensile behavior of tendons. *Connect. Tissue Res.* **18**, 293–305.

Diamant, J., Keller, A., Baer, E., Litt, M., and Arridge, R. G. C. (1972) Collagen: Ultrastructure and its relation to mechanical properties as a function of aging. *Proc. R. Soc. Lond. B* **180**, 293–315.

Eyre, D. R., Paz, M. A., and Gallop, P. A. (1984) Crosslinking in collagen and elastin. *Ann. Rev. Biochem.* **53**, 717–48.

Folkhard, W., Mosler, E., Gercken, W., Knörzer, E., Nemetschek-Gansler, H., Nemetschek, Th., and Koch, M. H. J. (1986) Quantitative analysis of the molecular sliding mechanism in native tendon collagen – Time resolved dynamic studies using synchrotron radiation. *Int. J. Biol. Macromol.* **9**, 169–75.

Fratzl, P., Fratzl-Zelman, N., and Klaushofer, K. (1993) Collagen packing and mineralization: An x-ray scattering investigation of turkey leg tendon. *Biophys. J.* **64**, 260–66.

Fratzl, P., Misof, K., Zizak, I., Rapp, G., Amenitsch, H., and Bernstorff, S. (1997) Fibrillar structure and mechanical properties of collagen. *J. Struct. Biol.* **122,** 119–22.

Hodge, A. J., and Petruska, J. A. (1963) Recent studies with the electron microscope on ordered aggregates of the tropocollagen molecule. In: *Aspects of Protein Structure,* G. N. Ramachandran (ed.). London, UK: Academic Press, pp. 289–300.

Hull, D. (1981) *An Introduction to Composite Materials.* Cambridge, UK: Cambridge University Press.

Hulmes, D. J. S., Wess, T. J., Prockop, D. J., and Fratzl, P. (1995) Radial packing, order and disorder in collagen fibrils. *Biophys. J.* **68,** 1661–70.

Kastelic, J., and Baer, E. (1980) Deformation in tendon collagen. *Proceedings of the 34th Symposium of the Society for Experimental Biology,* J. F. V. Vincent and J. D. Currey (eds.). Cambridge, UK: Cambridge University Press, pp. 397–435.

Lees, S., Eyre, D. R., and Barnard, S. M. (1990) β-APN dose dependence of mature crosslinking in bone matrix collagen of rabbit compact bone: Corresponding variation on sonic velocity and equatorial diffraction spacing. *Connect Tissue Res.* **24,** 95–105.

Light, N. D., and Bailey, A. J. (1979) Covalent crosslinks in collagen. Characterization and relationships to connective tissue disorders. In: *Fibrous Protein: Scientific, Industrial and Medical Aspects,* vol. 1, D. A. D. Perry and L. K. Creamer (eds.). London, UK: Academic Press, pp. 151–74.

Misof, K., Rapp, G., and Fratzl, P. (1997) A new molecular model for collagen elasticity based on synchrotron radiation evidence. *Biophys. J.* **72,** 1376–81.

Mosler, E., Folkhard, W., Knörzer, E., Nemetschek-Gansler, H., and Koch, M. H. J. (1985) Stress-induced molecular rearrangement in tendon collagen. *J. Mol. Biol.* **182,** 589–96.

Puxkandl, R., Zizak, I., Paris, O., Keckes, J., Tesch, W., Bernstorff, S., Purslow, P., and Fratzl, P. (2001) Viscoelastic properties of collagen—synchrotron radiation investigations and structural model. *Phil. Trans. R. Soc. Lond. B* **357,** 191–97.

Sasaki, N., and Odajima, S. (1996) Elongation mechanism of collagen fibrils and force-strain relations of tendon at each level of structural hierarchy. *J. Biomech.* **29,** 1131–36.

Sasaki, N., Shukunami, N., Matsushima, N., and Izumi, Y. (1999) Time resolved x-ray diffraction from tendon collagen during creep using synchrotron radiation. *J. Biomech.* **32,** 285–92.

Scott, J. E. (1991) Proteoglycan: Collagen interactions in connective tissues. Ultrastructural, biochemical, functional and evolutionary aspects. *Int. J. Biol. Macromol.* 13, 157–61.

Tang, S., Trackman, P. C., and Kagan, H. M. (1983) Reaction of aortic lysyl oxidase with β-aminopropionitrile. *J. Biol. Chem.* **10,** 4331–38.

Vincent, J. (1990) *Structural Biomaterials.* Princeton, NJ: Princeton University Press.

Wess, T. J., Hammersley, A. P., Wess, L., and Miller, A. (1998) A consensus model for molecular packing of type I collagen. *J. Struct. Biol.* **122,** 92–100.

TEN

Collagens with Elastin- and Silk-like Domains

J. Herbert Waite, Eleonora Vaccaro, Chengjun Sun,
and Jared M. Lucas

INTRODUCTION

'Without turbulence, there would be no life on earth' (Lugt, 1983). The full impact
of this particular perspective is nowhere more evident than in the high-energy
intertidal zone where mussels and other sessile organisms thrive amidst aeration, nutrient cycling, and waste removal ad libitum. The benefits of a turbulent
habitat are considerable, but there are also costs. A major cost for mussels is the
production of a robust holdfast or byssus. The byssus provides mussels with the
necessary tenacity to survive incessant buffeting by waves. In exchange, a mussel is committed to invest 10% or more of its energy and assimilated nitrogen
for making and maintaining a byssus (Jordan and Valiela, 1982).

Mussel byssus consists of a bundle of short threads that resemble tiny tendons (Figure 10.1). It is constructed entirely of extraorganismic, extracellular
connective tissue. In the *Mytilus* species (*M. edulis* and *M. galloprovincialis*),
each new thread has dimensions of a few centimetres in length and less than
0.1 mm in diameter and is made in about 5 minutes in the ventral groove of
the foot by a process akin to injection moulding (Waite, 1992). Like tendons,
the threads originate from the byssal retractor muscles by way of a rooted stem;
unlike tendons, byssal threads do not 'insert' onto foreign objects as tendons do
to bone (Currey, 1984). Instead, byssal adhesive plaques bond only to the surface
of exogenous substrata, such as stone (Tamarin et al., 1976). For morphological clarity, the byssus is divided into four parts (from proximal to distal): root,
stem, thread, and plaque or pad. Each thread is further divided into proximal
and distal portions according to appearance (i.e., smooth and stiff for the distal;
soft and crimped for the proximal). The formation and ultrastructure of mussel
byssus has been described in detail elsewhere (Bairati, 1991; Waite, 1992).

SHOCK-ABSORBING TETHERS

Based on generally accepted features of elastomers (Gibson and Ashby,
1997), byssal threads are ostensibly elastomeric: Young's modulus is low (range:

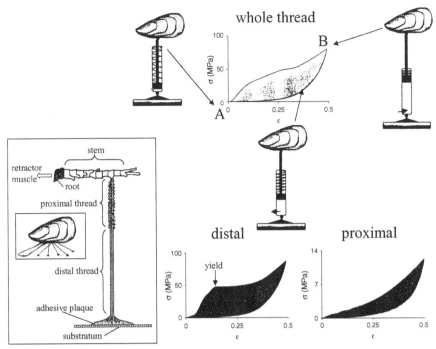

Figure 10.1: Stress-strain cycles of byssus from *M. galloprovincialis*. One thread is mod-elled as a spring-loaded shock absorber showing hysteresis with breakdown into distal and proximal contributions. Hysteresis is about 72% for the whole thread and distal portion, and a little lower at 47% for the proximal portion. All samples were extended to 0.5 (strain rate: $0.4-0.7$ s^{-1}). Inset at left shows mussel with typical byssal attachment and extended foot. Enlarged schematic shows dissected byssus with all threads but one removed for clarity. Thread diameter: 50–100 μm; length: 2–4 cm. Thread number per mussel in the field has a typical range of 20–200.

10–500 MPa), extensibility can be as high as 200%, and there is restorative re-coil. Like other protein elastomers – such as elastin, resilin, and abductin – byssal threads are quite tough (exhibit high-strain energy at break) (Table 10.1); how-ever, they are better than most at energy dissipation. Toughness and energy dissipation are both crucial properties for holdfasts. Energy dissipation in fibers subjected to cyclic stress-strain analysis is frequently normalized with respect to the total absorbed energy and reported as hysteresis or %hysteresis (Table 10.1). Hysteresis reflects a shock-damping capacity and is represented for one byssal thread by a confined spring and piston analogue (Figure 10.1). Enter a wave, the mussel is deflected by lift and drag forces. By restraining the mussel's deflection, a byssal thread assumes the load and stretches (curve A→B). As the wave passes, the load is relieved, and the thread relaxes (curve B→A) (Figure 10.1). Instant

Table 10.1. Comparative Tensile Properties for Structural Proteins Relevant to the Byssus of *M. galloprovincialis*

Tensile Element	Stiffness E_i, GPa	Strength, GPa	Extensibility	Hysteresis, %	Toughness, MJ/m^3	Source
Elastin	0.001	0.002	1.5	10	7.5	a
Periostracum	0.10	ND	0.9	ND	ND	b
Tendon collagen	1.5	0.15	0.12	7	5	a
Keratin, wool	0.5	0.2	0.5	45	20	c
Silk, viscid	0.003	0.5	2.7	65	150	a
Silk, MA	10	1.1	0.3	65	160	a
Titin (muscle)	0.002	ND	2	81	ND	d
Byssus, distal	0.50	0.15	0.8	72	50	b
proximal	0.05	0.02	1.6	47	15	b

Note: Strength or σ_{max} is the force to break divided by the cross-sectional area, extensibility or ε_{max} is the change in length (from initial to breaking) divided by the initial length, stiffness (E_i) is the initial slope of the σ versus ε curve, and the area under the σ versus ε curve gives the breaking strain energy or toughness. Hysteresis (%) is determined from cyclic stress-strain at non-breaking strains and represents the ratio of energy dissipated to the total energy absorbed. ND, not detected.

Sources: [a]Gosline et al. (1999); [b]our observations; [c]Wainwright et al., (1976), [d]Wang et al. (1993).

recoil is prevented by energy dissipation (viscous effects) and represented by friction in the piston element, thus sparing the mussel from being dashed against the rocks. The stress-strain cycle for one thread has been dissected into separate mechanical contributions for the distal and proximal portions of the thread (Smeathers and Vincent, 1979; Price, 1981; Bell and Gosline, 1996) (Figure 10.1). Of these, the distal is stronger, stiffer, and superior at damping (70% hysteresis), while the proximal portion is softer and weaker, with a lower but still significant hysteresis at 40%. Other protein structures show hysteresis ranging from 10% in elastin and tendon to a high of >80% for titin-derived passive tension in skeletal muscle (Wang et al., 1993). Major ampullate (MA) and viscid spider silks both have a hysteresis of 65%, which is appropriate given their role in the capture of flying insects (Gosline et al., 1999).

There is more to the mechanical behavior of byssal threads. Bell (as cited in Waite et al., 1998) first demonstrated that when strained beyond its yield point, the distal portion exhibited a dramatic stress softening (i.e., the initial modulus of the second cycle was reduced to about 20% of the modulus in the first cycle). Hysteresis was reduced as well; complete recovery to the modulus of the first cycle was slow (e.g., >24 hr), but significant partial recovery did occur within an hour (Bell, 1998; Vaccaro and Waite, 2001). Ironically, stress softening in individual threads may enhance the toughness in the entire byssus because it redistributes recurring loads. As illustrated in Figure 10.2, a roughly radial distribution of threads by the mussel subjected to uniaxial loading results in

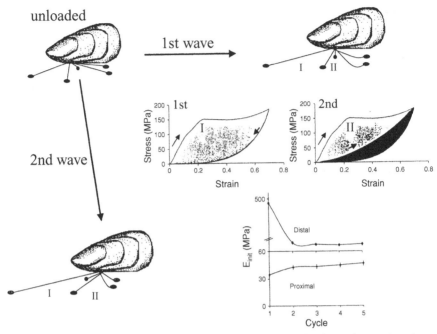

Figure 10.2: Stress-strain cycling of byssal threads for two cycles ('waves') to 70% strain. Stress softening shown for thread I in the second cycle leading to recruitment of thread II. Modulus matching in proximal and distal portions over five consecutive cycles appears to involve stress softening and strain stiffening in distal and proximal portions, respectively (bottom).

stress concentration at the leading thread (I). Suppose that thread I has been extended beyond its yield point by the first wave. Enter a second wave from the same direction. Due to stress softening, thread I's distal modulus has decreased from 500 to 80 MPa. Consequently, it extends with little resistance to loading. By so doing, threads II and III, which still have high initial stiffness, are recruited into action. Presumably, the process of stress softening followed by new recruitment continues until as many threads as possible contribute to load sharing. The proximal portion also shows a tendency to change stiffness with cyclic loading. In this case, there is strain stiffening from an initial modulus of 35 MPa to an asymptotic leveling at 50 MPa (Figure 10.2, bottom), an increase of about 40%.

INCREMENTAL MODULUS

In mediating attachment of mussel to stone, the byssus depends on a chain of morphological entities each with mechanical attributes that are distinct from those of adjoining structures. This is illustrated with respect to stiffness, for example [Figure 10.3(a)]. The modulus of skeletal muscle under passive tension

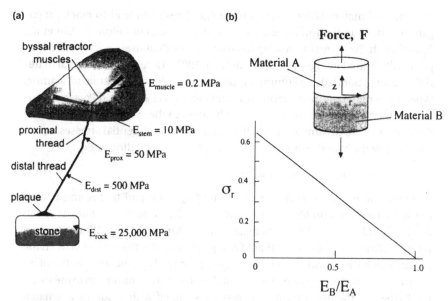

Figure 10.3: (a) Incremental steps in stiffness E_{init} in byssal attachment to stone. Initial moduli for different portions were determined from byssal sections taken from *M. galloprovincialis*. Strain rate was between 0.4 and 0.7 s^{-1}. (b) Stresses arising between two materials A and B with different modulus in a butt joint. Both A and B are assumed to have the same Poisson ratio (0.4), and the axial stress σ_z is assumed to be $= 1$. Residual stresses, σ_r and σ_θ, denote the radial and circumferential stresses, respectively, and are assumed to be the same. Problem looks at $E_B < E_A$.

is estimated at 0.2 MPa (Wang et al., 1993). The stem has a modulus of about 10 MPa (E. Vaccaro, unpublished data). Distinct proximal and distal portions within each thread differ in modulus by an order of magnitude [e.g., 50 (proximal) and 500 MPa (distal) in *M. galloprovincialis*]. Finally, each thread is attached to a rock or shell with a modulus of 25,000 MPa or more. Although the moduli appear to increase incrementally from the animal's soft tissues to the rock, many of these steps are nonetheless abrupt leaps in stiffness.

In manufactured materials, leaps in mechanical properties usually reflect structural discontinuities, such as joints and raise serious engineering concerns. When joining man-made materials, it is good design practice to ensure that critical material properties are well-matched. Depending on the application, critical properties might include Young's modulus, coefficient of expansion, and conductivity among others. Modulus is the most salient property in considering byssus. Without a good match in modulus, joined materials deform to different extents during loading (Suresh, 2001). Concentrated shear and axial stresses called residual stresses are generated where the interface between two mismatched materials intersects a free edge. These can lead to crack propagation parallel to the interface as well as other modes of failure (Rabin et al., 1995; Suresh, 2001). An interesting treatment of residual stresses in a model butt joint is illustrated in Figure 10.3(b) (Kinloch, 1987). Here, two cylindrical materials, A and B, bonded at a common interface, are subjected to uniaxial loading. When both materials have identical Young's moduli (E_B, E_A) and Poisson ratios (ν_B, ν_A), then the axial stress, σ_z, is the only stress in the system. However, when the materials are dissimilar, residual radial and circumferential stresses, σ_θ, σ_r, are generated where the interface meets a free edge according to Eq. (10.1):

$$\sigma_\theta = \sigma_r = [\nu_B - E_B\nu_A/E_A][\sigma_z/(1 - \nu_B)]. \tag{10.1}$$

The example in Figure 10.3(b) assumes that $\sigma_z = 1.0$ and that A and B have the same Poisson ratio (0.4) over a range of E_B/E_A. Note that σ_θ and/or σ_r can approach $2/3\,\sigma_z$ as E_B decreases relative to E_A (Kinloch, 1987). When material B has a stiffness that is a tenth that of A (e.g., such as the difference between the proximal and distal portions of thread), the predicted σ_r is nearly 3/5 that of σ_z.

In byssus, are the stepwise moduli correlated with discontinuities in the structure? How are the usual residual stresses associated with modulus mismatch avoided? The joints between the attachment plaque and stone and, above that, between the plaque and the distal thread do indeed exhibit abrupt structural discontinuities (Tamarin et al., 1976; Benedict and Waite, 1986). These are counterbalanced somewhat by large interfacial areas. Another discontinuity is apparent at the junction between the stem root and retractor muscles (Tamarin, 1975). The 'joint' between the distal and proximal portions of the thread has not been systematically scrutinized by transmission electron microscopy. This is surprising given the obvious morphological differences between the two (e.g., different diameters and smooth or corrugated surfaces on proximal and distal portions, respectively, evident upon examination by light or scanning electron microscopy) (Bairati, 1991; A. Bairati, personal observations, 2001). The proximal and distal portions of a mytilid thread, but not the area joining them, have been subjected to more detailed scrutiny by transmission electron microscopy than any other byssal structure (Vitellaro-Zuccarello et al., 1983; Benedict and Waite, 1986; Bairati, 1991). The distal portion consists of densely fused fibrils (diameter: 7–8 nm), all oriented parallel to the thread axis and dispersed with occasional voids. In transverse sections, the fibrils appear to be organised into hexagonal and pentagonal patterns with 20 nm center-to-center distances (Bairati and Vitellaro-Zuccarello, 1976). The proximal portion, in contrast, has some densely packed fibrils near the thread cuticle; however, solitary undulating fibrils (diameter: 7–8 nm) or small clusters thereof dominate the rest of the core. The undulations have an average pitch and diameter of about 70 and 100 nm, respectively (Bairati and Vitellaro-Zuccarello, 1976; Vitellaro-Zuccarello et al., 1983).

Transmission electron microscopy has yet to be fully exploited to investigate the distal-to-proximal transition, but a variety of other analytical techniques – including fiber X-ray diffraction (Rudall, 1955), light birefringence (Mercer, 1952; Gathercole and Keller, 1975; Bairati and Vitellaro-Zuccarello, 1974), amino acid composition (Mascolo and Waite, 1986), and protein extraction (Qin and Waite, 1995) – have detected the presence of gradients in byssal threads. Before embarking on the subject of gradients, however, the proteins of byssal threads deserve introduction.

PROTEINS IN BYSSAL THREADS

Composition

Freshly collected byssal threads are about 70% water by weight. Of the dry weight, about 96% is present as protein with carbohydrates and inorganic residue (ash) accounting for the rest (Table 10.2). Ash content depends on the extent of seawater removal. In threads washed with distilled water, heavy and transition metals predominate (Coombs and Keller, 1981). Collagen represents about 48% of the protein composition in distal and 33% in proximal portions. This agrees with an independent estimate of $\leq$50% collagen based on fiber X-ray diffraction of

Table 10.2. Compositional Analysis of *M. edulis* Byssal Threads

Byssal Thread Analysis	%	n	Method
Water content, % total wt.	70 ± 3	8	Microbalance[a]
Dry weight, % total wt.	30 ± 2	8	Lyophilisation[a]
Ash, % dry wt.	4.8	1	Combustion[a]
Cu/Fe/Zn, % dry wt.	0.1	—	Atomic absorption[b]
Other cations, % dry wt.	0.2	3	Atomic absorption[b]
Hexoses, % dry wt.	2.5	3	Anthrone[c]
Protein, % dry wt.	95 ± 3	5	Amino acid analysis[c,d]
Collagen, % protein, *Distal*	48 ± 2	4	Hydroxyproline ratio[e,d]
Proximal	33 ± 4	4	Hydroxyproline ratio[e,d]
PreCol, % protein, *Distal*	96		Col $\cong$ preCol/2[c]
Proximal	66		Col $\cong$ preCol/2[e]
Other protein, *Proximal*	34		—

Note: Water content was determined by weighing freshly collected blotted threads before and after freeze-drying. Ash content was estimated from the weight remaining after combustion at 530°C. The proportion of collagen in threads was estimated from the 4-hydroxyproline content in proximal and distal portions divided by the 4-hydroxyproline content of purified preCol-P and -D. Composition values are given in percent ± standard deviation when known and number of observations (*n*).

Sources: [a]Sun, unpublished data; [b]Coombs and Keller (1981); [c]Pikkarainen et al. (1968); [d]Mascolo and Waite (1986); [e]Qin and Waite (1995).

the distal portion (Rudall, 1955). Non-collagenous proteins typically account for about 34% of the proximal portion of thread, not including the non-collagenous domains in the byssal precursor collagens described later.

Byssal Proteins

The isolation of proteins from mytilid byssus is much hampered by the extremely tough and insoluble properties of the structure. In general, proteins are more extractable from younger threads than older threads; they are also more extractable from the proximal than distal portions. Perhaps, in the latter case, this is due to the smaller proportion of matrix in distal byssus. A higher cross-link density is another possible explanation. To isolate byssal proteins, we have used two extraction strategies, one for collagens (A) and the other for non–cross-linked proteins (B). Strategy A is the standard pepsinization method, whereby byssus is incubated for many hours at 5–10°C with pepsin at pH 3–5 (Qin and Waite, 1995). As native triple helical collagen is largely pepsin-resistant, this treatment releases soluble collagen domains (Col-P, Col-NG, and Col-D) and many smaller peptide fragments from a cross-linked matrix. Strategy B involves use of 8 M urea buffered at pH 3–4. This treatment releases most or all of the proteins not covalently cross-linked in the structure (unpublished observations).

The collagen domains released by strategy A appear to be homotrimers with monomeric α-chain masses of 40–50 kDa. The α-chains have been characterised by partial peptide mapping, mass spectrometry, and sequencing (Qin and Waite, 1995). Edman sequence data were used to make degenerate oligonucleotides that allowed preferential polymerase chain reaction amplification of a mussel foot cDNA library, and later, selection and amplification of cDNAs corresponding to complete α-chain sequences of the byssal collagens (Coyne et al., 1997; Qin et al., 1997). These collagens have been referred to as 'prepepsinized collagens' or preCols (Qin and Waite, 1995; Waite et al., 1998) and are aptly described as naturally occurring molecular 'chimaeras' and 'block copolymers' because of their linear array of structural motifs (Figure 10.4): (a) a central kinked collagen domain representing 52–58% of the sequence, (b) an acidic patch (C-side of collagen), (c) variable flanking domains (both N- and C-sides), and (d) amino- and carboxy-termini rich in histidine and 3,4-dihydroxyphenyl-alanine (DOPA) residues. The variable flanking domains in preCol-D have sequence motifs reminiscent of spider dragline and *Antheraea* cocoon silk (Guerette et al., 1996; Sezutsu and Yukuhiro, 2000). As shown in Table 10.3, there are six slightly substituted poly-alanine blocks, which in spun silks give rise to crystalline β-sheets (Simmons et al., 1996). PreCol NG has Gly-rich sequences (66% Gly) (Qin and Waite, 1998) for which several homologues are worth noting: mussel periostracum (about 55% Gly as per Waite et al., 1979), certain plant cell wall proteins (Sachette-Martins, 2000), and, of course, the amorphous regions of spider silk

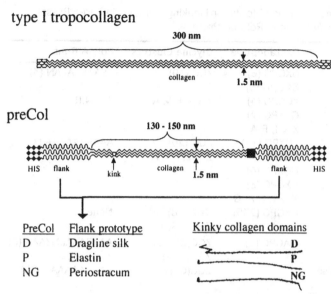

Figure 10.4: Schematic block structure of tropocollagen and the preCols-P, -D, and -NG. Collagenous domain of type I sea urchin collagen (Exposito et al., 1992) represents more than 95% of the tropocollagen sequence while the non-collagenous sequence is 5% (shown in scales). The closest homology of the flanking domains in each preCol is listed (left) and sequence motifs are given in Table 10.3. Central collagen domain occupies about 50% of the composition of each preCol. The effect of missing amino acids in each collagen domain is modelled using SYBIL (right). HIS, histidine.

(Lewis, 1992; Guerette et al., 1996). PreCol-P flanking sequences resemble the proline-containing pentapeptides of elastin (Brown-Augsburger et al., 1995) and flagelliform (viscid) silks (Hayashi and Lewis, 1998) (Table 10.3). Complete sequences for the α-chains of preCol-D, -NG, and -P are available in GenBank and have been discussed in a previous review (Waite et al., 1998). Using specific polyclonal antibodies and colloidal gold immunohistochemistry, preCol-P was shown to be associated with the undulating fibrils in the core of the proximal portion, while preCol-D was localized to the densely packed fibril bundles in the distal portion (Qin and Waite, 1995). Specific antibodies for preCol-NG are not available.

Strategy B liberated a variety of proteins particularly from proximal byssus. Notable among these was proximal thread matrix protein (PTMP-1), which was recently purified, characterized and cloned from two *Mytilus* species (Sun et al., 2002). PTMP-1 mass as determined by MALDI TOF mass spectrometry is 50 kDa, compared with the calculated 47.5 kDa based cDNA-deduced primary sequence. Carbohydrates probably account for most of the difference as concanavalin

Table 10.3. Some Sequence Motifs for Flanking Domains in Three Byssal PreCols from *M. edulis* and in Related Proteins

Protein	Pentapeptide	Glycine Clusters	Poly-A Runs
PreCol-P[a]	GXGPG (6)	ZGG (39)	ASAZAAAZAN (3)
M. edulis	GGGPG (5)		
	XGGPG (4)	Z = I, F, V, A	Z = N,R
	GGXPG (2)		
	X = I, F, A		
Elastin[b]	XGXPG (17)	ZGG (20)	AAAKAAAKAA (10)
Bos	GXXPG (8)	Z = L, I, V, A	
	GGXPG (4)		
	GXGPG (3)		
	X = V, F, A, G		
Viscid silk[c]	GXGPG (235)	GGZ (28)	None
Nephila	X = A, S, Y, V	Z = A, S, T	
PreCol-D[d]	GAGPG (2)	ZGG(G) (40)	AAAZAAA(A)ZAAA (6)
M. edulis		Z = L, A, F	Z = R, N, S
Spidroin1, MA silk[e]		ZGG(G) (100)	AAAAAAAAAA (25)
Nephila			
Wild silk[f]	None	GGY (117)	AAAAAAAAAAAA (80)
Antherea		GGZ (51)	
		Z = S, A, V	
PreCol-NG[g]	None	ZGG(G) (65)	ASAAARAAAAA (2)
M. edulis		Z = L, A, N, H	
Cell wall protein[h]	None	ZGGG(G)$_n$ (50)	None
Petunia		Z = A, V, L	

Note: Approximate number of repeats in parentheses.
Sources: [a]Coyne et al. (1997); [b]Brown-Augsburger et al. (1995); [c]Hayashi and Lewis (1998); [d]Qin et al. (1997); [e]Lewis (1992); [f]Sezutsu and Yukuhiro (2000); [g]Qin and Waite (1998); [h]Sachetto-Martins et al. (2000).

A-binding assays and analysis of neutral sugars show PTMP to be a glycoprotein (Sun et al., 2002). The structure of PTMP can be modelled by its strong homology to known proteins in the database (SwissProt). More than 80% of the primary sequence is present as two A-type domains of von Willebrand factor (also known as I-type integrin domains) (Pareti et al., 1987) [Figure 10.5(a)]. Each of these domains contains a metal-assisted collagen-binding sequence: DXSSS. Recent X-ray crystallographic analysis of the complex formed by the I domain of integrin, magnesium, and a collagenous peptide revealed that Mg^{2+} is octahedrally coordinated by an aspartate, two waters, and two serines in the DXSSS sequence and by a glutamate residue from the collagen (Bella and Berman, 2000). While formation of a similar complex is plausible between preCols and PTMP-1, it would be premature at this time to exclude other possibilities.

(a)

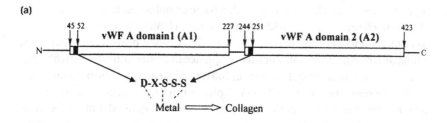

(b)

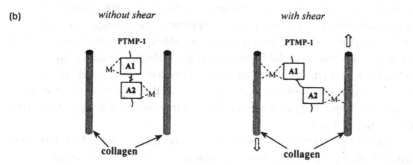

Figure 10.5: PTMP structure showing the prominence of two von Willebrand factor (vWF) domains (a). The demonstrated tendency of these domains to bind collagen under shear (Smith et al., 2000) is modelled to suggest matrix–fiber interactions in byssus during shear stiffening (b).

Collagen Domains and Sequence Transitions

Byssal preCols can be counted among a relatively small number of collagens that consist of fused linear blocks of disparate yet familiar structural domains. Two other striking examples are worthy of mention here: type I macrophage scavenger receptor protein (Kodama et al., 1990; Resnick et al., 1996) and *Hydra* minicollagen (Engel et al., 2001). The scavenger receptor is a trimer consisting of a membrane-anchored N-terminal domain, followed by a classic α-helical coiled coil (21 nm long), a triple helical collagen (17 nm) domain, and finally a globular C-terminal cap, where receptor binding occurs. The minicollagen from *Hydra* nematocysts is another intriguing block copolymer with a short central collagen domain (17 nm) that is flanked on both sides by poly-hydroxyproline sequences and capped by Cys-rich sequences at the N- and C-termini (Engel et al., 2001).

It would be instructive to observe how the transition from one secondary structural motif to another is negotiated in such block copolymers. The scavenger receptor protein and *Hydra* minicollagen both have sharp transitions, and the preCols follow this pattern. On the C-terminal side of the collagen domains in preCols, the acid-patch sequence punctuating the end of the Gly-X-Y repeats consists of a hexapeptide with a basic or polar residue, three nonpolar alkyl side chains, and two acidic residues [e.g., HIIVDD (preCol-NG), RIVIDD (preCol-P), and TLVIED (preCol-D)]. The significance of these hexapeptides, which also occur in the nonrepetitive sequence of spider and silkworm fibroins (Hayashi and Lewis, 1998; Datta et al., 2001; Sezutsu and Yukuhiro, 2000), is not known. Demarcation of the Gly-X-Y triplets from non-collagenous sequences on the N-terminal side of collagen domain is also well defined, but can be more subtle, in part because, like collagen, the other domains are also Gly-rich. Thus, in preCol-D, a repeated tripeptide sequence Gly-Gly-Z, switches suddenly from hydrophobic residues at Z to a persistently recurrent hydroxyproline (Qin et al., 1997). In preCol-P, an Ala-rich sequence, AAARANANGG, defines the end of the elastin-like domain and the beginning of the collagen domain (Coyne et al., 1997). While not typically associated with the elasticity of vertebrate elastins, alanine-rich clusters serve as host sequences for the lysyl residues involved in cross-links (Brown-Augsburger et al., 1995), a role not possible in preCols given the absence of lysines.

Structural peculiarities also occur on the collagenous side of the transition between collagen and the flanking sequences in preCols. As already noted, preCol-D has a string of more than 20 Gly-Gly-Y tripeptides clustered at the N-terminal edge of the collagen domain. Ordinarily, were it not for the presence of hydroxyproline in the Y position, the triple helical structure of collagen would be destabilised by such repeats, especially near the termini (Shah et al., 1997). This trend is not found in preCol-P and -NG, where the 7–10 Gly-Gly-Y sequences are dispersed throughout the collagen domain. In preCol-NG, however, two destabilizing Gly-X-Gly triplets and one Gly-Gly-Ser are clustered at the C-terminus of the collagen domain (Qin and Waite, 1998). Curiously, Gly-X-Gly triplets are also found near the C-terminal boundary of collagen in the scavenger receptor and minicollagen (Kodama et al., 1990; Engel et al., 2001). It is possible that the placement of destabilizing triplets near the boundaries 'softens' the collagen triple helix in anticipation of a less associative conformation in the three polypeptide chains.

IDENTIFICATION OF PROTEIN GRADIENTS

Mascolo (Mascolo and Waite, 1986) first identified biochemical gradients in byssal threads of *Mytilus*. After serially cutting byssal threads into 2 mm segments, she then hydrolyzed these and determined amino acid compositions. Glycine showed the greatest change from a high of 330 residues/1000 at the far-distal end to 240 residues/1000 at the far-proximal end; 4-hydroxyproline showed a similar trend, although the change was smaller. Taken together, the Gly and Hyp gradients suggest differences in the distribution of collagens. Qin and Waite (1995) identified gradient-specific collagens by isolating two pepsin-resistant collagen fragments, Col-P and Col-D, from serial byssal thread

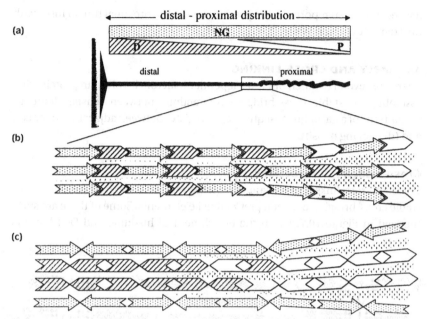

Figure 10.6: Gradients of chimaeric collagens preCol-D, -NG, and -P in byssal threads from *M. edulis*. Gradient shape is approximate as are relative proportions (a); model 1 of axial and lateral assembly of preCol dimers (poly [-NG-X]) in midbyssal threads where the gradient is thought to occur. All elements are assembled N→C with full lateral register (b). Model 2 of homopolymer axial assembly involving poly[-D] and poly[-NG]. PreCols are successively N-to-N and C-to-C (c). Interfibrous pattern denotes matrix protein. Arrow tails are N-terminal and heads are C-terminal.

sections and demonstrating that the distribution of these along the thread axis was graded. Graded distributions of two preCols (-D and -P) were also evident from western blots of protein extracted from successive serial sections of byssal threads and mussel feet using specific polyclonal antibodies for the collagenous domains of preCol-D and -P (Qin and Waite, 1995) as well as in situ hybridisation using preCol-specific probes (Qin et al., 1997; Qin and Waite, 1998). Because of the inevitable use of four to six tissue sections for analysis, these results do not specify a precise shape for the gradients, only a general trend. The distribution of preCol-D tapers off in a distal-to-proximal direction [Figure 10.6(a)], while preCol-P decreases in a proximal-to-distal direction reaching extinction before midthread in *M. edulis*. PreCol-NG ('no gradient'), in contrast, is uniformly distributed. Evidently, these gradients are produced prior to secretion because they are already detectable in expression libraries prepared from serial foot sections (Qin and Waite, 1998). Here, the relative abundance of each preCol mRNA can be precisely titrated using the rRNA EukA as an internal reference (Qin et al., 1997; Qin and Waite, 1998). Another gradient involving two variants of preCol-P has recently been reported to be present, where the proximal thread fuses with the stem (Coyne and Waite, 2000).

ASSEMBLY AND CROSS-LINKING
There are several distinct, but interdependent, factors involved in protein fiber assembly: (a) stable cross-bridging interactions between tensile elements, (b) patterns of axial or linear sequence, (c) register between adjacent sequences, and (d) packing density.

Cross-bridging Interactions
Covalent cross-links and chelate complexes represent the best candidates for stable cross-bridging between preCol tensile elements. Some of these are summarized in Figure 10.7. Given the enrichment of histidine and DOPA in the

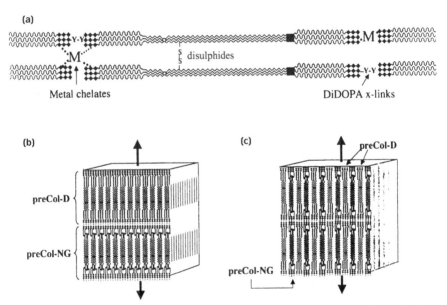

Figure 10.7: Models of preCol assembly in *Mytilus* byssus. (a) Summary of cross-links implicated in assembled preCols. DiDOPA cross-links have been detected by solid-state nuclear magnetic resonance (see text); disulphides are present in isolated preCols. Metal chelates are suggested by protein sequence and by the effect of EDTA on mechanical properties (Vaccaro and Waite, 2001). (b) View of tightly packed and fully aligned preCols in the distal portion in the equal stress or Reuss estimate of model 1. Loading is normal to the plane of each layer. Moduli and volume fractions for the collagen, silk-like, and Gly-rich domains are taken as 1.5, 10 and 0.1 GPa, and 0.5, 0.25, and 0.25, respectively. (c) Voigt estimate of model 2. Volume fractions and moduli are the same as previously stated.

amino- and carboxy-terminal sequences of all preCols, it was proposed that these residues might mediate end-to-end preCol binding as chelate complexes (Waite et al., 1998). Although chelate complexes are expected to provide axial interactions in isolated preCol polymer chains, they could also provide lateral bridges at higher packing densities. The chelated metals are likely to be zinc, copper, and iron: these three were detected in *Mytilus* byssus at up to 0.1% of dry weight (Table 10.2) (Coombs and Keller, 1981). If this amount was chelated entirely by preCols, it would represent about 3–4 metals per preCol in the distal portion. Many other metals are present at trace levels (Koide et al., 1978; Coombs and Keller, 1981). The stability constants of chelates formed between Fe(III) and DOPA-containing byssal precursors at pH 7.5 and 20°C were calculated to be more than 10^{40} in vitro (Taylor et al., 1996). Histidine in glycine-rich sequences (e.g., Gly-His-Gly) binds Cu(II) at 10^{20} (Smith et al., 1987); this tripeptide alone is repeated four times in the N-terminus of preCol-NG, and one or two times in the C-termini of -P and -D (Waite et al., 1998). The role of metal chelates in the cohesive assembly of preCols is also suggested by two recent observations: preCols were precipitated by micromolar amounts of Cu^{2+} (unpublished data), and EDTA treatment abolished the yield point in distal byssal threads (Vaccaro and Waite, 2001). Direct measurement of metal binding by preCols and, in particular by their His-rich termini, remains to be done.

End-to-end stabilization of preCols could be re-enforced by aryl coupling (Y-Y in Figure 10.7a). DOPA residues not engaged in chelate complexes are prone to auto-oxidation at slightly alkaline pH [e.g., seawater (Smith and Haskell, 2000)]. This oxidation may be catalyzed by significant catecholoxidase activity, which has been known for some time to be present in byssal threads (Waite, 1985). Oxidation of DOPA is typically by one or two electrons to DOPA semi-quinone or DOPA quinone, respectively. The chemical reactivity of quinones and semiquinones is known to be diverse and extensive in vitro (Waite, 1990); however, quinone-derived cross-links have not been isolated from mussel byssus, although somewhat better progress has been made with insect cuticle (Xu et al., 1997). Recently, Schaefer et al. (Merritt et al., 1996; McDowell et al., 1999) have pioneered the non-destructive solid-state analysis of scleroproteins by magic-angle spin nuclear magnetic resonance. In a study of byssus 'stressed' by conditions of high flow and aeration, the primary cross-link in ^{2}H- and ^{13}C-tyrosine-labelled byssal plaques was found to be 5,5'-diDOPA (McDowell et al., 1999). In the threads, aromatic amino acid content was too low to support the same conclusion at this time. DiDOPA was also formed in vitro during the enzyme-catalyzed oxidation of DOPA-containing byssal precursor peptides (Burzio and Waite, 2000). The reaction pathway was suggested as follows: DOPA undergoes a two-electron oxidation to DOPAquinone, which forms free radical semiquinones by reverse dismutation with unreacted DOPA. The free radicals couple preferentially with one another through the 5-C ring positions. While these results are tantalizing, they do not specifically implicate the involvement of preCol-linked DOPA residues at this time. The results also suggest that due to their inherent redox instability, diDopa cross-links may not be isolated intact from acid-hydrolysed byssus.

Another type of covalent interaction is the disulphide bond [—S—S—in Figure 10.7(a)]. As indicated, preCols are trimers. Each α-chain of preCol-D and -NG contains a single cysteine residue located toward the latter half of the collagen domain (Qin et al., 1997; Qin and Waite, 1998). PreCol-P has no cysteine. Disulphides form spontaneously from proximate pairs of cysteine residues by a one-electron oxidation. Disulphide formation between neighbouring preCols or preCol and a matrix protein would represent lateral interactions. PreCol-D and NG have a maximum capacity of three lateral disulphide interactions/molecule and a minimum of one (if the other two are involved in an intramolecular disulphide). There is no evidence for the aldimine- or aldol condensation-derived cross-links typical of collagen in byssus (van Ness et al., 1988).

Axial Sequence

The axial and lateral orders of tensile elements are at the root of understanding how the graded distribution of preCol-D and -P translates into a functional mechanical gradient. Concepts from polymer science bring to mind two paradigms of axial or linear sequence: homopolymers [those constructed from a single kind of repeat unit, e.g., poly[D] in which bracketed letters such as D refer to a particular preCol] and copolymers, with two or more types of ordered or random repeat units (e.g., poly[NG-P] or poly[D_x, -NG_y, -P_z], where subscripts denote relative proportions of each in the chain). Plausible models reconciling axial sequence with the stochastic distribution of preCols in byssal threads can be contrived for many combinations of homo- and/or copolymer chains in which spatial variability is controllable. Assuming the metal-induced 'stickiness' of the histidine-rich ends, even short chains would readily combine into longer ones upon addition of copper and/or zinc (see, e.g., Lilius et al., 1991).

Two slightly different models of axial assembly will be examined here. The axial model (model 1), proposed by Qin and Waite (1998), is based on the dimer NG-X (X is D or P) as the basic repeating unit. It assumes that preCols, like many collagens, assemble in an $N \rightarrow C$ orientation and that this assembly requires the coupling of specific sequences in the His-rich domains of adjacent preCols. Several such 'hand-shaking signatures' are evident: (1) preCol-D and -P have the tyrosine/DOPA-containing sequences, DY and KPGY, at their N- and C-termini, respectively (Waite et al., 1998). The reverse pattern holds for preCol-NG. (2) PreCol-D and -P both contain HAXSAG and HAHAH near the N-terminus and HGG (1–2xs) at the C-terminus. Again, the reverse pattern is true in preCol-NG.

Coupling preCols by way of such sequences may seem contrived, but it is reasonable here given that the local environment of histidine and DOPA functionalities has a pronounced effect on metal chelation and/or cross-linking ability (Smith et al., 1987; Burzio and Waite, 2000). Glycine and alanine are common spacers for histidine residues involved in metal chelation, but their effect on the affinity and specificity of histidine towards metals such as zinc, nickel, and copper, for example, is different (Smith et al., 1987). What emerges from this model is a copolymer of two specific dimers (poly[NG-D] and poly[NG-P]), in which the proportion of the two can be gradually changed. Thus, beginning at the distal end, preCol-NG would be required as a linker between the C-terminus of one preCol-D and the N-terminus of the following preCol-D, and, eventually upon approaching the proximal portion, preCol-P [Figure 10.6(b)].

Another interesting model (model 2) to reconcile the distributions of preCol,-D, -NG, and -P in byssal threads involves spatially variable homopolymers. With this one, we discard the assumption that axial orientation necessarily be N→C or head-to-tail. Both head-to-head and tail-to-tail orientations are present in the peculiar kinked collagens of dogfish eggcase (Knupp et al., 1999). Interactions such as these would allow a linear sequence, as pictured in Figure 10.6c, with preCols in N-to-N-terminal and C-to-C-terminal couplings. Note that this still accommodates the 'hand-shaking' signatures listed previously. PreCol-NG would form uniform homopolymers (poly[-NG]) from the distal-to-proximal end of the thread, while the poly[-D] chains formed distally would show an increasing tendency to substitution by preCol-P as they pass from thread midlength into the proximal portion [Figure 10.6(c)].

Register and Density

Register refers to the nature of alignment of a linear tensile element with its lateral neighbours. In models of fiber-forming interstitial type I collagens, axial assembly (such as that described previously) is prevented by a gap region about 36 nm wide separating each collagen from its linear neighbours. A consequence of this gap is that fiber formation (also known as fibrillogenesis) in collagens is driven by lateral rather than axial interactions (Prockop and Fertala, 1998). Lateral interactions between neighbouring tropocollagens are uniformly shifted by a quarter-length or about 75 nm (Prockop et al., 1979). Quarter-stagger periodicity is not observed in byssus and is difficult to envision in byssal preCols for the following reasons: (1) collagen domains cover only 50% of the preCol molecule (compared with 95% in tropocollagens), (2) preCol collagen domains are markedly kinked by one or more missing glycine residues, and (3) the His-rich termini are large extended secondary structures. Predicted metal chelation between the His-rich N- and C-terminal domains would prevent gaps between preCols, such as those present in type I collagen. We suspect that densely packed distal preCols are more likely to be in register than the loosely packed ones of the proximal portion: Unique kinks, stacking of β-sheet crystals (See Figure 4 in Waite et al., 1998), and the formation of lateral disulphides would all tend to favor a tight register between packed poly[D] segments.

MECHANICAL MODELS

Perhaps future work will reveal new facts that disqualify these two models. Models are meant to challenge the data, not to validate them. Models 1 and 2 attempt to reconcile preCol assembly with the intricacies of available sequence, domain structure, preferred orientation of collagens, and metal-binding environments, as well as the stochastic preCol distributions. But are the models consistent with known mechanical properties of byssal threads? The answer is affirmative if certain assumptions are made, the main one being that the stiffness of the silk-like (D), elastin-like (P), and glycine-rich flanking domains can be deduced from the stiffness of spider MA silk, elastin, and periostracum, respectively (Table 10.1). The preCol collagen domain is also presumed to be like tendon collagen, although the complexity of tendon structure and the unknown consequence of missing Gly residues must add some uncertainty to this (Sasaki and Odajima, 1996).

Mechanical models have been designed to assess the effect of serial or parallel components in composite materials (Harris, 1980). Of the two portions, the distal is better disposed to modelling since, by composition, about 90% of it consists of densely fused chains of collagen, silk, and Gly-rich segments. If neighbouring silk segments are in tight register, perhaps with stacked β-sheets (Figure 10.7b), then the array would resemble a smectic liquid crystal and conform well to the requirements for a series of alternating hard and soft layers under normal uniaxial tension (model 1). The Reuss estimate (10.2) for calculating the composite modulus for such a structure is

$$1/E_c = V_1/E_1 + V_2/E_2 + \text{etc.} \tag{10.2}$$

Using volume fractions (V_n) of 0.5 for collagen and 0.25 for silk and periostracum-like domains and moduli (E_n) of 1.5 GPa, 10 GPa, and 0.1 GPa for collagen, silk, and periostracum, respectively, a composite E_c of 350 MPa is calculated. This is well within the range of measured moduli for the distal portion of byssus from *Mytilus* species for which we have byssal protein sequence (range: 150–500 MPa – Price, 1981; Bell and Gosline, 1996; Smeathers and Vincent, 1979; Vaccaro and Waite, 2001). If one segregates the preCol-D and -NG elements into separate linear fibrils [model 2; Figure 10.7(c)] with tight packing and disulphide bonds, then the Voigt estimate can be applied (10.3):

$$E_c = V_1 E_1 + V_2 E_2 + \text{etc.} \tag{10.3}$$

This time, the composite modulus E_c is estimated to be 1.4 GPa. Although this modulus is two to three times higher than the observed stiffness, our model is not compromised by the presence of microvoids as distal threads commonly are (Bariati and Vitellaro-Zuccarello, 1994). A Reuss estimate for the proximal portion yields an E_c of 8 MPa, which is less than half the observed E of 20–50 MPa for proximal threads. Noninclusion of the very significant proportion of noncollagenous matrix (34%) in proximal threads must contribute to the underestimation.

Neither model specifically addresses the nonlinear peculiarities of mechanical behavior. These include yield and stress softening in the distal portion, as well as strain stiffening in the proximal portion. Model 2, however, does offer a more plausible explanation for yield and recovery in the distal portion: at yield stress, metal chelates in the stiff poly[-D] chains might fail and transfer their load to intact poly[-NG] chains. These would extend with little initial resistance and, upon relaxation, recover their initial length (see stress-strain curve for distal portion in Figure 10.1). Future studies will examine this possibility in detail. Recent identification of von Willebrand A domains in PTMP and the shear-induced binding of these to type I collagen (Smith et al., 2000) suggest that this protein may contribute to the reversible strain-induced stiffening observed in proximal threads [Figure 10.5(b)] (Sun et al., 2001).

CONCLUSIONS

Byssus is essentially an array of elastomerically graded fibers that mussels produce to attach to hard surfaces in their habitat and depend upon to moderate the forces buffeting them. Gradients are most evident in the span of a byssal thread between the attachment plaque and the stem. Here, stiffness drops by a factor of 10. Correlated to this mechanical ramp-down is the graded distribution of two chimaeric collagens: preCol-D, and -P and a nongraded third, -NG. All share similar collagen and terminal domains. The flanking domains wedged between these, however, exhibit a distinct resemblance to three well-known structural paradigms: the crystalline β-sheets (D) and amorphous domains (NG) of spider silk, as well as the β-spirals of elastic pentapeptides (P). The axial and lateral self-assembly of preCols during byssal thread formation results in a distribution in which preCols with β-sheets predominate distally, while those with β-spirals prevail proximally.

Living organisms produce a great variety of graded structures. Structures with graded porosities, for example, are particularly common in nature (Currey, 1984; Gibson and Ashby, 1997; Amada, 1995). Graded structures and the pathways for producing them probably persist because of the adaptive advantages they offer. Given the complexity of biological structures, however, information about structure – function relationships remains difficult to assess at a molecular level.

Byssal threads may represent a singular opportunity for gaining insights into the molecular nature of structural gradients. Such insights are also emerging from industry: functionally graded materials are being increasingly utilized for their superior performance in the next generation of electronic, energy conversion, and aeronautic materials (Kaysser and Ilschner, 1995; Koizumi and Niino, 1995). In these technologies, it is generally appreciated that gradient structures smooth stress distribution, reduce stress concentrations, eliminate stress singularities, and in most cases improve bonding strength and fracture toughness (Erdogan, 1995). Many of these advantages are manifestly suggested in the structure and performance of mussel byssus.

ACKNOWLEDGMENTS
We thank E. Carrington, B. Gaylord, J. Gosline, and S. Rapoport for comments and stimulating discussion.

REFERENCES
Amada, S. (1995) Hierarchical functionally gradient structures of bamboo, barley, and corn. *MRS Bull.* **20,** 35–36.

Bairati, A., and Vitellaro Zuccarello, L. (1974) The ultrastructure of the byssal apparatus of *Mytilus galloprovincialis*. Analysis of byssus by polarized light microscopy. *J. Submicrosc. Cytol.* **6,** 367–79.

Bairati, A., and Vitellaro Zuccarello, L. (1976) The ultrastructure of the byssal apparatus of *Mytilus galloprovincialis*. IV. Observations by transmission electron microscopy. *Cell Tiss. Res.* **166,** 219–34.

Bairati, A. (1991) The byssus of the mussel *Mytilus* from the molecules to the organ: Functional performance resides in the ultrastructural assemby. In: *Form and Function in Zoology*, (G. Lanzavecchia and R. Valvassori (eds.). Modena, Italy: Mucchi, pp. 163–77.

Bell, E. C., and Gosline, J. M. (1996) Mechanical design of mussel byssus: Material yield enhances attachment strength. *J. Exp. Biol.* **199,** 1005–17.

Bell, E. C. (1998) Mechanical design of mussel byssus. Presented at the annual meeting of the Society of Experimental Biology, Edinburgh, UK.

Bella, J., and Berman, H. M. (2000) Integrin-collagen complex: A metal-glutamate handshake. *Structure* **8,** R121–26.

Benedict, C.V., and Waite, J. H. (1986) Composition and ultrastructure of the byssus of *Mytilus edulis*. *J. Morph.* **189,** 261–70.

Brown-Augsburger, P., Tisdale, C., Broekelmann, T., Sloan, C., and Mecham, R. P. (1995) Identification of an elastin cross-linking domain that joins three peptide chains. *J. Biol. Chem.* **270:** 17778–83.

Burzio, L. A., and Waite, J. H. (2000) Cross-linking in adhesive quinoproteins: Studies with model decapeptides. *Biochem.* **39,** 11147–53.

Coombs, T. L., and Keller, P. J. (1981) *Mytilus* byssal threads as an environmental marker for metals. *Aquat. Toxicol.* **1,** 291–300.

Coyne, K. J., Qin, X. X., and Waite, J. H. (1997) Extensible collagen in mussel byssus: A natural block copolymer. *Science* **277**, 1830–32.

Coyne, K. J., and Waite, J. H. (2000) In search of molecular dovetails in mussel byssus: From the threads to the stem. *J. Exp. Biol.* **203**, 1425–31.

Currey, J. (1984) *The Mechanical Adaptations of Bones.* Princeton, NJ: Princeton University Press.

Datta, A., Ghosh, A. K., and Kundu, S. C. (2001) Purification and characterization of fibroin from the tropical saturniid silkworm, *Antheraea mylitta. Insect Biochem Mol. Biol.* **31**, 1013–18.

Engel, U., Pertz, O., Fauser, C., Engel, J., David, C. N., and Holstein, T. W. (2001) A switch in disulfide linkage during minicollagen assembly in *Hydra* nematocysts. *EMBO J.* **20**, 3063–73.

Erdogan, F. (1995) Fracture mechanics of functionally graded materials. *MRS Bull.* **20**, 43–44.

Exposito, J. Y., D'Alessio, M., Solursh, M., and Ramirez, R. (1992) Sea urchin collagen evolutionarily homologous to vertebrate pro-$\alpha2$(I) collagen. *J. Biol. Chem.* **267**, 15559–64.

Gathercole, L. J., and Keller, A. (1975) Light microscopic waveforms in collagenous tissues and their structural implications. In: *Structure of Fibrous Biopolymers*, E. D. T. Atkins and A. Keller (eds.). London, UK: Butterworths, pp. 153–75.

Gibson, L. J., and Ashby, M. F. (1997) *Cellular Solids,* 2nd ed. Cambridge, UK: Cambridge University Press.

Gosline, J. M., Guerette, P. A., Ortlepp, C. S., and Savage, K. N. (1999) The mechanical design of spider silks: From fibroin sequence to mechanical function. *J. Exp. Biol.* **202**, 3292–3305.

Guerette, P. A., Ginzinger, D. G., Weber, B. H. F., and Gosline, J. M. (1996) Silk properties determined by gland-specific expression of a spider fibroin gene family. *Science* **272**, 112–15.

Harris, B. (1980) The mechanical behaviour of composite materials. *Symp. Soc. Exp. Biol.* **34**, 37–74.

Hayashi, C. Y., and Lewis, R. V. (1998) Evidence from flagelliform cDNA for the structural basis of elasticity and modular nature of spider silks. *J. Mol. Biol.* **275**, 773–84.

Jordan, T. E., and Valiela, I. (1982) A nitrogen budget for the ribbed mussel and its significance in nitrogen flow in a New England salt marsh. *Limnol. Oceanogr.* **27**, 75–90.

Kaysser, W. A., and Ilschner, B. (1995) FGM research activities in Europe. *MRS Bull.* **20**, 22–25.

Kinloch, A. J. (1987) *Adhesion and Adhesives: Science and Technology.* London, UK: Chapman & Hall.

Koide, M., Lee, D. S., and Goldberg, E. D. (1982) Metal and transuranic records in mussel shell, byssal threads and tissues. *Estuar. Coast. Shelf Sci.* **15**, 679–95.

Kodama, T., Freeman, M., Rohrer, L., Zabrecky, J., Matsudaira, P., and Krieger, M. (1990) Type one macrophage scavenger receptor contains α-helical and collagen-like coiled coils. *Nature* **343**, 531–35.

Koizumi, M., and Niino, M. (1995) Overview of FGM research in Japan. *MRS Bull.* **20**, 19–21.

Knupp, C., Luther, P. K., Morris, E. P., Knight, D. P., and Squire, J. M. (1999) Partially systematic molecular packing in the hexagonal columnar phase of dogfish egg case collagen. *J. Struct. Biol.* **126,** 121–30.

Lewis, R. V. (1992) Spider silk: The unraveling of a mystery. *Acc. Chem. Res.* **25,** 392–98.

Lilius, G., Persson, M., Bülow, L., and Mosbach, K. (1991) Metal affinity precipitation of proteins carrying genetically attached polyhistidine tails. *Eur. J. Biochem.* **198,** 499–504.

Lugt, H. J. (1983) *Vortex Flow in Nature and Technology.* New York: Wiley.

Mascolo, J. M., and Waite, J. H. (1986) Protein gradients in byssal threads of some marine bivalve molluscs. *J. Exp. Zool.* **240,** 1–7.

McDowell, L. M., Burzio, L. A., Waite, J. H., and Schaefer, J. (1999) REDOR detection of cross-links formed in mussel byssus under high flow stress. *J. Biol. Chem.* **274,** 20293–95.

Mercer, E. H. (1952) Observations on the molecular structure of byssus fibres. *Austral. Mar. Freshwater Res.* **3,** 199–204.

Merritt, M. E., Christensen, A. M., Kramer, K. J., Hopkins, T. L., and Schaefer, J. (1996) Detection of intercatechol cross-links in insect cuticle by solid-state carbon-13 and nitrogen-15 NMR. *J. Am. Chem. Soc.* **118,** 11278–82.

Pareti, F. I., Niiya, K., McPherson, J. M., Ruggeri, Z. M. (1987) Isolation and characterization of two domains of human von Willebrand factor that interact with fibrillar collagen types I and III. *J. Biol. Chem.* **262,** 13835–41.

Pikkarainen, J., Rantanen, J., Vastamäki, M., Lampiaho, K., and Kulonen, E. (1968) On collagens of invertebrates with special reference to *Mytilus edulis. Eur. J. Biochem.* **4,** 555–60.

Price, H. A. (1981) Byssus thread strength in the mussel, *Mytilus edulis. J. Zool. Lond.* **194,** 245–55.

Prockop, D. J., Kivirrikko, K. I., Tuderman, L., and Guzman, N. (1979) The biosynthesis of collagen and its disorders. *N. Engl. J. Med.* **301,** 13–23.

Prockop, D. J., and Fertala, A. (1998) The collagen fibril: The almost crystalline structure. *J. Struct. Biol.* **122,** 111–18.

Qin, X. X., and Waite, J. H. (1995) Exotic collagen gradients in the byssus of the mussel *Mytilus edulis. J. Exp. Biol.* **198,** 633–44.

Qin, X. X., Coyne, K. J., and Waite, J. H. (1997) Tough tendons: Mussel byssus has collagen with silk-like domains. *J. Biol. Chem.* **272,** 32623–27.

Qin, X. X., and Waite, J. H. (1998) A collagenous precursor that may mediate block copolymer gradients in mussel byssal threads. *Proc. Natl. Acad. Sci. U.S.A.* **95,** 10517–22.

Rabin, B. H., Williamson, R. L., and Suresh, S. (1995) Fundamentals of residual stresses in joints between dissimilar materials. *MRS Bull.* **20,** 37–39.

Resnick, D., Chatterton, J. E., Schwartz, K., Slayter, H., and Krieger, M. (1996) Structures of class A macrophage scavenger receptors. *J. Biol. Chem.* **271,** 26924–30.

Rudall, K. M. (1955) The distribution of collagen and chitin. *Symp. Soc. Exp. Biol.* **9,** 49–71.

Sachetto-Martins, G., Franco, L., Oliveira, D. E. (2000) Plant glycine rich proteins: A family or just proteins with a common motif? *Biochim. Biophys. Acta* **1492,** 1–14.

Sasaki, N., and Odajima, S. (1996) Stress strain curve and Young's modulus as determined by the x-ray diffraction technique. *J. Biomech.* **29,** 655–58.

Sezutsu, H., and Yukuhiro, K. (2000) Dynamic rearrangement within the *Antheraea pernyi* gene is associated with four types of repetitive units. *J. Mol. Evol.* **51,** 329–38.

Shah, N. K., Sharma, M., Kirkpatrick, A., Ranshaw, J. A. M., and Brodsky, B. (1997) Gly-Gly containing triplets of low stability adjacent to a type III collagen epitope. *Biochem.* **36,** 5878–83.

Simmons, A. H., Michal, C. A., and Jelinski, L. W. (1996) Molecular orientation and two-component nature of the crystalline fraction of spider dragline silk. *Science* **271,** 84–87.

Smeathers, J. E., and Vincent, J. F. V. (1979) Mechanical properties of mussel byssus threads. *J. Moll. Stud.* **49,** 219–30.

Smith, C., Estavillo, D., Emsley, J., Bankston, L. A., Liddington, R. C., and Cruz, M. A. (2000) Mapping the collagen-binding site in the domain of the glycoprotein Ia/IIa (integrin $\alpha 2\beta 1$). *J. Biol. Chem.* **275,** 4205–9.

Smith, G. J., and Haskell, T. J. (2000) The fluorescent oxidation products of dihydroxyphenylalanine and its esters. *J. Photochem. Photobiol. B* **55,** 103–8.

Smith, M. C., Furman, T. C., and Pidgeon, C. (1987) Immobilized iminodiacetic acid metal peptide complexes. Identification of chelating peptide purification handles for recombinant proteins. *Inorg. Chem.* **26,** 1965–69.

Steinert, P. M., and Marekov, L. N. (1995) The proteins elafin, filaggrin, keratin intermediate filaments, loricrin, and small proline rich proteins 1 and 2 are isodipeptide cross-linked components of the epidermal cornified cell envelop. *J. Biol. Chem.* **270,** 17702–11.

Sun, C.-J., Vaccaro, E., and Waite, J. H. (2001) Oxidative stress and the mechanical properties of naturally occurring chimeric collagen-containing fibers. *Biophys. J.* **81,** 3590–95.

Sun, C. J., Lucas, J. M., and Waite, J. H. (2002) Collagen-binding matrix proteins from elastomeric extraorganismic byssal fibers. *Biomacromolecules* **3,** 1240–48.

Suresh, S. (2001) Graded materials for resistance to contact deformation and damage. *Science* **292,** 2447–51.

Tamarin, A. (1975) An ultrastructural study of byssus stem formation in *Mytilus californianus. J. Morphol.* **145,** 151–77.

Tamarin, A., Lewis, P., and Askey, J. (1976) The structure and formation of the byssus adhesive plaque in *Mytilus. J. Morphol.* **149,** 199–221.

Taylor, S. W., Chase, D. B., Emptage, M. H., Nelson, M. J., and Waite, J. H. (1996) Ferric ion complexes of a DOPA-containing adhesive protein from *Mytilus edulis. Inorg. Chem.* **35,** 7572–77.

Vitellaro-Zuccarello, L., De Biasi, S., and Bairati, A. (1983) Ultrastructure of the byssal apparatus of a mussel. V. Localization of the collagenic and elastic components in the threads. *Tiss. Cell* **15,** 547–54.

Vaccaro, E., and Waite, J. H. (2001) Yield and post-yield behavior of mussel thread: A self healing biomolecular material. *Biomacromolecules,* **2,** 906–911.

van Ness, K. P., Koob, T. J., and Eyre, D. R. (1988) Collagen cross-linking: Distribution of hydroxypyridinium cross-links among invertebrate phyla and tissues. *Comp. Biochem. Physiol.* **91B,** 531–34.

Waite, J. H., Saleuddin, A. S. M., and Andersen, S. O. (1979) Periostracin—A soluble precursor of sclerotized periostracum in *Mytilus edulis* L. *J. Comp. Physiol.* **130,** 301–7.

Waite, J. H. (1985) Catecholoxidase in the byssus of the common mussel *Mytilus edulis. J. Mar. Biol. Assoc. U.K.* **65,** 359–71.

Waite, J. H. (1990) The phylogeny and chemical diversity of quinone-tanned glues and varnishes. *Comp. Biochem. Physiol.* **97B,** 19–29.

Waite, J. H. (1992) The formation of mussel byssus: Anatomy of a natural manufacturing process. *Results Prob. Cell Differ.* **19,** 27–54.

Waite, J. H., Qin, X. X., and Coyne, K. J. (1998) The peculiar collagens of mussel byssus. *Matrix Biol.* **17,** 93–106.

Wainwright, S. A., Biggs, W. D., Currey, J. D., and Gosline, J. M. (1976) *Mechanical Design in Organisms.* Princeton, NJ: Princeton University Press.

Wang, K., McCarter, R., Wright, J., Beverly, J., and Ramirez-Mitchell, R. (1993) Viscoelasticity of the sarcomere matrix of skeletal muscles. *Biophys. J.* **64,** 1161–77.

Xu, R. D., Xin, H., Hopkins, T. L., and Kramer, K. J. (1997) Catecholamine and histidyl protein cross-linked structures in sclerotized insect cuticle. *Insect Biochem. Mol. Biol.* **27,** 101–8.

ELEVEN

Conformational Compliance of Spectrins in Membrane Deformation, Morphogenesis, and Signalling

Graham H. Thomas and Dennis E. Discher

INTRODUCTION

The spectrin-based membrane skeleton (SBMS) is a remarkably diverse and multifunctional molecular scaffold. This structure was first identified as the primary cytoskeletal element in the erythrocyte, where it is responsible for maintaining cell shape and plasma membrane integrity (Mohandas and Evans, 1994; Tse and Lux, 1999). Now recognized as a ubiquitous structure in metazoan cells, the actual and postulated processes in which the SBMS participates additionally include the generation of specialised membrane domains; polarization of specific proteins in cells; protein sorting; vesicle transport; endocytosis; morphogenesis; and even yet to be characterized nuclear function(s) (De Matteis and Morrow, 2000; Bennett and Baines, 2001; Thomas, 2001; Tse et al., 2001).

Spectrin in the erythrocyte is a heterotetrameric protein comprised of two α- and two β-chains. Each chain is a long, rope-like molecule consisting largely of 106 amino acid triple α-helical repeat units (spectrin repeats), but with specialized protein modules within and at either end of this array (itemized in Figure 11.1). Dimerisation of α- and β-spectrins occurs in an antiparallel fashion and dimer pairs interact via the 'head-to-head' interaction to generate tetramers. This arrangement leaves an actin-binding domain at either end of the tetramer, and a major activity of spectrin is to cross-link actin into highly branched networks associated with membrane surfaces (Figure 11.2; Bennett and Baines, 2001). The association between the SBMS and a membrane is mediated by direct, or indirect, association with integral membrane proteins and by direct binding to phospholipids via the pleckstrin homology (PH) domain (Lemmon et al., 2002) at the C-terminus of most non-erythroid isoforms (see later).

Spectrins are part of a larger superfamily of proteins that include α-actinins and dystrophins, and are all thought to have evolved from a common α-actinin-like ancestor (Byers et al., 1989; Thomas et al., 1997). If one sets aside the differences in oligomeric form and the variations in length, the proteins in this family all exhibit the same general organisation (Figure 11.1). They all have an

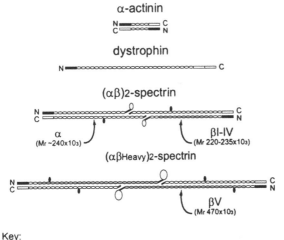

Figure 11.1: Spectrin superfamily of proteins.

N-terminal actin-binding domain based on two calponin homology domains, a tandem series or array of spectrin repeats, and a C-terminal EF-hand type Ca^{2+} binding region. These features have been strikingly conserved since before the arthropod/vertebrate split (for an extended discussion on the evolution of this gene family, see Thomas et al., 1997). The spectrin repeat array forms a flexible extended rod in all cases, the specific length of which is clearly dictated by the number of spectrin repeats. Thus, the shortest member, α-actinin, has four repeats and exhibits a contour length of 35–45 nm in rotary-shadowed images (Bennett et al., 1984; Meyer and Aebi, 1990), while the longest tetramer $(\alpha\beta_H)_2$-spectrin has a total of 47 repeats and an extended contour length of 250 nm (Dubreuil et al., 1990). In addition to establishing length, specialised spectrin repeats near the N-terminus of β-spectrins and the C-terminus of α-spectrins mediate the antiparallel dimerisation of these two chains (Viel and Branton, 1994; Begg et al., 2000). In addition, incomplete spectrin repeats in α- and β-spectrin mediate tetramerisation via complementary head-to-head interactions (Tse et al., 1990; Figure 11.1).

Vertebrate genomes contain two α-spectrin genes and five β-spectrin genes, and alternative splicing creates further diversity. The number of potential combinations is huge and the rules governing which pairs of α- and β-isoforms can bind together have not been fully elucidated. In invertebrates, the situation is somewhat simpler. There is one α-spectrin gene and two β-spectrin genes making the choice of β-chain the primary determinant of the SBMS. The contribution of alternative splicing to invertebrate spectrin diversity is currently unknown. In both vertebrates and invertebrates, the β-chains are categorized into 'conventional' β and 'heavy' $β_H$ isoforms which differ primarily in the number of repeats (17 and 30, respectively; Figure 11.1).

TARGETING SPECTRIN TO THE MEMBRANE

Most of the better characterized functions of spectrin occur while the SBMS is anchored to a fluid, but incompressible, lipid bilayer (Evans and Skalak, 1979a, 1979b; Discher et al., 1994), and it is important to bear this in mind when considering the biophysical properties and extensibility of these large proteins. It is also clear that the membrane domains occupied by different spectrin isoforms seldom overlap. For example, in invertebrates, the two β-isoforms show a clear segregation along the apicobasal axis in almost all epithelial cells. $β_H$ is confined to the apical domain, while conventional β-spectrin is found exclusively in the basolateral domain (e.g., Thomas and Williams, 1999). This is consistent with the notion that different spectrin isoforms have membrane domain-specific functions and also indicates that there must be mechanisms present to effect such segregation.

Although a tetrameric form may be required for stable membrane association (Lee et al., 1997; Zarnescu and Thomas, 1999) and the SH3 domain on α-spectrin will autonomously localize to the membrane (Meriläinen et al., 1993), the α-subunit appears to play a relatively minor role in specific membrane targeting (reviewed in De Matteis and Morrow, 2000). β-Spectrins on the other hand appear to contribute the most potent and specific membrane associations and can associate with cellular membranes in several ways along their length.

N-terminal Actin-Binding Domain

The N-terminus of β-spectrin contains two calponin homology domains that mediate binding to F-actin. This binding is facilitated by other proteins which can, in turn, bind to integral membrane proteins. For example, members of the protein 4.1 family typically bind to short juxtamembrane regions of integral membrane proteins while also binding to spectrin near their N-terminus (e.g., Discher et al., 1995; Marfatia et al., 1997). A similar role may be played by α-catenin in coupling spectrin to cadherin-based adhesive complexes (Pradhan et al., 2001).

Central Spectrin Repeat Region

Ankyrin

Perhaps the best-characterised membrane association is via the adapter molecule ankyrin which binds to the 15th/16th repeat of conventional β-spectrin.

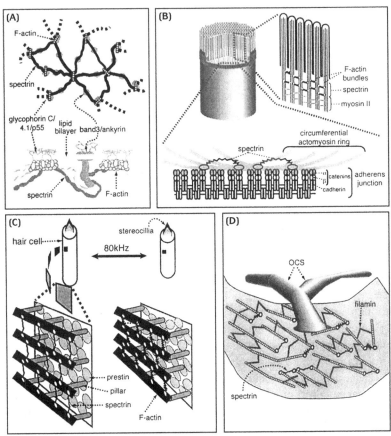

Figure 11.2: A selection of spectrin networks. (A) The spectrin membrane skeleton in the erythrocyte. *Upper illustration* is a schematic drawing of the typical appearance of a red cell SBMS spread visualized after rotary-shadowed staining in the electron microscope (see Elgsaeter et al., 1986 for example). Note the multiple rope-like spectrin tetramers interconnecting F-actin nodes. The network is retained at the membrane through indirect binding to integral membrane proteins via the adapter proteins 4.1 and ankyrin. This structure is shown in cross-section in the lower illustration. (B) Two spectrin-associated microfilament populations at the apical pole of an epithelial cell (upper left) are illustrated. The upper right illustration shows how parallel actin bundles supporting microvilli protrude into the apical cytoplasm and are cross-linked by myosin II and spectrin in distinct zones (see Bement and Mooseker, 1996, for review). The lower illustration shows a hypothetical arrangement for spectrin in association with the apical contractile ring inspired by the phenotypes of spectrin mutants and the demonstrated association of at least one β-spectrin with α-catenin (Pradhan et al., 2001). Here, microfilaments (gray arrows) are anchored to the junction in ternary complexes involving the β-spectrin actin binding domain and α-catenin N-terminus. (C) The spectrin membrane skeleton associated with the auditory OHC amplifier. This cell undergoes reversible changes in height driven by reversible changes in the conformation of prestin (gray circles/ellipses) in response to

Ankyrins are encoded by a multigene family, gain additional diversity by alternative splicing, and mediate the interaction with many integral membrane proteins (Bennett and Chen, 2001). Ankyrin binding also seems to be a primary pathway for the association with or organisation of some specific membrane domains. Thus, in mice, the organisation of a βIV-spectrin domain in the axon initial segments in Purkinje neurons is dependent on ankyrin G (Jenkins and Bennett, 2001). Similarly, in *Drosophila*, the ankyrin-mediated association of β-spectrin with the cell adhesion molecule neuroglian is sufficient to drive the assembly of domains containing this spectrin isoform in response to cell contact (Dubreuil and Grushko, 1999). In contrast, two observations have emerged which indicate that β_H isoforms are dependent exclusively on ankyrin-independent pathways for specific membrane association. First, sequence analysis has indicated that β_H isoforms do not contain the well-conserved ankyrin binding site found in the conventional β's (Thomas et al., 1997; Stabach and Morrow, 2000). Secondly, colocalization experiments between β_H and ankyrin show no overlap in the distribution of these two proteins (Lee et al., 1997).

Ankyrin-Independent Membrane Association

The central repeat region also appears to mediate a modest level of ankyrin-independent association with membranes (Davis and Bennett, 1994; Lombardo et al., 1994). However, the binding sites for membrane targets in this region are not well characterized (reviewed in De Matteis and Morrow, 2000).

Spectrin C-terminus

The C-terminus of β-spectrin is a region of considerable interest. All β-spectrin genes encode a phospholipid-binding PH domain region that is variably incorporated at the C-terminus through alternative splicing. Most non-erythroid isoforms incorporate the PH domain conferring upon them a direct site of association with the membrane in addition to the more stereotypical integral membrane protein anchors (see De Matteis and Morrow, 2000). In competition experiments, this region can account for up to 80% of the membrane-binding

---

cyclical depolarization. Here, roughly vertical, soft spectrin tetramers interconnect perpendicular stiff microfilaments producing a network that prefers to deform in the vertical axis (Tolomeo et al., 1996). The identity of the membrane anchors is unknown. (D) Illustration shows a simplified three-dimensional schematic of part of the resting platelet cell focusing on the membrane skeleton. This membrane has tubular infolding to an extensive internal membrane network known as the open canalicular system (OCS). This may form a reservoir of membrane that facilitates rapid cell shape changes in conjunction with exocytosis during platelet activation. Only the membrane-associated microfilaments and actin cross-linking proteins are illustrated. Cell shape changes during activation are facilitated by actin severing and later microfilament growth (Hartwig et al., 1999).

activity of purified spectrin tetramers, making this a major membrane association domain (Lombardo et al., 1994).

PH domains are found in more than 100 proteins and can have widely varying phospholipid-binding properties. Some are exquisitely specific while others are quite promiscuous, and the structural basis for such discrimination is becoming increasingly well characterised (Lietzke et al., 2000; Ferguson et al., 2000). The vertebrate β-spectrin PH domain is structurally well understood and is known to bind to phosphatidyl inositol 4,5-bisphosphate (PI45P_2) and PI345P_3 with equally modest affinity (Rameh et al., 1997), as it lacks the ability to discriminate phosphorylation at the 3' position on the inositol ring. PI45P_2, being by far the more abundant of these two lipids in the membrane, is thus thought to dominate the spectrin PH domain–membrane interactions. In whole cell assays, PI45P_2 plays a central role in the adhesion between the cell membrane and the cytoskeleton, and it seems likely that spectrin contributes to at least some of this (Sheetz, 2001). However, the only functional data with regard to phospholipid-mediated membrane association by a spectrin is at the Golgi complex, where the binding of βIII-spectrin to the Golgi is directly regulated by PI45P_2 levels (Godi et al., 1999).

The PH domain sequences may also participate in protein–protein interactions at their C-terminal side. Notably, the examples of $G_{\beta\gamma}$ binding in this region, such as that by the β-adrenergic receptor kinase PH domain, raise the exciting possibility that the β-spectrin C-terminus may be directly involved in anchoring or nucleating signalling complexes. The sequences surrounding the PH domain at the end of each β-spectrin are very diverse, varying significantly in sequence and extent between orthologs and paralogs, suggesting that each C-terminus has specific functions and is a site of rapid evolution. Beyond acting as a membrane association domain (Lombardo et al., 1994), no specific function has yet been ascribed to the spectrin C-terminus.

ROLES OF THE SPECTRIN-BASED MEMBRANE SKELETON

The list of integral and peripheral membrane proteins that bind to the SBMS is long (see list in De Matteis and Morrow, 2000) and certainly incomplete. Such interactions serve not only to anchor the SBMS to the membrane, but may also stabilize the protein in the membrane by greatly decreasing its turnover rate (e.g., Hammerton et al., 1991). Juxtaposition of the SBMS against the plasma membrane leads to corralling of integral membrane proteins, as well as steric hindrance of lateral diffusion (Sako and Kusumi, 1995). It is also likely to influence the protein composition of a membrane in more subtle ways through the suppression of endocytosis (Kamal et al., 1998). Furthermore, SBMS are not limited to the plasma membrane, but are also found on internal membrane compartments. Here they may play a role in shepherding some proteins through the secretory pathway by facilitating protein sorting at the endoplasmic reticulum/Golgi (reviewed in De Matteis and Morrow, 2000), and/or by attaching the motor protein dynein to cargo vesicles via the dynactin complex (Muresan et al., 2001).

Through its diverse and manifold interactions, the SBMS thus has the potential to impact significantly on the protein composition of the membrane domain that it defines or facilitates transport to. This, coupled with the observation that different spectrin isoforms are typically found in mutually exclusive regions on a membrane, led to the prediction that the SBMS would be an important player in the production of cell polarity, most notably apicobasal polarity in epithelia (Yeaman et al., 1999). However, recent genetic tests of this hypothesis in suitable model organisms have indicated that an essential role in the development and maintenance of epithelial apicobasal polarity per se is unlikely. Analysis of the phenotypic consequences of mutations in either the apical or basolateral membrane skeletons (or both) in *Drosophila* and *Caenorhabditis* has clearly shown that the SBMS is an essential structure (Lee et al., 1997; McKeown et al., 1998; Thomas et al., 1998; Moorthy et al., 2000; Dubreuil et al., 2000), but that polarized epithelia are in all cases established and exhibit other phenotypes, such as epithelial instability (Lee et al., 1993) or problems during morphogenesis (McKeown et al., 1998; Zarnescu and Thomas, 1999). Nonetheless, it is important to emphasize that while the apicobasal axis is intact in such cases, the polarisation/stability of many proteins in the membrane may still be affected. de Matteis and Morrow itemize ~50 known interacting proteins (De Matteis and Morrow, 2000), and more limited examples of cell polarity which are dependent upon the SBMS are accumulating (e.g., Jenkins and Bennett, 2001).

Not surprisingly, large effects on cellular processes result from the absence of spectrin, and genetic studies have been influential in defining cellular roles for the SBMS. The role of this network in the erythrocyte is particularly well understood through the study of numerous viable hereditary anaemias extant in the human population. These phenotypes dramatically illustrate the crucial role of the SBMS in membrane resilience and sustained circulation of the red cell. In the absence of an intact SBMS, erythrocytes lose their biconcave morphology and exhibit spontaneous endovesiculation (Mohandas and Evans, 1994; Tse and Lux, 1999). However, the general relationship between the SBMS and red cell shape and stability is complex. Mutations in different components have widely differing effects on cell shape, with some cells remaining discoidal and others becoming spherocytic, elliptocytic, ovalocytic, or echinocytic (Tse and Lux, 1999). Such behaviour, along with the tendency to lose membrane by vesiculation, cannot easily be predicted because the shape and stability of the membrane are a collaboration between the SBMS itself and the proteins to which it is anchored. Secondary effects are also difficult to model or predict (Elgsaeter et al., 1986; Van Dort et al., 2001).

The determination of cell shape and membrane support are both roles which may be applicable in the non-erythroid SBMS; however, the non-erythroid membrane skeleton appears to have many more functions that are only now being fully appreciated as genetic models are beginning to yield their results (Bennett and Baines, 2001). The phenotypic characterization of mutants in the α, β, and β_H genes of *Drosophila* and *Caenorhabditis* has demonstrated the importance of this structure, with severe lethal and/or morphogenetic consequences in each case (Lee et al., 1993; McKeown et al., 1998; Thomas et al., 1998; Dubreuil et al., 2000; Moorthy et al., 2000). Not surprisingly, the most severe effects arise from the mutations in the lone α-spectrin gene, since this is the binding partner for both β-isoforms and leaves no part of the SBMS untouched. Characterized fly α-spectrin mutant phenotypes include late embryonic lethality (Lee et al., 1993), failure to acidify the larval midgut (Dubreuil et al., 1998), as well as defects in oogenesis that include cell germline cell divisions (de Cuevas et al., 1996) and epithelial instability (Lee et al., 1997). In the worm RNAi-induced depletion of α-spectrin encoded by the *spc-1* locus is similarly devastating, with lethality accompanied by conspicuous morphogenetic defects (Moorthy et al., 2000).

Selective mutation of the β or β_H SBMS leads to differing phenotypes emphasizing the functional diversity conferred by the β-chain in the network. Mutations in β-spectrin have recently been isolated in the fly (β-*spec*) and worm (*bgs1/unc-70*). Both are lethal, with defects in midgut structure seen in the fly (Dubreuil et al., 2000) and both neuronal outgrowth as well as muscle structure in the worm (Hammarlund et al., 2000; Moorthy et al., 2000).

In the fly and worm mutations in the β_H isoform (*karst* and *sma-1* mutants, respectively) lead to extensive, but not complete lethality in the former, while *sma-1* worms are viable (McKeown et al., 1998; Thomas et al., 1998; Moorthy et al., 2000). In both cases, surviving individuals exhibit significant developmental abnormalities, for which there is a striking parallel to the phenotypes in the two organisms: A defect in apical contractility in epithelia may underlie the morphogenetic defects seen, leading to a failure in worm elongation and follicle cell morphogenesis in the fly. Interestingly, spectrin is often associated with myosin II (reviewed in Thomas, 2001), a relationship with particular relevance in the context of this review, since the unfolding of spectrin repeats requires forces commensurate with the activity of this motor protein (see later).

In the mouse, knockout technology has mostly been applied towards the erythroid SBMS; however, elimination of ankyrin B or G function causes a variety of neural defects (Scotland et al., 1998; Zhou et al., 1998). Significantly Ank B−/− mice resemble mutants lacking the ankyrin-associated neural cell adhesion molecule L1-CAM, demonstrating the importance of the linkage of this molecule to the SBMS for its function. The mouse *quivering* mutations – which also cause major neurological defects that include tremoring, paralysis, and deafness – were recently located to the βIV gene (Parkinson et al., 2001). In humans, the Providence mutant allele of the βI gene is common to both its erythroid and non-erythroid spliceoforms and is lethal in the neonatal period (Weed et al., 1996). It is currently unclear whether this lethality arises primarily from an erythroid or a non-erythroid defect, or both. Together, the previous studies emphasize the importance of the SBMS for organismal development and viability.

The non-erythroid SBMS also appears to be a significant target in a number of postnatal pathologies. For example, spectrin is a major target for the calcium-activated protease calpain during ischemic injury (e.g., White et al., 2000). In the heart, such disruption of the SBMS has recently been correlated with membrane blebbing, osmotic fragility, and the transition from reversible to irreversible ischemic injury (Armstrong et al., 2001). In addition to these examples, two recent and provocative results suggest a potentially direct involvement of spectrin in genome stability. First, αII spectrin has recently been found to complex with a number of Fanconi anaemia proteins in the nucleus (McMahon et al., 1999) that are linked to BRCA-mediated DNA repair processes (Garcia-Higuera et al., 2001). Second, a recently isolated binding partner of the human α-spectrin SH3 domain is associated with loss of heterozygosity in some prostatic adenocarcinomas (Macoska et al., 2001). Were this protein to be significantly destabilized as a consequence of SBMS mutations, there could again be proliferative consequences.

Lastly, in keeping with the observation that spectrin is involved in the secretory pathway (De Matteis and Morrow, 2000), recent experiments have isolated spectrin as part of protein complexes that bind to both microfilaments and microtubules (Sisson et al., 2000). Moreover, the phenotype of the zebrafish mutation *reisling* (β-spectrin) hints at a possible functional connection since it perturbs the microtubule marginal band in erythrocytes (Liao et al., 2000). Such results indicate that spectrin may form a novel functional bridge between the microfilament and microtubule cytoskeletons.

Other members of the spectrin superfamily are also thought to have important structural roles, as well as specialized functions based on specific protein domains. Thus, lateral association of α-actinin chains–with only four homologous repeats plus N-terminal actin binding domains–allows for cross-linking of F-actin, imparting stability to structures that range from focal adhesions to Z-lines of myotubes (Kreis and Vale, 1993). In contrast, dystrophin and utrophin are hypothesized to impart stability to diverse membranes as monomeric actin-binding proteins that anchor adhesion complexes (Kreis and Vale, 1993). Length and extensibility are deemed central to function of all these proteins, making an understanding of their biophysical behaviour of paramount importance. To this end, we present a current view of the most conspicuous elastic component of this family: the spectrin repeat.

THE SPECTRIN REPEAT

The defining structural feature of the spectrin superfamily, as well as the central basis of spectrin length and extensibility, lies in the tandemly repeated triple-helical spectrin repeats (Figures 11.1 and 11.3). A typical spectrin repeat is a 5- to 6-nm long triple α-helix of about 106 amino acids in length. Tandem arrays of repeats are thought to have arisen by intragenic duplication of this motif from an ancestral α–actinin-like molecule (Byers et al., 1989; Thomas et al., 1997). Neighbouring repeats are less well conserved than the equivalent repeat in different isoforms, indicating that such duplications have not occurred since before the arthropod/vertebrate split (see Thomas et al., 1997, for discussion). Overall, conservation is modest and is probably dictated primarily by structural requirements, with a few noted exceptions where specialized protein–protein interactions occur (e.g. ankyrin binding). The helices of each repeat are referred to as *a*, *b*, and *c* in N-terminal to C-terminal order, with sequence conservation being greatest in the *a* helix and declining steadily towards *c* (Muse et al., 1997).

Most spectrin repeats do not have an assigned function, and yet a characteristic repeat array length for each family member has been maintained during

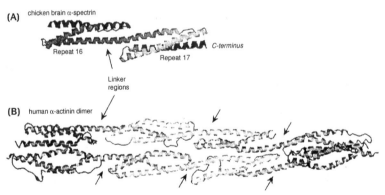

Figure 11.3: Spectrin tandem repeat structures. (A) Ribbon structure of the 16th and 17th repeating units of chicken brain α-spectrin, based on one of the five crystal structures of four cloned spectrin fragments (Grum et al., 1999). As in all five structures reported, the two triple-helical, coiled-coil repeats are connected by an α-helical linker. (B) Ribbon structure of a dimer of human α-actinin's 4 repeat rod domain (Ylanne et al., 2001). Again, all coiled-coil repeats are connected by an α-helical linker. Identity with red cell spectrin's α18–21 and β1–4 repeats–which nucleate spectrin dimerization–ranges from about 10– 50% per domain.

evolution. This suggests that either the length of the overall molecule or the biophysical properties generated by its constituent repeats may impose a selective constraint on these proteins. In two cases, absolute length appears to be flexible. First, the addition of one spectrin repeat into the middle of the α-actinin 4 repeat array had no affect on the ability of this protein to rescue α-actinin mutants in *Drosophila* (Dubreuil and Wang, 2000). Second, in-frame deletions of repeats in dystrophin produce the relatively mild Becker muscular dystrophy phenotype instead of the more lethal Duchennes form of the trait. In one striking example, a patient with loss of 16/24 repeats was still ambulatory at 61 years of age (England et al., 1990). These examples would suggest that, at least in these superfamily members, array length per se is less important than other features of the repeats, such as their extensibility.

Key structural features of the three-helix spectrin repeat were first proposed by Speicher and Marchesi (1984). Although this proved to be an early example of successful tertiary structure prediction based on primary sequence, it took a decade before the first crystal structure was obtained for a *Drosophila* spectrin repeat (Yan et al., 1993). More recently, the tandem repeats of chicken brain αII-spectrin (Figure 11.3A; Grum et al., 1999) and the complete four-repeat rod domain of α-actinin (Figure 11.3B; Ylanne et al., 2001) have been determined, motivating deeper studies of extensibility. Unstructured linkers between each repeat were originally assumed, but a unifying surprise has been the appearance of contiguous helices between adjacent domains. Just how many linkers form such contiguous helices is open to question. Dystrophin is notable here for having four distinctive proline-rich 'hinge' regions (Kreis and Vale, 1993). Electron microscopy has indicated that different spectrin forms have different stiffnesses or persistence lengths (Coleman et al., 1989), and at least for the spectrins of the highly resilient red cell, chain flexibility suggests less ordered linkers (e.g. Discher et al., 1998). Nonetheless, a contiguous helix between at least some tandem repeats is intriguing in its implications for spectrin family elasticity, especially domain unfolding. Recent denaturation studies demonstrate that tandem domains unfold in concert as a single unit, just as isolated repeats do (MacDonald and Pozharski, 2001). Tensile stresses directed from the N- to C-terminus of a spectrin chain could cause similar conformational changes, while also being more relevant to spectrins physiological function. New methods of single molecule manipulation make such tests of extensibility accessible.

SPECTRIN'S CONFORMATIONAL COMPLIANCE IN ISOLATION

Direct stretching of individual spectrin chains under the action of an atomic force microscope (AFM) tip [Figure 11.4(A)] reveals considerable complexity. Initially applied to the multidomain muscle protein titin (both native and recombinant; Rief et al., 1997), this groundbreaking approach with AFM has now

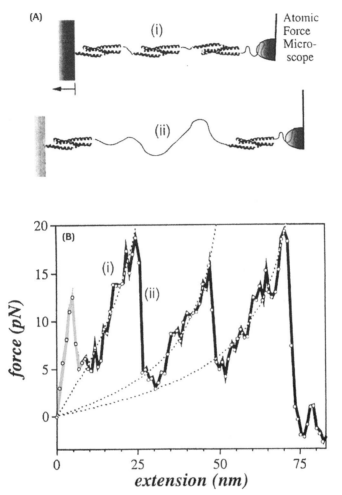

Figure 11.4: Mechanical response of recombinant β1–4 spectrin when extended under force. (A) Illustrated action of an Atomic Force Microscope cantilever as it might pull and unfold a chain of spectrin repeats. (B) Force versus extension curve of a single spectrin chain (Carl et al., submitted for publication). Unfolding of the spectrin domains leads to a sawtooth pattern of very regular spacing.

been used to show that spectrins also respond to extension as highly non-linear, even *unfoldable* quasi-elastic springs. Depending on the rate of forced extension, the distending force increases disproportionately with the fractional chain extension, x, defined relative to the 'maximum' end-to-end length: $x = L/L_{max}$. Such AFM experiments begin by adsorbing protein to substrates ranging from mica to aminosilanized glass or gold and then, at random, contacting the surface with the probing AFM tip, in the hopes of forming a second contact. At low protein concentrations, a single protein or protein loop is believed to span the gap between AFM tip and substrate [Figure 11.4(A)]. Results such as those in Figure 11.4(B) show that, after a small initial 'desorption' peak, a β-spectrin construct consisting of just the first four repeats is extended off of the substrate to a significant fraction of its folded length (~25 nm). In this process, the extensional force increases almost linearly up to about 20 pN. However, after this initial extension, there is a dramatic sawtooth pattern of force peaks and valleys which is understood to reflect domain unfolding.

Based on the initial phase of extension alone, an effective spring constant of $k_{sp} \approx 1$ pN/nm is readily calculated, even though it is probably not physiological. This apparent stiffness could represent a range of processes, including the work of forced desorption from the surface layer of spectrin chains. Alternatively, the simplest linear elasticity theory for an entropically flexible polymer equates k_{sp} to several structural parameters: namely, $k_{sp} = 3k_B T/(2b\, L_{max})$, where T is the bath temperature that drives thermal motions of the chain on length scales greater than the persistence length b, but less than some limiting contour length, L_{max}. For an unfolded polypeptide chain, $b \approx 0.4$ nm; a folded spectrin repeat ought to have $b \geq 5$ nm based on the approximate length of a single repeat. Given these numbers, the only self-consistent polymer elasticity model for the illustrated experiment is that an unfolded segment of chain of total unfolded length $L_{max} \approx 10$ nm (about one unfolded helix) connects the remaining folded repeats to the tip or substrate. This is consistent with the idea that unfolded regions of the chain at either the tip or the substrate augment the chain's elasticity or even dominate it (Evans and Ritchie, 1999), making the chain appear stiffer or softer than natural. A third alternative is that the measured stiffness genuinely reflects a more enthalpic rather than entropic basis for elasticity. For example, the multiple structures reported for brain α-spectrin (Grum et al., 1999) have been interpreted as an indication that particular residues can exist part of the time in one helix and part of the time in an adjacent loop – at a likely cost of breaking hydrogen bonds and hydrating hydrophobic residues. However, a value of $k_{sp} \approx 1$ pN/nm is at least 2 orders of magnitude too high, compared with a broad array of measurements of network elasticity (see later). The initial phase of spectrin extension by the AFM thus seems less likely to offer significant insight into the elasticity of red cell spectrin.

The most remarkable feature of the single molecule extension curves that Figure 11.4B typifies is the finding that the extensional force curve has both peaks and valleys. As currently interpreted, a peak and decay in the force curve reflects a sudden slackening upon domain unfolding. Additional data on the periods of the sawtooth patterns, for αI-spectrin as well as βI-spectrin, reveal bimodal distributions of peak-to-peak unfolding lengths (Carl et al., submitted for publication). Importantly, the higher of the two averages is exactly twice the lower average and yet it occurs at nearly the same force, though at lower frequency. In addition, the higher average is considerably greater than the average contour length of the 106–120 amino acid repeats probed, and the lower of the two average values is approximately two-thirds of a spectrin domain's fully stretched contour length. The latter likely corresponds to the unfolding of most of a full spectrin repeat, while the higher peak in the bimodal distribution represents *cooperative* unfolding of tandem repeats. The process is viewed as cooperative since the force scale for unfolding two repeats is the same as that for one repeat. Regardless, it is clear that these triple helical repeats of spectrins can unfold in single or tandem repeat events under forces of only tens of pNs at typical rates of AFM extension (~1 nm/ms). For comparative reference: (i) individual myosin molecules (perhaps the red cells as well) exert peak forces of ~5 pN, and (ii) current estimates of maximum spectrin forces in simulations of membrane deformation are 5–10 pN (Discher et al., 1998).

Qualitatively similar observations of multistate unfolding have likewise been reported for one repeat of chicken brain αII-spectrin flexibly linked in series (Lenne et al., 2000). However, full unfolding was said to be complemented by a more rare partial unfolding event even though the first report of spectrin unfolding by Rief et al. (1999) definitively stated that 'no intermediates can be seen'. More work appears needed. Initial results, for example, on the effects of α-β associations suggest that unfolding is certainly no harder in a dimer than a monomer. In addition, with all measurements done to date at room temperature, 37°C may hold some surprises, as this is halfway to the melting temperature of typical spectrin domains. Finally, as most thoroughly demonstrated for titin unfolding (Rief et al., 1997), unfolding forces increase logarithmically with chain extension rate. Thus, the maximum forces illustrated in Figure 11.4B are likely to be at least several-fold higher than those needed for unfolding spectrin at slower, more physiological rates of extension relevant as well to in vitro processes, such as micropipette aspiration of cells. However, it is clear that fluctuations in force and extension are intrinsic to spectrin's mechanical response at the most basic single molecule level.

SPECTRIN EXTENSIBILITY AT THE RED CELL MEMBRANE
The red cell membrane [Figure 11.2(A)] is probably the very best characterized of all cellular membranes – organelle membranes included – and spectrin is well known to be a central contributor to the erythrocyte membrane's resilience (see above). Beyond AFM, recent molecular scale insights into the properties and behaviour of erythrocyte spectrin have been obtained from all atom simulations of thermal and mechanical stability (Paci and Karplus, 2000; Zhang et al., 2001). Normally, in cross-linking short actin filaments at the plasma membrane, αβ-spectrin oligomers form a remarkably ordered and seemingly

stable triangulated network. How spectrin and this network sustain stresses and strains both globally and locally, down to the atomic scale, has been the concentrated focus of significant current effort.

The considerable extensibility of spectrin evident in AFM has now also been demonstrated to a modest degree in the intact red cell network. Earlier fluorescence imaged microdeformation results (Discher et al., 1994) had suggested sustainable network extensions in dilation of at least $\lambda \sim 2$ relative to the unstrained or natural state of the membrane network. Fluorescence-patterned photobleaching of the network has revealed, in an even more explicit fashion, a distorted or sheared state with sustainable extension ratios in the pipette-axis direction of magnitude $\lambda \sim 2$–3 (Figure 11.5; Lee et al., 1999). The strains are imposed not on single molecules directly but on a brush-like, semidisordered spectrin network of initial mesh size ~ 55–70 nm and a dehydrated height or thickness of perhaps 10 nm (Takeuchi et al., 1998). Of course, hydration as well as Brownian fluctuations – elaborated below – tend to expand the network into the third, unconstrained dimension of the cell interior. Indeed, very recent measurements of network thickness and compressibility (Heinrich et al., 2001) generally confirm this picture and also provide further support for a very soft, polymer-like network.

THERMAL FLUCTUATIONS AND THEIR DEFORMATION ENHANCEMENT

Elastically constrained thermal motions of network components are hallmarks of entropically elastic structures. Recent polymer and tether simulations have assumed this in taking an explicit statistical physics approach to network structure-function (Boey et al., 1998; Picart et al., 2000). Only recently have high spatial resolution experiments on intact membranes begun to reveal the network's thermal motions. It was first shown that nanoparticles attached to either single spectrin or 'immobile' band 3 molecules (Tomishige et al., 1998) could be forcibly displaced with laser tweezers, locally deforming the network. At small or vanishing forces, the thermal fluctuations of these same components about a fixed mean position were readily detected and appeared to be in the range $<X^2> \approx (25$–50 nm$)^2$. Similar results have now been found with bead-labelled F-actin (Figure 11.6) and glycophorin C. Analogous to a Stokes–Einstein relationship between mean-squared displacement of a particle and the ambient viscosity, a fundamental relation for the equipartitioning of thermal energy allows one to extract an effective network elasticity from the fluctuation amplitude:

$$\tfrac{1}{2}k_{sp} <X^2> = \tfrac{1}{2}k_B T. \tag{11.1}$$

More rigorous theories (for either enthalpic or entropic springs) show k_{sp} approximates the theoretical expectation for the network shear modulus μ

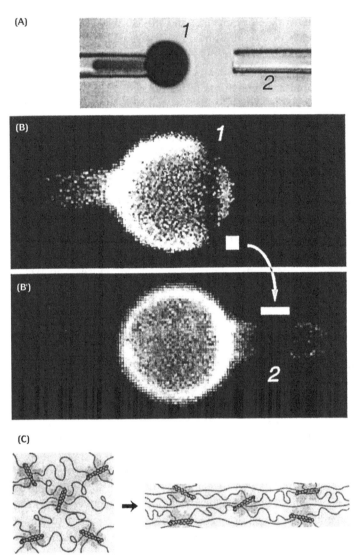

Figure 11.5: Membrane skeleton deformation as quantitated by laser-patterned photo-bleaching of rhodamine-phalloidin-labeled actin (Lee et al., 1999). A single cell is aspirated into a ~1-μm diameter glass micropipette (A), patterned, and imaged in both an unstrained state (B) and a strained state (B′). (C) Scaled schematic showing actin protofilaments interconnected by convoluted spectrin so that, when reversibly deformed, spectrin chains are either extended or compressed.

(A)

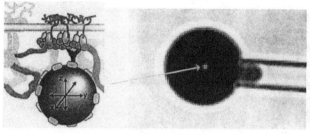

(B)

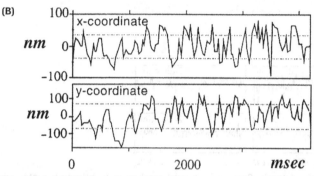

Figure 11.6: Attachment of a single fluorescent nanoparticle (~40 nm diameter) to either actin (A) or glycophorin allows tracking of the spectrin-constrained thermal motions of the actin nodes (B). The significant amplitude of the localized mean-squared displacements reveals the network's considerable elasticity (Lee and Discher, 2001).

(Lee and Discher, 2001). Indeed, the magnitude of $k_{sp} \approx 0.002$–0.007 pN/nm, although broad, generally agrees with measurements for μ obtained by a long list of techniques which assume little to nothing about molecular scale motions (Henon et al., 1999; Mohandas and Evans, 1994). The small value of k_{sp} also reveals the essentially soft character of the network's microelasticity.

In strained regions of the network on the micropipette-aspirated projection (see Figure 11.5), fluctuations of nanoparticles are even more intriguing. In regions and directions of the network that are most strongly stretched, nanoparticle fluctuations are *increased* (Figure 11.7). Equation (11.1) then implies that the effective spring constant of the network is reduced or softened at these points. Indeed, when a number of particles are tracked and their fluctuations related to the known network extensions, the softened response is shown to be nonlinear [Figure 11.8(A)], with a lower than expected force required to increasingly strain or extend the network. The same form of softened force versus extension response can be shown to be well approximated in shape by the unfolding-refolding equilibrium of spectrin-like repeats [Figure 11.8(B)]. Essentially all of

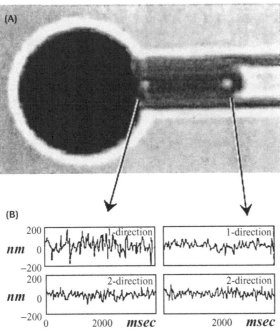

Figure 11.7: Motions of 40-nm fluorescent beads attached to F-actin and located on the aspirated projection of a micropipette-deformed red cell ghost (Lee and Discher, 2001). (A) Beads were tracked at various positions along the projection, displaying (B) fluctuations that are dependent in magnitude on both direction and position along the aspirated projection. In particular, near the micropipette entrance, fluctuations in the axial, 1-direction are almost twice those near the aspirated projection's tip. Fluctuations in the circumferential 2-direction, again near the micropipette entrance, are more weakly affected, but appear slightly suppressed compared with those near the tip.

the sawtooth peaks in the unfolding pattern of Figure 11.4 are smoothed out to give a hyperbolic-appearing force versus extension curve. Given the $\sim 10^7$ spectrin repeats in the red cell, a few localized spectrin domains are conjectured to unfold under the regionally sustained stresses and strains of aspiration.

OTHER SPECTRIN NETWORKS

The red cell network is crucial to membrane resilience and sustained circulation, and is by far the best characterized and understood of the spectrin networks. However, a number of other SBMS are sufficiently well understood to be worth discussing in light of what we know about the erythroid network.

The Terminal Web

A spectrin network is clearly required for epithelial cell integrity and epithelial sheet morphogenesis. Two distinct populations of microfilaments are known

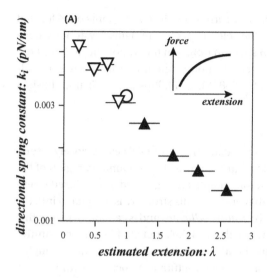

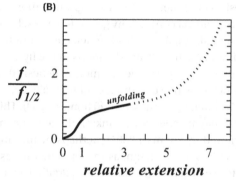

Figure 11.8: Softening of the response in deformation. (A) Measured softening of the network in micropipette aspiration as obtained by nanoparticle tracking on aspirated cells. (B) Single spectrin chain model of the unfolding-refolding equilibrium as a function of force, f, normalized by the force to unfold half of the domains (for details, see Lee et al., 2001).

to exist at the apical end of epithelial cells. The first is found in cells manifesting a brush border – a high density of microvilli, each supported by a parallel bundle of microfilaments [Figure 11.2(B)]. The actin bundles in a brush border protrude into the apical cytoplasm, where they are densely cross-linked in an ill-understood region called the terminal web. At least two cross-linking proteins are present here and appear to be spatially distinct. Myosin II is found in the more basal regions, while spectrin is found in the more apical regions (reviewed in Bement and Mooseker, 1996). The function of spectrin in this region is unknown. The terminal web is assembled as one of the final steps in brush border formation, correlates with the flattening out of the apical cell surface (Fath and

Burgess, 1995), and exhibits dynamic turnover of its components once formed
(Stidwill et al., 1984). However, the role of spectrin remains mysterious since
elimination of the terminal web spectrin $(\alpha\beta_H)_2$ in the fly appears to have little
or no perceptible effect on the organization of the microvilli, even at the ultrastructural level (Thomas et al., 1998; M. D. Phillips and GHT unpublished
data).

The Apical Contractile Ring

The second apical microfilament population in epithelial cells is closely associated with the junctional complexes at the apical/basal boundary region of the
lateral membrane. Here, the microfilaments are organized into a mixed orientation contractile ring along with myosin II. This structure is integrated into the
junctional complexes, notably the *zonula adherens* and is capable of generating
tension (Burgess, 1982; Hirokawa et al., 1983; Keller et al., 1985). A constitutive
level of tension probably keeps the apical surface of most epithelia 'taught';
however, localized up-regulation of this structure is associated with bending
of epithelia during morphogenesis (Gumbiner, 1992). Strikingly, mutations in
the apical β_H isoform cause failure in apical contractility generated by a related
structure in the worm (McKeown et al., 1998) and lead to detachment of the
actin from the membrane (Praitis and Austin, 2000). Mutations in the fly gene
for β_H exhibit a morphogenetic defect during follicle cell morphogenesis that
is easily rationalized if a similar defect is present and are additionally correlated with disruption of the *zonula adherens* (Zarnescu and Thomas, 1999). This
has led to the suggestion that β_H may be specifically linking the apical contractile ring to the junction (reviewed in Thomas, 2001), perhaps by forming
a tertiary complex along with F-actin and α-catenin, as seems to be the case
for at least one conventional vertebrate β-chain [Figure 11.2(B); Pradhan et al.,
2001]. Spectrin tetramers in this region could function simply to attach microfilaments to the junction and as scaffolds for other proteins or as elastic elements
that facilitate some aspect of the contractile process itself. Just how extended
spectrin in this region is in a resting or contracting epithelial cell remains to be
determined.

The Outer Hair Cell Lateral Membrane

A third spectrin network that is biophysically well-characterized resides in the
cortex of the cochlear outer hair cell [OHC; Figure 11.2(C)]. This cell is particularly intriguing because of its role as an amplifier in auditory perception.
Sound waves travelling in the overlying fluid trigger depolarization via mechano-
chemical opening of K^+ channels in the apical stereocillial bundle. Depolarization triggers a substantial conformational change in the integral membrane
protein prestin that is rapidly reversed on repolarization. In this way, the prestin

electro-mechanical motor likely causes oscillations in the projected membrane surface area (Adachi and Iwasa, 1999). Both in vivo and in isolated OHCs, this change in surface area is translated exclusively into an oscillation in cell length rather than circumference via the associated spectrin membrane skeleton.

The OHC is cylindrical and contains cortical microfilaments much longer than the 35 nm 'protofilaments' of the red cell. The longer microfilaments are juxtaposed to the basolateral membrane and cross-linked by spectrin in a roughly orthogonal orientation [Figure 11.2(C)]. This SBMS is present in patches and, while each of these may exhibit some tilt relative to the apicobasal axis, the mean connected orientation of the microfilaments is circumferential (Holley et al., 1992). Within this SBMS, preferred deformation of the soft spectrin chains over the microfilaments is thought to direct the membrane deformations along the vertical axis of the cell (Tolomeo et al., 1996). Although overall strains tend to be small in the OHC during excitation, localized strains might be large and would tend to modulate spectrin domain conformations.

Regional resonance is a key property in discriminating between different pitches in the cochlea, and multiple cells and structures in addition to the OHC contribute to this tuning (Holley, 2000). This regionalization takes the form of a continuous gradient of resonant frequency responses along the length of the cochlea, such that OHC resident at any given location will exhibit maximal amplification at a characteristic sound frequency. It appears plausible that the sensitivity of the OHC electro-mechanical response may reflect changes in spectrin conformation such that the SBMS is part of the tuning apparatus. Disruption of spectrin certainly reduces force generation by the OHC (Raphael et al., 2000).

Platelet Plasma Membrane

The fourth relatively well described non-erythroid SBMS is associated with the resting platelet plasma membrane. Superficially resembling the red cell structure in electron micrographs, this is clearly more elaborate and includes the actin cross-linking protein ABP280/filamin1, as well as utrophin/dsytrophin-related protein 1 [Figure 11.2(D)]. This structure is of interest in the context of this review since the platelet undergoes a major cell shape change upon activation, assuming first a spherical morphology, subsequent actin-driven protrusions and ultimately spreading (reviewed in Hartwig et al., 1999; Fox, 2001). Much of the membrane for the initial expansion may be derived from pre-existing tubular infoldings of the plasma membrane referred to as the open canalicular system. Platelet activation is accompanied by massive rearrangements of the actin cytoskeleton that includes severing of microfilaments by gelsolin, the release of spectrin from actin and calpain-mediated cleavage of many proteins, including spectrin and ABP280/filamin 1. The membrane skeleton in this case may be necessary for restraining expansion in the resting state; however, it is currently unknown if spectrin or any of the other proteins is under tension in the resting platelet. Curiously, activation and membrane skeleton rearrangement are also correlated with the release of membrane microvesicles (Fox et al., 1991), something that is also seen in mutant erythrocytes lacking SBMS components, underscoring the importance of such structures in stabilizing the membrane.

PERSPECTIVE

Many key questions remain unanswered about the behaviour of spectrin and the spectrin-based membrane skeleton during developmental processes, where cell shape changes are of widespread importance. For example, we have an increasingly clear picture of spectrin under tension, but studies visualizing spectrin in situ by quick-freeze deep etch techniques (Ursitti et al., 1991) reveal contour lengths that are significantly shorter than those seen in rotary-shadowed images of purified spectrin tetramers (e.g., Coleman et al., 1989). While the red cell is relentlessly pounded by the circulatory system, many of these other cell types perform more sedentary functions, and such images therefore suggest that non-erythroid spectrin repeat arrays may spend much of their time folded beyond the repeat level and possibly even compressed (see Bloch and Pumplin, 1992). Under physiological conditions, the erythrocyte SBMS may be somewhat compressed by the lipid bilayer (Discher et al., 1998), and expansion from a compressed state is clearly evident at subphysiological ionic strengths (Svoboda et al., 1992). This expansive tendency on lowering the ionic strength is presumably due to electrostatic repulsion between closely opposed charges and is thus consistent with the notion that the spectrin rod is folded beyond the repeat level in vivo. Whether the spectrin repeat array in any given isoform actually behaves in this way or escapes compression by deforming towards the cell interior as discussed above remains to be seen. One therefore wonders if the cell might not take advantage of this potential and depend upon spectrin expansion under some circumstances. For example, if a membrane is undergoing expansion through vesicle delivery, then the accompanying cytoskeleton must also expand to prevent blebbing. Significant growth is likely to be facilitated by the simultaneous delivery or recruitment of new SBMS proteins; however, smaller scale or more transient events could utilize the expandability of spectrin in the extant membrane skeleton to contribute to a membrane tension which regulates membrane delivery.

The physical consequence of changing the relationship between the spectrin/actin network and its membrane anchors is another unknown. First, the role of protein anchors has been only partially defined. The SBMS has been characterized as a series of linked mosaics, wherein heterogeneous groupings of integral membrane proteins form units of limited mobility constrained by the behaviour of the underlying spectrin network (De Matteis and Morrow, 2000). Some measurements of the mobility of such constrained proteins in the red cell have been reviewed here; but it remains unknown to what extent this is affected by the choice of spectrin isoform, or how complexes involving different proteins might affect the dynamics of any given isoform. Secondly, the direct association of spectrin with phospholipids requires further definition. In at least one case binding to $PI45P_2$ via the PH domain appears to mediate association of spectrin with the Golgi compartment in a switch-like fashion (Godi et al., 1999). However, this may not be the case for all spectrin isoforms and any co-requirement for association with an integral or peripheral membrane protein remains undefined. Thus, it remains possible that modulations in the abundance of phospholipids could reversibly dissociate this one domain of the β-chain from the membrane modulating the behaviour of the underlying network. Defining such properties and their modulation may impose limits on the dynamic behaviour of spectrin in vivo.

One area that needs to be addressed before we can fully comprehend the biophysical properties of the SBMS during development is the role of network rearrangement. It is clear that a network per se is required to maintain epithelial cells in the fly, since conditional mutations discretely affecting the α-spectrin tetramerization site and thus network formation cause massive tissue destabilization at non-permissive temperatures (Deng et al., 1995). On the other hand, it is equally clear that dynamic rearrangement of the SBMS occurs during the development of this organism (Thomas and Kiehart, 1994; Thomas and Williams, 1999). The balance between network rearrangement versus elastic compliance is unclear during development. Which morphogenetic process correlates with one mechanism or the other is equally unclear.

Finally, the results summarized here include some of the first demonstrations of stress-induced conformational changes in spectrin repeat structures. When compared with a growing list of forcibly unfolded protein domains – primarily stressed by AFM methods (Rief et al., 1997) – the triple helical fold of spectrin already appears to be among the most facile of folds. In addition, the initial suggestions of cooperativity in forced unfolding also poise these molecules to respond in complex ways to stress. A particularly intriguing notion is that unfolding may modulate the interaction with other proteins. The ankyrin binding site is a specialized repeat making this an interaction that may well be modulated mechanically. Alternatively unfolding of spectrin might attract new binding partners, ranging from proteases to chaperones to new membrane associations. Similar phenomena might also modulate the potential signalling activity of spectrin's SH3, PH, or Ca^{2+}-calmodulin binding domains leading to mechanical modulation of cell signalling. Such issues will undoubtedly be the challenge for much future effort.

REFERENCES

Adachi, M., and Iwasa, K. H. (1999) Electrically driven motor in the outer hair cell: Effect of a mechanical constraint. *Proc. Natl. Acad. Sci. U.S.A.* **96**, 7244–49.

Armstrong, S. C., Latham, C. A., Shivell, C. L., and Ganote, C. E. (2001) Ischemic loss of sarcolemmal dystrophin and spectrin: Correlation with myocardial injury. *J. Mol. Cell. Cardiol.* **33**, 1165–79.

Begg, G. E., Harper, S. L., Morris, M. B., and Speicher, D. W. (2000) Initiation of spectrin dimerization involves complementary electrostatic interactions between paired triple-helical bundles. *J. Biol. Chem.* **275**, 3279–87.

Bement, W. M., and Mooseker, M. S. (1996) The cytoskeleton of the intestinal epithelium. In: *The Cytoskeleton*, vol. 3. J. E. Hesketh, I. F. Pryme, (eds.). U.S.A.: Greenwhich JAI Press, Inc., pp. 359–404.

Bennett, J. P., Zaner, K. S., and Stossel, T. P. (1984) Isolation and some properties of macrophage alpha-actinin: Evidence that it is not an actin gelling protein. *Biochemistry* **23**, 5081–86.

Bennett, V., and Baines, A. J. (2001) Spectrin and ankyrin-based pathways: Metazoan inventions for integrating cells into tissues. *Physiol. Rev.* **81**, 1353–92.

Bennett, V., and Chen, L. (2001) Ankyrins and cellular targeting of diverse membrane proteins to physiological sites. *Curr. Opin. Cell. Biol.* **13**, 61–67.

Bloch, R. J., and Pumplin, D. W. (1992) A model of spectrin as a concertina in the erythrocyte membrane skeleton. *Trends Cell Biol.* **2**, 186–89.

Boey, S. K., Boal, D. H., and Discher, D. E. (1998) Simulations of the erythrocyte cytoskeleton at large deformation. I. Microscopic models. *Biophys. J.* **75**, 1573–83.

Burgess, D. R. (1982) Reactivation of intestinal epithelial cell brush border motility: ATP-dependent contraction via a terminal web contractile ring. *Cell Biol.* **95**, 853–63.

Byers, T. J., Husain-Chishti, A., Dubreuil, R. R., Branton, D., and Goldstein, L. S. (1989) Sequence similarity of the amino-terminal domain of Drosophila beta spectrin to alpha actinin and dystrophin. *J. Cell Biol.* **109**, 1633–41.

Coleman, T. R., Fishkind, D. J., Mooseker, M. S., and Morrow, J. S. (1989) Contributions of the β-subunit to spectrin structure and function. *Cell Motil. Cytoskel.* **12**, 248–63.

Davis, L. H., and Bennett, V. (1994) Identification of two regions of β_G spectrin that bind to distinct sites in brain membranes. *J. Biol. Chem.* **269**, 4409–16.

de Cuevas, M., Lee, J. K., and Spradling, A. C. (1996) Alpha-spectrin is required for germline cell division and differentiation in the Drosophila ovary. *Development* **122**, 3959–68.

De Matteis, M. A., and Morrow, J. S. (2000) Spectrin tethers and mesh in the biosynthetic pathway. *J. Cell Sci.* **113**, 2331–43.

Deng, H., Lee, J. K., Goldstein, L. S., and Branton, D. (1995) Drosophila development requires spectrin network formation. *J. Cell Biol.* **128**, 71–79.

Discher, D. E., Mohandas, N., and Evans, E. A. (1994) Molecular maps of red cell deformation: Hidden elasticity and in situ connectivity. *Science* **266**, 1032–35.

Discher, D. E., Winardi, R., Schischmanoff, P. O., Parra, M., Conboy, J. G., and Mohandas, N. (1995) Mechanochemistry of protein 4.1's spectrin-actin-binding domain: Ternary complex interactions, membrane binding, network integration, structural strengthening. *J. Cell Biol.* **130**, 897–907.

Discher, D. E., Boal, D. H., and Boey, S. K. (1998) Simulations of the erythrocyte cytoskeleton at large deformation. II. Micropipette aspiration. *Biophys. J.*, **75**, 1584–97.

Dubreuil, R. R., Byers, T. J., Stewart, C. T., and Kiehart, D. P. (1990) A beta-spectrin isoform from Drosophila (beta H) is similar in size to vertebrate dystrophin. *J. Cell Biol.* **111**, 1849–58.

Dubreuil, R. R., Frankel, J., Wang, P., Howrylak, J., Kappil, M., and Grushko, T. A. (1998) Mutations of alpha spectrin and labial block cuprophilic cell differentiation and acid secretion in the middle midgut of Drosophila larvae. *Dev. Biol.* **194**, 1–11.

Dubreuil, R. R., and Grushko, T. (1999) Neuroglian and DE-cadherin activate independent cytoskeleton assembly pathways in Drosophila S2 cells. *Biochem. Biophys. Res. Commun.* **265**, 372–75.

Dubreuil, R. R., and Wang, P. (2000) Genetic analysis of the requirements for alpha-actinin function. *J. Muscle Res. Cell. Motil.* **21**, 705–13.

Dubreuil, R. R., Wang, P., Dahl, S., Lee, J., and Goldstein, L. S. (2000) Drosophila beta spectrin functions independently of alpha spectrin to polarize the Na,K ATPase in epithelial cells. *J. Cell Biol.* **149**, 647–56.

Elgsaeter, A., Stokke, B. T., Mikkelsen, A., and Branton, D. (1986) The molecular basis of erythrocyte shape. *Science* **234**, 1217–23.

England, S. B., Nicholson, L. V. B., Johnson, M. A., Forrest, S. M., Love, D. R., Zubrzycka-Gaarn, E. E., Bulman, D. E., Harris, J. B., and Davies, K. E. (1990) Very mild muscular dystrophy associated with the deletino of 46% of dystrophin. *Nature* **343**, 180–82.

Evans, E. A., and Skalak, R. (1979a) Mechanics and thermodynamics of biomembranes. Part 1. *CRC Crit. Rev. Bioeng.* **3**, 181–330.

Evans, E. A., and Skalak, R. (1979b) Mechanics and thermodynamics of biomembranes. Part 2. *CRC Crit. Rev. Bioeng.* **3**, 331–418.

Evans, E., and Ritchie, K. (1999) Strength of a weak bond connecting flexible polymer chains. *Biophys. J.* **76**, 2439–47.

Fath, K. R., and Burgess, D. R. (1995) Not actin alone. *Curr. Biol.* **5**, 591–93.

Ferguson, K. M., Kavran, J. M., Sankaran, V. G., Fournier, E., Isakoff, S. J., Skolnik, E. Y., and Lemmon, M. A. (2000) Structural basis for discrimination of 3-phosphoinositides by pleckstrin homology domains. *Mol. Cell* **6**, 373–84.

Fox, J. E., Austin, C. D., Reynolds, C. C., and Steffen, P. K. (1991) Evidence that agonist-induced activation of calpain causes the shedding of procoagulant-containing microvesicles from the membrane of aggregating platelets. *J. Biol. Chem.* **266**, 13289–95.

Fox, J. E. (2001) Cytoskeletal proteins and platelet signaling. *Thromb. Haemost.* **86**, 198–213.

Garcia-Higuera, I., Taniguchi, T., Ganesan, S., Meyn, M. S., Timmers, C., Hejna, J., Grompe, M., and D'Andrea, A. D. (2001) Interaction of the Fanconi anemia proteins and BRCA1 in a common pathway. *Mol. Cell.* **7**, 249–62.

Godi, A., Santone, I., Pertile, P., Marra, P., Di Tullio, G., Luini, A., Corda, D., and De Matteis, M. A. (1999) ADP-ribosylation factor regulates spectrin skeleton assembly on the Golgi complex by stimulating phosphatidylinositol 4,5-bisphosphate synthesis. *Biochem. Soc. Trans.* **27**, 638–42.

Grum, V. L., Li, D., MacDonald, R. I., and Mondragon, A. (1999) Structures of two repeats of spectrin suggest models of flexibility. *Cell* **98**, 523–35.

Gumbiner, B. M. (1992) Epithelial morphogenesis. *Cell* **69**, 385–87.

Hammarlund, M., Davis, W. S., and Jorgensen, E. M. (2000) Mutations in beta-spectrin disrupt axon outgrowth and sarcomere structure. *J. Cell Biol.* **149**, 931–42.

Hammerton, R. W., Krzeminski, K. A., Mays, R. W., Ryan, T. A., Wollner, D. A., and Nelson, J. W. (1991) Mechanism for regulating cell surface distribution of Na^+, K^+-ATPase in polarized epithelial cells. *Science*, **254**, 847–50.

Hartwig, J. H., Barkalow, K., Azim, A., and Italiano, J. (1999) The elegant platelet: Signals controlling actin assembly. *Thromb. Haemost.* **82**, 392–98.

Heinrich, V., Ritchie, K., Mohandas, N., and Evans, E. (2001) Elastic thickness compressibilty of the red cell membrane. *Biophys. J.* **81**, 1452–63.

Henon, S., Lenormand, G., Richert, A., and Gallet, F. (1999) A new determination of the shear modulus of the human erythrocyte membrane using optical tweezers. *Biophys. J.* **76**, 1145–51.

Hirokawa, N., Keller, T. S., III, Chasan, R., and Mooseker, M. (1983) Mechanism of brush border contractility studied by the quick-freeze, deep-etch method. *J. Cell Biol.* **96**, 1325–36.

Holley, M. (2000) Hearing. Tuning in with motor proteins [news; comment]. *Nature* **405**, 130–31, 133.

Holley, M. C., Kalinec, F., and Kachar, B. (1992) Structure of the cortical cytoskeleton in mammalian outer hair cells. *J. Cell Sci.* **102**, 569–80.

Jenkins, S. M., and Bennett, V. (2001) Ankyrin-G coordinates assembly of the spectrin-based membrane skeleton, voltage-gated sodium channels, and L1 CAMs at Purkinje neuron initial segments. *J. Cell Biol.* **155**, 739–46.

Kamal, A., Ying, Y., and Anderson, R. G. (1998) Annexin VI-mediated loss of spectrin during coated pit budding is coupled to delivery of LDL to lysosomes. *J. Cell Biol.* **142**, 937–47.

Keller, T. C. S., III, Conzelman, K. A., Chasan, R., and Mooseker, M. A. (1985) Role of myosin in terminal web contraction in isolated intestinal epithelial brush borders. *J. Cell Biol.* **100**, 1647–55.

Kreis, T., and Vale, R. (1993) *Guidebook to the Cytoskeletal and Motor Proteins*. Oxford, UK: Oxford University Press.

Lee, J. C., Bermudez, H., Discher, B. M., Sheehan, M. A., Won, Y. Y., Bates, F. S., and Discher, D. E. (2001) Preparation, stability, and in vitro performance of vesicles made with diblock copolymers. *Biotechnol. Bioeng.* **73**, 135–45.

Lee, J. C., and Discher, D. E. (2001) Deformation-enhanced fluctuations in the red cell skeleton with theoretical relations to elasticity, connectivity, and spectrin unfolding. *Biophys. J.* **81**, 3178–92.

Lee, J. C., Wong, D. T., and Discher, D. E. (1999) Direct measures of large, anisotropic strains in deformation of the erythrocyte cytoskeleton. *Biophys. J.* **77**, 853–64.

Lee, J. K., Coyne, R. S., Dubreuil, R. R., Goldstein, L. S., and Branton, D. (1993) Cell shape and interaction defects in alpha-spectrin mutants of *Drosophila melanogaster*. *J. Cell Biol.* **123**, 1797–1809.

Lee, J. K., Brandin, E., Branton, D., and Goldstein, L. S. B. (1997) α-spectrin is required for ovarian follicle monolayer integrity in *Drosophila melanogaster*. *Development* **124**, 353–62.

Lemmon, M. A., Ferguson, K. M., and Abrams, C. S. (2002) Pleckstrin homology domains and the cytoskeleton. *FEBS Lett.* **513**, 71–76.

Lenne, P. F., Raae, A. J., Altmann, S. M., Saraste, M., and Horber, J. K. (2000) States and transitions during forced unfolding of a single spectrin repeat. *FEBS Lett.* **476**, 124–28.

Liao, E. C., Paw, B. H., Peters, L. L., Zapata, A., Pratt, S. J., Do, C. P., Lieschke, G., and Zon, L. I. (2000) Hereditary spherocytosis in zebrafish riesling illustrates evolution of erythroid beta-spectrin structure, and function in red cell morphogenesis and membrane stability. *Development* **127**, 5123–32.

Lietzke, S. E., Bose, S., Cronin, T., Klarlund, J., Chawla, A., Czech, M. P., and Lambright, D. G. (2000) Structural basis of 3-phosphoinositide recognition by pleckstrin homology domains. *Mol. Cell* **6**, 385–94.

Lombardo, C. R., Weed, S. A., Kennedy, S. P., Forget, B. G., and Morrow, J. S. (1994) βII-spectrin (fodrin) and βIΣ2-spectrin (muscle) contain NH_2- and COOH-terminal membrane association domains (MAD1 and MAD2). *J. Biol. Chem.* **269**, 29212–19.

MacDonald, R. I., and Pozharski, E. V. (2001) Free energies of urea and of thermal unfolding show that two tandem repeats of spectrin are thermodynamically more stable than a single repeat. *Biochemistry* **40**, 3974–84.

Macoska, J. A., Xu, J., Ziemnicka, D., Schwab, T. S., Rubin, M. A., and Kotula, L. (2001) Loss of expression of human spectrin src homology domain binding protein 1 is associated with 10p loss in human prostatic adenocarcinoma. *Neoplasia* **3**, 99–104.

Marfatia, S. M., Morais-Cabral, J. H., Kim, A. C., Byron, O., and Chishti, A. H. (1997) The PDZ domain of human erythrocyte p55 mediates its binding to the cytoplasmic carboxyl terminus of glycophorin C. Analysis of the binding interface by in vitro mutagenesis. *J. Biol. Chem.* **272**, 24191–97.

McKeown, C., Pratis, V., and Austin, J. (1998) The *sma-1* gene encodes a $β_{Heavy}$-spectrin required for *C. elegans* morphogenesis. *Development* **125**, 2087–98.

McMahon, L. W., Walsh, C. E., and Lambert, M. W. (1999) Human alpha spectrin II and the Fanconi anemia proteins FANCA and FANCC interact to form a nuclear complex. *J. Biol. Chem.* **274**, 32904–8.

Meriläinen, J., Palovuori, R., Sormunen, R., Wasenius, V.-M., and Lehto, V.-P. (1993) Binding of the α-fodrin SH3 domain to the leading lamellae of locomoting chicken fibroblasts. *J. Cell Sci.* **105**, 647–54.

Meyer, R. K., and Aebi, U. (1990) Bundling of actin filaments by alpha-actinin depends on its molecular length. *J. Cell Biol.* **110**, 2013–24.

Mohandas, N., and Evans, E. (1994) Mechanical properties of the red cell membrane in relation to molecular structure and genetic defects. *Annu. Rev. Biophys. Biomol. Struct.* **23**, 787–818.

Moorthy, S., Chen, L., and Bennett, V. (2000) *Caenorhabditis elegans* beta-G spectrin is dispensable for establishment of epithelial polarity, but essential for muscular and neuronal function. *J. Cell Biol.* **149**, 915–30.

Muresan, V., Stankewich, M. C., Steffen, W., Morrow, J. S., Holzbaur, E. L., and Schnapp, B. J. (2001) Dynactin-dependent, dynein-driven vesicle transport in the absence of membrane proteins: A role for spectrin and acidic phospholipids. *Mol. Cell* **7**, 173–83.

Muse, S. V., Clark, A. G., and Thomas, G. H. (1997) Comparisons of the nucleotide substitution process among repetitive segments of the alpha- and beta-spectrin genes. *J. Mol. Evol.* **44**, 492–500.

Paci, E., and Karplus, M. (2000) Unfolding proteins by external forces and temperature: The importance of topology and energetics. *Proc. Natl. Acad. Sci. U.S.A.* **97**, 6521–26.

Parkinson, N. J., Olsson, C. L., Hallows, J. L., McKee-Johnson, J., Keogh, B. P., Noben-Trauth, K., Kujawa, S. G., and Tempel, B. L. (2001) Mutant beta-spectrin 4 causes auditory and motor neuropathies in quivering mice. *Nat. Genet.* **29**, 61–65.

Picart, C., Dalhaimer, P., and Discher, D. E. (2000) Actin protofilament orientation in deformation of the erythrocyte membrane skeleton. *Biophys. J.* **79**, 2987–3000.

Pradhan, D., Lombardo, C. R., Roe, S., Rimm, D. L., and Morrow, J. S. (2001) alpha-Catenin binds directly to spectrin and facilitates spectrin-membrane assembly in vivo. *J. Biol. Chem.* **276,** 4175–81.

Praitis, V. A., and Austin, J. (2000) SMA-1 spectrin is required for embryonic morphogenesis in *C. elegans. Mol. Biol. Cell* **11,** 550a.

Rameh, L. E., Arvidsson, A., Carraway, K. L., 3rd, Couvillon, A. D., Rathbun, G., Crompton, A., VanRenterghem, B., Czech, M. P., Ravichandran, K. S., Burakoff, S. J., Wang, D. S., Chen, C. S., and Cantley, L. C. (1997) A comparative analysis of the phosphoinositide binding specificity of pleckstrin homology domains. *J. Biol. Chem.* **272,** 22059–66.

Raphael, R. M., Popel, A. S., and Brownell, W. E. (2000) A membrane bending model of outer hair cell electromotility. *Biophys. J.* **78,** 2844–62.

Rief, M., Gautel, M., Oesterhelt, F., Fernandez, J. M., and Gaub, H. E. (1997) Reversible unfolding of individual titin immunoglobulin domains by AFM. *Science* **276,** 1109–12.

Rief, M., Pascual, J., Saraste, M., and Gaub, H. E. (1999) Single molecule force spectroscopy of spectrin repeats: Low unfolding forces in helix bundles. *J. Mol. Biol.* **286,** 553–61.

Sako, Y., and Kusumi, A. (1995) Barriers for lateral diffusion of transferrin receptor in the plasma membrane as characterised by receptor dragging by laser tweezers: Fence versus tether. *J. Cell Biol.* **129,** 1559–74.

Scotland, P., Zhou, D., Benveniste, H., and Bennett, V. (1998) Nervous system defects of AnkyrinB (−/−) mice suggest functional overlap between the cell adhesion molecule L1 and 440-kD AnkyrinB in premyelinated axons. *J. Cell Biol.* **143,** 1305–15.

Sheetz, M. P. (2001) Cell control by membrane-cytoskeleton adhesion. *Nat. Rev. Mol. Cell Biol.* **2,** 392–96.

Sisson, J. C., Field, C., Ventura, R., Royou, A., and Sullivan, W. (2000) Lava lamp, a novel peripheral Golgi protein, is required for *Drosophila melanogaster* cellularization. *J. Cell Biol.* **151,** 905–18.

Speicher, D. W., and Marchesi, V. T. (1984) Erythrocyte spectrin is comprised of many homologous triple helical segments. *Nature* **311,** 177–80.

Stabach, P. R., and Morrow, J. S. (2000) Identification and characterization of beta V spectrin, a mammalian ortholog of Drosophila beta H spectrin. *J. Biol. Chem.* **275,** 21385–95.

Stidwill, R. P., Wysolmerski, T., and Burgess, D. R. (1984) The brush border cytoskeleton is not static: In vivo turnover of proteins. *J. Cell Biol.* **98,** 641–45.

Svoboda, K., Schmidt, C. F., Branton, D., and Block, S. M. (1992) Conformation and elasticity of the isolated red blood cell membrane skeleton. *Biophys. J.* **63,** 784–93.

Takeuchi, M., Miyamoto, H., Sako, Y., Komizu, H., and Kusumi, A. (1998) Structure of the erythrocyte membrane skeleton as observed by atomic force microscopy. *Biophys. J.* **74,** 2171–83.

Thomas, G. H. (2001) Spectrin: The ghost in the machine. *Bioessays* **23,** 152–60.

Thomas, G. H., and Kiehart, D. P. (1994) Beta heavy-spectrin has a restricted tissue and subcellular distribution during Drosophila embryogenesis. *Development* **120,** 2039–50.

Thomas, G. H., Newbern, E. C., Korte, C. C., Bales, M. A., Muse, S. V., Clark, A. G., and Kiehart, D. P. (1997) Intragenic duplication and divergence in the spectrin superfamily of proteins. *Mol. Biol. Evol.* **14,** 1285–95.

Thomas, G. H., Zarnescu, D. C., Juedes, A. E., Bales, M. A., Londergan, A., Korte, C. C., and Kiehart, D. P. (1998) Drosophila beta heavy-spectrin is essential for development and contributes to specific cell fates in the eye. *Development* **125,** 2125–34.

Thomas, G. H., and Williams, J. A. (1999) Dynamic rearrangement of the spectrin membrane skeleton during the generation of epithelial polarity in Drosophila. *J. Cell Sci.* **112,**
2843–52.

Tolomeo, J. A., Steele, C. R., and Holley, M. C. (1996) Mechanical properties of the lateral
cortex of mammalian auditory outer hair cells. *Biophys. J.* **71,** 421–29.

Tomishige, M., Sako, Y., and Kusumi, A. (1998) Regulation mechanism of the lateral diffusion of band 3 in erythrocyte membranes by the membrane skeleton. *J. Cell Biol.* **142,**
989–1000.

Tse, W. T., Lecomte, M. C., Costa, F. F., Garbarz, M., Feo, C., Boivin, P., Dhermy, D., and
Forget, B. G. (1990) Point mutation in the beta-spectrin gene associated with alpha
I/74 hereditary elliptocytosis. Implications for the mechanism of spectrin dimer self-
association. *J. Clin. Invest.* **86,** 909–16.

Tse, W. T., and Lux, S. E. (1999) Red blood cell membrane disorders. *Br. J. Haematol.* **104,**
2–13.

Tse, W. T., Tang, J., Jin, O., Korsgren, C., John, K. M., Kung, A. L., Gwynn, B., Peters, L. L., and
Lux, S. E. (2001) A new spectrin, beta IV, has a major truncated isoform that associates
with promyelocytic leukemia protein nuclear bodies and the nuclear matrix. *J. Biol.
Chem.* **276,** 23974–85.

Ursitti, J. A., Pumplin, D. W., Wade, J. B., and Bloch, R. J. (1991) Ultrastructure of the human
erythrocyte cytoskeleton and its attachment to the membrane. *Cell. Motil. Cytoskel.* **19,**
227–43.

Van Dort, H. M., Knowles, D. W., Chasis, J. A., Lee, G., Mohandas, N., and Low, P. S. (2001)
Analysis of integral membrane protein contributions to the deformability and stability
of the human erythrocyte membrane. *J. Biol. Chem.* **276,** 46968–74.

Viel, A., and Branton, D. (1994) Interchain binding at the tail end of the Drosophila spectrin
molecule. *Proc. Natl. Acad. Sci. U.S.A.* **91,** 10839–43.

Weed, S. A., Stabach, P. R., Oyer, C. E., Gallagher, P. G., and Morrow, J. S. (1996) The lethal
hemolytic mutation in beta I sigma 2 spectrin Providence yields a null phenotype in
neonatal skeletal muscle. *Lab. Invest.* **74,** 1117–29.

White, B. C., Sullivan, J. M., DeGracia, D. J., O'Neil, B. J., Neumar, R. W., Grossman,
L. I., Rafols, J. A., and Krause, G. S. (2000) Brain ischemia and reperfusion: molecular
mechanisms of neuronal injury. *J. Neurol. Sci.* **179,** 1–33.

Yan, Y., Winograd, E., Viel, A., Cronin, T., Harrison, S. C., and Branton, D. (1993) Crystal
structure of the repetitive segments of spectrin. *Science* **262,** 2027–30.

Yeaman, C., Grindstaff, K., and Nelson, W. J. (1999) New perspectives on mechanisms
involved in generating epithelial cell polarity. *Physiol. Rev.* **79,** 73–98.

Ylanne, J., Scheffzek, K., Young, P., and Saraste, M. (2001) Crystal structure of the alpha-
actinin rod reveals an extensive torsional twist. *Structure (Cambr.)* **9,** 597–604.

Zarnescu, D. C., and Thomas, G. H. (1999) Apical spectrin is essential for epithelial morphogenesis but not apicobasal polarity in Drosophila. *J. Cell Biol.* **146,** 1075–86.

Zhang, Z., Weed, S. A., Gallagher, P. G., and Morrow, J. S. (2001) Dynamic molecular modeling of pathogenic mutations in the spectrin self-association domain. *Blood* **98,** 1645–53.

Zhou, D., Lambert, S., Malen, P. L., Carpenter, S., Boland, L. M., and Bennett, V. (1998)
AnkyrinG is required for clustering of voltage-gated Na channels at axon initial segments
and for normal action potential firing. *J. Cell Biol.* **143,** 1295–1304.

TWELVE

Giant Protein Titin: Structural and Functional Aspects

J. Trinick and L. Tskhovrebova

INTRODUCTION

The spatial organization of acto-myosin contractile machine varies widely in different muscle cells. Each of the major muscle types – smooth, oblique-, and cross-striated – has versions differing in myosin/actin ratio, myofilament size, accessory proteins, and overall level of structural order. These variations relate directly to functional differences of the muscles, especially their speed and strength of contraction. The greatest structural order is in cross-striated muscles, and highest myosin/actin ratio is in striated muscles of vertebrates. These have exactly structured myofilaments integrated into highly ordered contractile units: the sarcomeres. This beautifully controlled architecture provides the basis for highly coordinated acto-myosin interactions and for the rapid production of macroscopic levels of force and displacement.

The giant protein titin is thought to play a pivotal role in defining and maintaining structural order in vertebrate striated muscle. The properties of the titin molecule allow it not only to be a centrepiece of contractile protein assembly, but also to control the level of mechanical sensitivity and elasticity of the sarcomere. This chapter describes some of the recent ideas concerning the relationship between structural and mechanical properties of single titin molecules and how these properties integrate to determine muscle function (see also reviews by Trinick, 1994, 1996; Wang, 1996; Linke and Granzier, 1998; Gautel et al., 1999; Trinick and Tskhovrebova, 1999; Gregorio and Antin, 2000).

The titin molecule (Figure 12.1) is more than 1-μm long and is a single polypeptide chain with molecular weight ranging from approximately 3 to 4 MDa, depending on isoform (Freiburg et al., 2000). Sequence analysis indicates that most of the polypeptide is folded into ~300 domains similar to I-set immunoglobulins (Ig) or type III fibronectins (Fn) arranged in tandem (Labeit and Kolmerer, 1995), and this predicted, multidomain structure correlates well with the beaded appearance of about 4 nm periodicity observed in the electron microscope (Trinick et al., 1984). The structures of several Ig and Fn domains

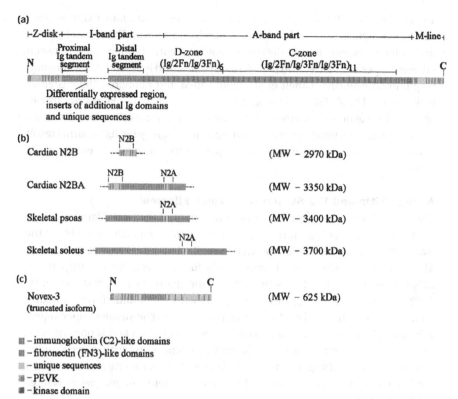

- ▦ – immunoglobulin (C2)-like domains
- ▦ – fibronectin (FN3)-like domains
- ▦ – unique sequences
- ▦ – PEVK
- ▪ – kinase domain

Figure 12.1: Domain structure of the titin molecule and isoform-related differences in the I-band part. (a) Conserved part of titin. This includes the A-band segment and the C- and N-terminal portions of the I-band segment. The central region of the I-band, as well as the Z- and M-line regions, are isoform-specific. (b) Examples of differentially expressed regions of the I-band part of titin molecule. These include the unique sequences and the variable part of the proximal poly-Ig segment. (c) Novex-3, truncated titin isoform. The N-terminal part is similar to the full length molecule and contains the Z-disk region (except Z-repeats) and the constitutively expressed proximal Ig tandem segment. The C-terminal part is isoform specific, and is formed by a giant unique sequence with a few insertions of Ig domains. In muscle, novex-3 is located within and near the Z-disk region. MW, molecular weight. (See color section.)

representative for different regions of the molecule have been solved to atomic resolution by means of nuclear magnetic resonance and X-ray diffraction (Pfuhl and Pastore, 1995; Improta et al., 1996; Muhle-Goll et al., 1998; Mayans et al., 2001).

In situ, the titin molecule spans half the sarcomere, with its C-terminus in the M-line and its N-terminus in the Z-disk (Furst et al., 1988; Labeit et al., 1992). The structures of the domains and their patterns within the molecule are adapted for different functions in the sarcomere, and four major regions can be distinguished: Z-disk, I-band, A-band, and M-line. The A-band part of the molecule reveals a high degree of sequence conservation, both between different types of muscles and in different species, and is essentially of constant length in all isoforms (Labeit et al., 1990; Labeit and Kolmerer, 1995; Greaser et al., 1996). The Z-disk, M-line, and especially the I-band regions are tissue-specific. Unlike other sarcomere components, such as myosin whose isoforms result from several genes, titin is the product of a single gene. Its isoforms result from differential splicing of the I-band part of the molecule (Freiburg et al., 2000).

A-Band Titin and the Structure of Thick Filament

The A-band part of titin is an integral component of the thick filament, which led to the suggestion that it may act as a template to regulate assembly of the exactly 294 myosin molecules known to form the filament (Whiting et al., 1989). Throughout most of this region, the Ig and Fn domains are arranged into patterns or 'super repeats' that reflect periodicities in the filament structure (Labeit et al., 1992). There are two types of super-repeats with seven and eleven domains, which are 25–30 nm and ~43 nm long, respectively. The smaller super-repeat is located closer to the N-terminal end of the A-band titin and is present in six copies. Immediately adjacent is the larger super-repeat present in eleven copies. This arrangement of super-repeats correlates with the structure and size of what are known as the D- and C-zones of the thick filament (see reviews by Gautel, 1996; Trinick, 1996).

The packing of titin within the thick filament is poorly understood, and it is not known if it forms the core of the filament or is located at a higher radius. On the basis of in situ structural studies it has been suggested that the molecules are aligned parallel to the filament long axis, rather than following helical tracks around the filament (Squire et al., 1998). The presence of six molecules in each half of the thick filament (Cazorla et al., 2000; Liversage et al., 2001) correlates with the trigonal symmetry of the filament (Luther and Squire, 1978) and suggests that there are two titin molecules associated with each of the three subfilaments making up the thick filament (Maw and Rowe, 1980).

The major proteins interacting with titin in thick filaments are myosin, C-protein, and M-line proteins (Labeit et al., 1992; Soteriou et al., 1993b; Obermann et al., 1997). In vitro, titin interacts with the light meromyosin part of the myosin molecule, mainly at a site located at about 20 nm from the end of the myosin tail (Houmeida et al., 1995). The interaction with myosin is thought to specifically involve the titin Fn domains (Labeit et al., 1992; Bennett and Gautel, 1996; Gautel et al., 1999). This conclusion is based on several considerations, including location of Fn domains in only the A-band part of the molecule, correlation between the number of Fn domain groups and the number of levels of myosin heads in each half of the thick filament, and correlation between myosin-binding affinity of recombinant titin fragments and the number of Fn domains present (Labeit et al., 1992). Studies with expressed constructs also suggest a single binding site for C-protein in the 11-domain super-repeat of A-band titin (Freiburg and Gautel, 1996), which explains the attachment of C-protein to the filament at 43 nm intervals (Bennett et al., 1986).

I-Band Titin and Mechanism of Muscle Elasticity

The I-band part of titin forms flexible connections between the end of the thick filament and Z-disk. These connections centre thick filaments in the sarcomere, and they form part of the main routes of mechanical connectivity through relaxed sarcomeres (Horowits et al., 1986). Compared with the A-band part, I-band titin is more complex and is highly tissue-specific (Labeit and Kolmerer, 1995; Freiburg et al., 2000; Bang et al., 2001). Only the N-terminal (or 'proximal') and C-terminal (or 'distal') segments of this region, with 15 and 22 Ig domains, respectively, are conserved between isoforms. The part between these segments varies greatly in structure and size (Figure 12.1). The major subregions here are a variable part of the proximal segment of Ig domains; the novex-N2 region consisting of alternating Ig domains and unique sequences; and the PEVK region, a unique sequence named from its preponderance of these aminoacid residues. The N2 region also has two versions differing in sequence: skeletal titins contain only so called N2A isoforms, whereas cardiac titins contain either the N2B isoforms, or both N2A and N2B (the N2BA isoforms). Both types of muscle also express a truncated version of titin (novex-3) that does not have the A-band part (Bang et al., 2001). The tertiary structures of the unique sequences cannot be predicted from database comparisons and remain unknown.

Information about the arrangement of I-band titin in situ is very limited and current schemes are largely hypothetical (Tskhovrebova and Trinick, 2000). It is likely that there are two major sections separated by the PEVK region. Near the A/I-junction, the six titin molecules emerging from the thick filament are likely to remain bound together in a single bundle. This idea is supported by the observations of the 'end-filaments' at the ends of thick filaments, both in situ (Bennett et al., 1997) and in separate filaments (Trinick, 1981; Liversage et al., 2001). At the PEVK region, the titin bundles are likely to branch into thinner bundles or separate molecules extending to the Z-disk. A requirement for branching is implied by necessity for titin to adjust its hexagonal organization in the A-band to the tetragonal symmetry of Z-disk (Liversage et al., 2001), due to the likely interaction with thin filaments near Z-disk (Linke et al., 1997; Trombitas and Granzier, 1997). The relatively high negative charge of the PEVK region could also favour branching at this site.

When muscle is stretched, stress is transmitted to the I-band part of titin which then extends. The following estimates illustrate the likely magnitude of changes in contour length of I-band titin during normal muscle function (see also Goulding et al., 1997). The operating ranges of sarcomere lengths are estimated to be 1.9–2.5 μm in cardiac muscles, 2.0–3.4 μm in psoas muscles and 2.2–3.8 μm in soleus muscles (Millman, 1998). This gives I-band length ranges between 0.05–0.35 μm, 0.12–0.82 μm, and 0.21–1.01 μm, respectively. The mean contour lengths of I-band titin isoforms estimated from sequence are 0.2 μm, 0.37 μm, and 0.49 μm. Thus, sarcomere operation over the maximum range of lengths requires the I-band titin to undergo at least a 2-fold compression and more than 2-fold elongation relative to the mean contour length in each muscle. Estimates based on mechanical measurements on muscle fibres and myofibrils indicate that the extension of titin beyond its contour length produces forces in the range from a few to tens of picoNewtons per molecule (Wang et al., 1991; Granzier and Irving, 1995; Linke et al., 1996).

To explain this remarkable extensibility of titin molecules, which allows them to remain functional over a large range of sarcomere lengths, reversible unfolding of the polypeptide was proposed and a mechanism based on estimates of domain folding energy was put forward (Soteriou et al., 1993a). Erickson (1994) extended these ideas to consideration of the critical forces necessary to denature the domains and break intermolecular bonds. Subsequent in vitro and in situ studies confirmed these predictions and also showed how non-uniform mechanical properties and unfolding characteristics are likely to derive directly from variations in structure along the molecule.

Titin Extensibility In Vitro

Electron microscopy of isolated titin molecules at high ionic strength, which favours solubility, reveals a highly flexible configuration (Maruyama et al., 1984; Trinick et al., 1984; Wang et al., 1984). The 'persistence length' of the molecule, which is a quantitative measure of its stiffness, has an average value of 13–15 nm and the radius of gyration is about 70 nm, based on dynamic light scattering (Fujime and Higuchi, 1993) and electron microscopy (Tskhovrebova and Trinick, 2001). Electron microscopy (Wang et al., 1984; Nave et al., 1989; Tskhovrebova and Trinick, 1997) also illustrates how the equilibrium coiled configuration can be easily deformed by small mechanical forces, resulting in straightened or even partial unfolded molecules (Figure 12.2).

Recent years have seen the exciting development of methods that allow mechanical experiments with single molecules. This is an important breakthrough that has opened the way to a new information that would be masked in an ensemble average. The many advantages of titin as a model system have resulted in it becoming a paradigm for this type of approach. Using optical tweezers

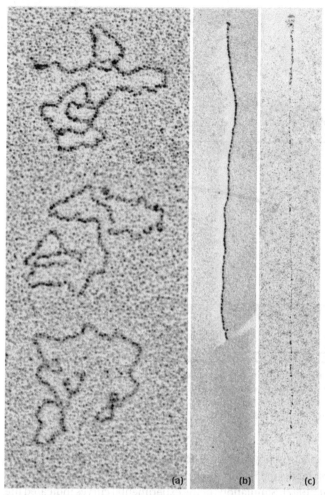

Figure 12.2: Electron micrographs illustrating titin molecules purified from rabbit back muscle, and metal-shadowed after drying on a mica substrate in the coiled (a), straightened (b), and unfolded (c) configurations. The coiled configuration closely reflects the equilibrium state adopted by the isolated molecule at high ionic strength. The equilibrium configuration in most cases is distorted by hydrodynamic forces related to drying. These forces are responsible for straightening or 'molecular combing' of the molecules. At relatively high forces, straightening is accompanied by unfolding of the polypeptide. The micrograph in (c) illustrates partially unfolded A-band part of titin molecule comprising mainly Ig and Fn domains. Diameter of the unfolded molecule is less than half of the native (b), while the length significantly exceeds the native length.

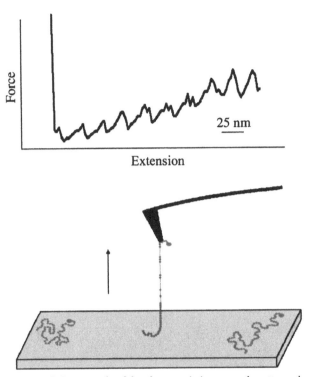

Figure 12.3: Example of the characteristic sawtooth pattern observed during stretching titin by atomic force spectroscope cantilever. The distance between peaks is about 25 nm, close to the length of unfolded Ig and Fn domains. This indicates that the sawtooth pattern reflects sequential unfolding of the domains.

and atomic force spectroscopy, the force–extension relationship of single titin molecules was first measured in vitro (Kellermayer et al., 1997; Rief et al., 1997; Tskhovrebova et al., 1997). Simulations of the experimental curves indicated the probable presence of two serially linked springs, which were identified as the Ig-Fn and PEVK regions. The data showed that extension of the molecule up to 1.5-fold did not affect the native fold of the Ig-Fn domains, and the increase of contour length was due mainly to the unfolded PEVK polypeptide.

Reversible unfolding of titin domains at high extensions and forces was also directly illustrated (Rief et al., 1997; Tskhovrebova et al., 1997). Unfolding of the domains occurs sequentially, leading to a characteristic sawtooth-like pattern with each peak resulting from a single domain unfolding (Figure 12.3). Computer simulations based on the structure of I27 (corresponds to I91 in the latest numeration of domains (Freiburg et al., 2000)), one of the few domains solved to atomic resolution (Improta et al., 1996), indicate that the force peak arises due to a potential barrier associated with the native state of the domain (Lu et al., 1998; Marszalek et al., 1999; Lu and Schulten, 2000). The barrier reflects the necessity (after an initial low-force stretch from original 4 nm to 7–10 nm) to break simultaneously a set of hydrogen bonds. As soon as this barrier is overcome, the polypeptide unravels as a worm-like chain. The measured fold-breaking force varies between domains reflecting small variations in their bonding patterns (Rief et al., 1998; Li et al., 2000). It also depends critically on the rate of domain extension (Rief et al., 1998) explaining the variability in results obtained in different types of experiments. Almost complete release of tension is required to collapse the polypeptide chain and allow refolding of the domains. Biochemical studies indicate refolding times of single domains vary over a wide range, from tens of milliseconds to tens of seconds, whereas mechanical experiments suggest that refolding of the polypeptide with tethered ends is slower in comparison with a free chain in solution (Carrion-Vazquez et al., 1999; Scott et al., 2002). The refolding properties of the titin molecule can be strikingly demonstrated in the electron microscope when, after a complete chemical denaturation-renaturation cycle, the renatured molecules appear indistinguishable from native (Tskhovrebova and Trinick, 2000).

The conclusions made from the in vitro studies can be summarised as follows: under mechanical stress the end-to-end distance of the titin molecule can undergo relatively fast and reversible changes from less than the contour length to substantially longer. This capacity derives both from the bending properties of the native protein and from folding-unfolding transitions of the polypeptide. Unfolding will start from regions lower mechanical stability, such as the unique sequences, and will then continue in regions of higher mechanical stability, such as in the beta-sandwich domains.

The differences in mechanical stability between the PEVK and Ig domains almost certainly arise from the differences in their tertiary structures. In the PEVK region, studies of sequence and the folding properties of expressed fragments indicate the presence of 28-residue long 'polyproline helix-coil' motif as the likely structural repeat (Greaser, 2001; Gutierrez-Cruz et al., 2001; Ma et al., 2001). This would be radically different from the beta-sandwich domains and could result in lower potential energy barrier and lower resistance to mechanical stress.

Comparison of In Vitro and In Situ Titin Extensibility

The conclusions from in vitro studies are in qualitative agreement with in situ observations (Gautel and Goulding, 1996; Granzier et al., 1996; Linke et al., 1996). According to the current consensus, sarcomere extension beyond rest length first causes straightening of the I-band part of the titin molecule from a resting, coiled state. Further extension induces local unfolding of the PEVK region. High speeds of extension and/or high forces are likely to lead to unfolding of Ig domains (see, e.g., Minajeva et al., 2001; Li et al., 2002).

This is a general scheme and the latest data indicate that extensibility may be more finely tuned and dependent on fibre type than previously thought. Thus, in cardiac muscles, the unique sequence of the N2B region is also involved in the multiphase extension of the molecule (Linke et al., 1999). The presence of more than one titin isoforms in a fibre may further contribute to delicate modulation of its extensible properties (Trombitas et al., 2001; Bang et al., 2001).

Quantitative comparisons of data from in vitro and in situ mechanical experiments is complicated by several factors that still need to be resolved. One such major factor is the arrangement of titin in the sarcomere. The approach used in single molecule experiments for studying titin extensibility (Kellermayer et al., 1997; Rief et al., 1997; Tskhovrebova et al., 1997) is of special interest, because it mimics the situation in muscle by applying mechanical force to the molecule tethered at both ends. An essential difference, however, is that in vivo titin in I-band is located within the ordered thin filament lattice of sarcomere. Thus, all conformational changes occur within the confined lattice space.

Figure 12.4 illustrates schematically the likely conformation of titin in the I-band at slack sarcomere length. It is obvious that the molecule cannot adopt an equilibrium conformation similar to the isolated molecule in solution. Based on its persistence length, the isolated I-band part of titin in vitro can be approximated as a sphere of radius R_G between ~15 and ~45 nm for different isoforms. However, in situ, the average conformational space per molecule is restricted by the elongated and narrow volume with an approximate radius of 8.5 nm and a length of about 50 nm, 120 nm, and 210 nm in cardiac, psoas, and soleus muscles, respectively. Thus, the configurational deformation resulting from changes in sarcomere length may not occur as a coil-to-rod transition, as in solution, but rather in a more ordered way maintaining the elongated shape of the conformational space. As a consequence of this confinement, compression of the molecule during muscle contraction may involve higher viscosity, and its extension during muscle stretch may have different dynamics compared with a free chain.

These estimates are based on the assumption that there are six titin molecules extending from each end of thick filament (Cazorla et al., 2000; Liversage et al., 2001). If the molecules do not extend as monomers, but as side-to-side dimers or trimers, this will change quantitative estimates. However, the effects of confinement on conformation and dynamics of titin in the I-band will remain.

Additional factors that may modulate extensibility in vivo and in situ are the reported interactions of titin in the I-band with thin filaments and Ca^{2+} ions. Evidence of three actin binding sites within the molecule has been claimed. One

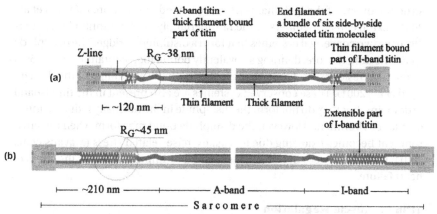

Figure 12.4: Effect of confinement on the configuration of the titin I-band part within (a) psoas, and (b) soleus muscles at slack sarcomere lengths. Scheme is drawn at the scale proportional to the known dimensions and relative positions of major contractile filaments. The I-band lengths in slack sarcomeres are about 120 and 210 nm in psoas and soleus muscles, respectively, and contour lengths of the I-band parts of corresponding titin isoforms are 320 and 440 nm (excluding actin-bound portions). The isolated I-band titin segment in solution will occupy a volume of approximately spherical shape with a radius of 38–45 nm, as can be judged from estimates of radius of gyration. Within the sarcomere the average space per single I-band segment is restricted by a volume of elongated shape with length equal to the I-band and radius 8–10 nm. The latter value is based on a thick filament spacing of 43 nm (Millman, 1998) and the estimate of six titin molecules per half thick filament (Cazorla et al., 2000; Liversage et al., 2001). Thus, there is a discrepancy between the space required and that actually available. This implies that titin strands in the I-band at slack sarcomere length either are arranged in a more ordered way than can be predicted from solution observations or are highly interwired. (See color section.)

Ca^{2+}-independent actin binding site is in the proximal Ig segment and includes domains I2–I3 (Freiburg et al., 2000). Another such site covers titin domains I47–I49 near the I/A band junction (Jin, 1995). The PEVK region is thought to contain a third actin binding site, which is proposed to be Ca^{2+}-calmodulin (skeletal muscle; Gutierrez-Cruz et al., 2001), or Ca^{2+}-S100A1 (cardiac muscle; Yamasaki et al., 2001) regulated. Ca^{2+}-binding properties of high-affinity ($K_a \sim 10^7$ M^{-1}) with resultant changes in the secondary structure of the protein were found in a 400-kDa proteolytic fragment of the I-band titin. The fragment is likely to include the distal Ig segment and the I/A junction part of titin, as well as part of the PEVK region (Tatsumi et al., 2001). Titin-actin interactions may play a role in the Ca^{2+}-sensitivity of stiffness (Stuyvers et al., 1998) and modulation of contractility (Linke et al., 2002) of cardiac muscles.

It has also been proposed that changes in the mechanical properties of titin may be induced by exercise. This was suggested by analysis of the atomic structure of one of the Ig domains from the I-band part of titin. (Mayans et al., 2001). Extracellular members of the Ig superfamily, such as antibodies, generally have two conserved cysteines that form a disulphide bridge. The disulphide spans the domain core, defining a covalently bound loop containing about 50% of the total residues, and thus changes the mechanical stability of the domain (Carl et al., 2001). These conserved cysteines are also present in the titin I-band Ig domains, but they do not form the disulphide in the normally reducing intracellular environment. However, the disulphide bond may form when the environment becomes oxidizing due to, e.g., exercise-related stress (Mayans et al., 2001). If substantiated, this mechanism could be important under the stress of heart failure.

Titin in Muscle Regulation

Stiffness of muscles is known to correlate with size of the I-band part of the titin isoform present (Labeit and Kolmerer, 1995; Linke et al., 1996). In skeletal muscles, longer isoforms are found in more compliant muscles and shorter isoforms in stiffer ones. Cardiac muscles are generally stiffer than skeletal ones, and this correlates with their smaller isoforms. It seems plausible that, by varying the length of the I-band part of titin, muscle controls its average operating range of sarcomere lengths. Through control of the sarcomere length, titin is also likely to influence the optimal level of active force and the speed of force production by the actin–myosin interaction. Furthermore, strain accumulated in I-band titin during contraction may generate the sarcomere rest-length restoring force. A version of this mechanism where titin functions as a bidirectional spring was suggested for cardiac muscles (Helmes et al., 1996).

Recently, the possibility of involvement of titin in stretch-related regulatory mechanisms of muscle has also been discussed, particularly in Ca^{2+}-signalling (Furukawa et al., 2001) and in stretch activation (Cazorla et al., 2001; Fukuda et al., 2001; Muhle-Goll et al., 2001). The functional importance of stretch activation and its relation to the Frank-Starling law of the heart, relating contractile force with venous filling pressure (Allen and Kentish, 1985; Lakatta, 1987), have stimulated attempts to uncover the underlying molecular mechanisms. Currently, the dominant hypothesis is that it is a stretch-related decrease in myofilament separation, rather than increased sarcomere length by itself, that causes the observed increase in Ca^{2+}-sensitivity of force production (see, e.g., review by Solaro and Rarick, 1998). According to the proposed mechanism, a decrease in myofilament spacing increases the probability of actin–myosin interactions and thus switches on the myosin-mediated activation of thin filaments. Titin as the length sensor is suggested to play a major role in controlling myofilament spacing. The tension in stretched I-band titin, which is transmitted directly to thick filaments, may not be purely axial and its radial component could force them to move closer to thin filaments stimulating binding and activation (Cazorla et al., 2001).

The importance of titin for normal functioning of muscle is also supported by observations of abnormalities in its expression and organisation in diseased heart (Hein et al., 1994; Morano et al., 1994). Loss of the stretch-induced increase in Ca^{2+}-sensitivity and the consequent inability of failing heart to use the Frank-Starling mechanism correlates with the absence of titin (Schwinger et al., 1994). Ischemia was shown to be characterised by non-elastic deformation of sarcomeres, resulting in increased sarcomere length and decreased interfilament distance, both of which are likely to be controlled by titin (Lu et al., 1999). Alterations of contractile properties in cardiac muscle observed in tachycardia also correlate with changes in the expressed titin isoforms (Bell et al., 2000).

Finally, the proposals that titin has major roles in controlling and maintaining the highly ordered structure of striated muscle sarcomeres are supported by observations of developmental processes during myofibrillogenesis (see for recent reviews: Fulton, 1999; Gautel et al., 1999; Gregorio and Antin, 2000). These studies show that titin is one of the first sarcomeric proteins to be synthesised, and its signalling and interactive properties are likely to be of critical importance for both the correct assembly of myofilaments and their integration into sarcomere structure (e.g., van der Ven et al., 1999, 2000). These conclusions support the idea that titin acts as a 'molecular ruler' for the thick filament formation, explaining their exactly sized structure in vertebrate striated muscles (Whiting et al., 1989; Trinick, 1994). In striated muscles of invertebrates, control of thick filament organisation is presumably based on alternative mechanisms, since these muscles apparently lack the single polypeptide protein spanning half the filament length. In vertebrates, the correlation of subtle differences between titin isoforms and the sarcomeres where they are present, suggest a close relationship between the structure and function of titin molecule and the structure and function of muscle.

ACKNOWLEDGEMENTS

We thank the British Heart Foundation for its financial support.

REFERENCES

Allen, D. G., and Kentish, J. C. (1985) The cellular basis of the length-tension relation in cardiac muscle. *J. Mol. Cell. Cardiol.* **17,** 821–40.

Bang, M.-L., Centner, T., Fornoff, F., Geach, A. J., Gotthardt, M., McNabb, M., Witt, C. C., Labeit, D., Gregorio, C. C., Granzier, H., and Labeit, S. (2001) The complete gene sequence of titin: Expression of an unusual ~700-kDa titin isoform, and its interaction with obscurin identify a novel Z-line to I-band linking system. *Circ. Res.* **89,** 1065–72.

Bell, P. B., Nyland, L., Tischler, M. D., McNabb, M., Granzier, H., and LeWinter, M. M. (2000) Alterations in the determinants of diastolic suction during pacing tachycardia. *Circ. Res.* **87,** 235–40.

Bennett, P., Craig, R., Starr, R., and Offer, G. (1986) The ultrastructural location of C-protein, X-protein and H-protein in rabbit muscle. *J. Muscle Res. Cell Motil.* **7,** 550–67.

Bennett, P. M., and Gautel, M. (1996) Titin domain patterns correlate with the axial disposition of myosin at the end of the thick filament. *J. Mol. Biol.* **259,** 896–903.

Bennett, P. M., Hodkin, T. E., and Hawkins, C. (1997) Evidence that the tandem Ig domains near the end of the muscle thick filament form an inelastic part of the I-band titin. *J. Struct. Biol.* **120,** 93–104.

Carl, P., Kwok, C. H., Manderson, G., Speicher, D. W., and Dicher, D. E. (2001) Forced infolding modulated by disulphide bonds in the Ig domains of a cell adhesion molecule. *Proc. Natl. Acad. Sci. U.S.A.* **98,** 1565–70.

Carrion-Vazquez, M., Oberhauser, A. F., Fowler, S. B., Marszalek, P. E., Broedel, S. E., Clarke, J., and Fernandez, J. M. (1999) Mechanical and chemical unfolding of a single protein: a comparison. *Proc. Natl. Acad. Sci. U.S.A.* **96,** 3694–9.

Cazorla, O., Freiburg, A., Helmes, M., Centner, T., McNabb, M., Wu, Y., Trombitás, K., Labeit, S., and Granzier, H. (2000) Differential expression of cardiac titin isoforms and modulation of cellular stiffness. *Circ. Res.* **86,** 59–67.

Cazorla, O., Wu, Y., Irving, T. C., and Granzier, H. (2001) Titin-based modulation of calcium sensitivity of active force in mouse skinned cardiac myocytes. *Circ. Res.* **88,** 1028–35.

Erickson, H. P. (1994) Reversible unfolding of fibronectin type III and immunoglobulin domains provides the structural basis for stretch and elasticity of titin and fibronectin. *Proc. Natl. Acad. Sci. U.S.A.* **91,** 10114–8.

Freiburg, A., and Gautel, M. (1996) A molecular map of the interactions between titin and myosin-binding protein C. Implications for sarcomeric assembly in familial hypertrophic cardiomyopathy. *Eur. J. Biochem.* **235,** 317–23.

Freiburg, A., Trombitás, K., Hell, W., Cazorla, O., Fougerousse, F., Centner, T., Kolmerer, B., Witt, C., Beckmann, J. S., Gregorio, C. C., Granzier, H., and Labeit, S. (2000) Series of exon-skipping events in the elastic spring region of titin as the structural basis for myofibrillar elastic diversity. *Circ. Res.* **86,** 1114–21.

Fujime, S., and Higuchi, H. (1993) An analysis of the dynamic light scattering spectra of worm-like chains: β-connectin from striated muscle. *Macromolecules* **26,** 6261–6.

Fukuda, N., Sasaki, D., Ishiwata, S., and Kurihara, S. (2001) Length dependence of tension generation in rat skinned cardiac muscle. Role of titin in the Frank-Starling mechanism of the heart. *Circulation* **104,** 1639–45.

Fulton, A. B. (1999) The elastic filament system in myogenesis. In: *Reviews of Physiology Biochemistry and Pharmacology,* vol. 138, D. Pette and D. Furst (eds.). Heidelberg, Germany: Springer, pp. 139–61.

Furst, D. O., Osborn, M., Nave, R., and Weber, K. (1988) The organization of titin filaments in the half-sarcomere revealed by monoclonal antibodies in immuno-electron microscopy – a map of 10 nonrepetative epitopes starting at the Z-line extends close to the M-line. *J. Cell Biol.* **106,** 1563–72.

Furukawa, T., Ono, Y., Tsuchiya, H., Katayama, Y., Bang, M.-L., Labeit, D., Labeit, S., Inagaki, N., and Gregorio, C. C. (2001) Specific interaction of the potassium channel β-subunit minK with the sarcomeric protein T-cap suggests a T-tubule-myofibril linking system. *J. Mol. Biol.* **313,** 775–84.

Gautel, M. (1996) The super-repeats of titin/connectin and their interactions: glimpses at sarcomeric assembly. In: *Advances in Biophysics*, vol. 33, S. Ebashi (ed.). Tokyo, Japan: Elsevier, pp. 27–37.

Gautel, M., and Goulding, D. (1996) A molecular map of titin/connectin elasticity reveals two different mechanisms acting in series. *FEBS Lett.* **385**, 11–4.

Gautel, M., Mues, A., and Young, P. (1999) Control of sarcomeric assembly: the flow of information on titin. In: *Reviews of Physiology Biochemistry and Pharmacology*, vol. 138, D. Pette and D. Furst (eds.). Heidelberg, Germany: Springer, pp. 97–137.

Goulding, D., Bullard, B., and Gautel, M. (1997) A survey of in situ sarcomere extension in mouse skeletal muscle. *J. Muscle Res. Cell Motil.* **18**, 465–72.

Granzier, H. L., and Irving, T. C. (1995) Passive tension in cardiac muscle: contribution of collagen, titin, microtubules, and intermediate filaments. *Biophys. J.* **68**, 1027–44.

Granzier, H., Helmes, M., and Trombitás, K. (1996) Nonuniform elasticity of titin in cardiac myocytes: A study using immunoelectron microscopy and cellular mechanics. *Biophys. J.* **70**, 430–42.

Greaser, M. (2001) Identification of new repeating motifs in titin. *Proteins* **43**, 145–9.

Greaser, M. L., Sebestyen, M. G., Fritz, J. D., and Wolff, J. A. (1996) cDNA sequence of rabbit cardiac titin/connectin. In: *Advances in Biophysics*, vol. 33, S. Ebashi (ed.). Tokyo, Japan: Elsevier, pp. 13–25.

Gregorio, C. C., and Antin, P. B. (2000) To the heart of myofibril assembly. *Trends Cell Biol.* **10**, 355–62.

Gutierrez-Cruz, G., van Heerden, A. H., and Wang, K. (2001) Modular motif, structural folds and affinity profiles of the PEVK segment of human fetal skeletal muscle titin. *J. Biol. Chem.* **276**, 7442–9.

Hein, S., Scholz, D., Fujitani, N., Rennollet, H., Brand, T., Friedl, A., and Schaper, J. (1994) Altered expression of titin and contractile proteins in failing human myocardium. *J. Mol. Cell. Cardiol.* **26**, 1291–306.

Helmes, M., Trombitás, K., and Granzier, H. (1996) Titin develops restoring force in rat cardiac myocytes. *Circ. Res.* **79**, 619–26.

Horowits, R., Kempner, E. S., Bisher, M. E., and Podolsky, R. J. (1986) A physiological role for titin and nebulin in skeletal muscle. *Nature* **323**, 160–4.

Houmeida, A., Holt, J., Tskhovrebova, L., and Trinick, J. (1995) Studies of the interaction between titin and myosin. *J. Cell Biol.* **131**, 1471–81.

Improta, S., Politou, A. S., and Pastore, A. (1996) Immunoglobulin-like modules from titin I-band: extensible components of muscle elasticity. *Structure* **4**, 323–37.

Jin, J.-P. (1995) Cloned rat cardiac titin class-I and class-II motives – expression, purification, characterization, and interaction with F-actin. *J. Biol. Chem.* **270**, 6908–16.

Kellermayer, M. S. Z., Smith, S. B., Granzier, H. L., and Bustamante, C. (1997) Folding-unfolding transitions in single titin molecules characterized with laser tweezers. *Science* **276**, 1112–6.

Labeit, S., Barlow, D. P., Gautel, M., Gibson, T., Holt, J., Hsieh, C. L., Francke, U., Leonard, K., Wardale, J., Whiting, A., and Trinick, J. (1990) A regular pattern of two types of 100-residue motif in the sequence of titin. *Nature* **345**, 273–6.

Labeit, S., Gautel, M., Lakey, A., and Trinick, J. (1992) Towards a molecular understanding of titin. *EMBO J.* **11**, 1711–6.

Labeit, S., and Kolmerer, B. (1995) Titins: giant proteins in charge of muscle ultrastructure and elasticity. *Science* **270**, 293–6.

Lakatta, E. G. (1987) Starling law of the heart is explained by an intimate interaction of muscle length and myofilament calcium activation. *J. Amer. College Cardiol.* **10,** 1157–64.

Li, H., Oberhauser, A. F., Fowler, S. B., Clarke, J., and Fernandez, J. M. (2000) Atomic force microscopy reveals the mechanical design of a modular protein. *Proc. Natl. Acad. Sci. U.S.A.* **97,** 6527–31.

Li, H., Linke, W. A., Oberhauser, A. F., Carrion-Vazquez, M., Kerkvliet, J. G., Lu, H., Marszalek, P. E., and Fernandez, J. M. (2002) Reverse engineering of the giant muscle protein titin. *Nature* **418,** 998–1002.

Linke, W. A., Ivemeyer, M., Olivieri, N., Kolmerer, B., Ruegg, J. C., and Labeit, S. (1996) Towards a molecular understanding of the elasticity of titin. *J. Mol. Biol.* **261,** 62–71.

Linke, W. A., Ivemeyer, M., Labeit, S., Hinssen, H., Ruegg, J. C., and Gautel, M. (1997) Actin-titin interaction in cardiac myofibrils: Probing a physiological role. *Biophys. J.* **73,** 905–19.

Linke, W. A., and Granzier, H. (1998) A spring tale: New facts on titin elasticity. *Biophys. J.* **75,** 2613–4.

Linke, W. A., Rudy, D. E., Centner, T., Gautel, M., Witt, C., Labeit, S., and Gregorio, C. (1999) I-band titin in cardiac muscle is a three-element molecular spring and is critical for maintaining thin filament structure. *J. Cell Biol.* **146,** 631–44.

Linke, W. A., Kulke, M., Li, H., Fujita-Becker, S., Neagoe, C., Manstein, D. J., Gautel, M., and Fernandez, J. M. (2002) PEVK domain of titin: An entropic spring with actin-binding properties. *J. Struct. Biol.* **137,** 194–205.

Liversage, A. D., Holmes, D., Knight, P. J., Tskhovrebova, L., and Trinick, J. (2001) Titin and the sarcomere symmetry paradox. *J. Mol. Biol.* **305,** 401–9.

Lu, H., Isralewitz, B., Krammer, A., Vogel, V., and Schulten, K. (1998) Unfolding of titin immunoglobulin domains by steered molecular dynamics simulation. *Biophys. J.* **75,** 662–71.

Lu, L., Greyson, C. R., Ursell, P. C., and Schwartz, G. G. (1999) Non-elastic deformation of myocardium in low-flow ischemia and reperfusion: ultrastructure-function relations. *J. Mol. Cell. Cardiol.* **31,** 1157–69.

Lu, H., and Schulten, K. (2000) The key event in force-induced unfolding of titin's immunoglobulin domains. *Biophys. J.* **79,** 51–65.

Luther, P., and Squire, J. (1978) Three-dimensional structure of the vertebrate muscle M-region. *J. Mol. Biol.* **125,** 313–24.

Ma, K., Kan, L. S., and Wang, K. (2001) Polyproline II helix is a key structural motif of the elastic PEVK segment of titin. *Biochemistry* **40,** 3427–38.

Marszalek, P. E., Lu, H., Li, H. B., Carrion-Vazquez, M., Oberhauser, A. F., Schulten, K., and Fernandez, J. (1999). Mechanical unfolding intermediates in titin modules. *Nature* **402,** 100–3.

Maruyama, K., Kimura, S., Yoshidomi, H., Sawada, H., and Kikuchi, M. (1984) Molecular size and shape of beta-connectin, an elastic protein of striated muscle. *J. Biochem.* (Tokyo) **95,** 1423–33.

Maw, M. C., and Rowe, A. J. (1980) Fraying of A-filaments into three subfilaments. *Nature* **286,** 412–4.

Mayans, O., Wuerges, J., Canela, S., Gautel, M., and Wilmanns, M. (2001) Structural evidence for a possible role of reversible disulphide bridge formation in the elasticity of the muscle protein titin. *Structure* **9,** 331–40.

Millman, B. M. (1998) The filament lattice of striated muscle. *Physiol. Rev.* **78,** 359–91.

Minajeva, A., Kulke, M., Fernandez, J. M., and Linke, W. A. (2001) Unfolding of titin domains explains the viscoelastic behavior of skeletal myofibrils. *Biophys. J.* **80,** 1442–51.

Morano, I., Hadicke, K., Grom, S., Koch, A., Schwinger, R. H., Bohm, M., Bartel, S., Erdmann, E., and Krause, E. G. (1994) Titin, myosin light chains and C-protein in the developing and failing human heart. *J. Mol. Cell. Cardiol.* **26,** 361–8.

Muhle-Goll, C., Pastore, A., and Nilges, M. (1998) The 3D structure of a type I module from titin: a prototype of intracellular fibronectin type III domains. *Structure* **6,** 1291–1302.

Muhle-Goll, C., Habeck, M., Cazorla, O., Nilges, M., Labeit, S., and Granzier, H. (2001) Structural and functional studies of titin's fn3 modules reveal conserved surface patterns and binding to myosin S1 – a possible role in the Frank-Starling mechanism of the heart. *J. Mol. Biol.* **313,** 431–47.

Nave, R., Furst, D. O., and Weber, K. (1989) Visualization of the polarity of isolated titin molecules – a single globular head on a long thin rod as the M-band anchoring domain. *J. Cell Biol.* **109,** 2177–87.

Obermann, W. M., Gautel, M., Weber, K., and Furst, D. O. (1997) Molecular structure of the sarcomeric M band: mapping of titin and myosin binding domains in myomesin and the identification of a potential regulatory phosphorylation site in myomesin. *EMBO J.* **16,** 211–20.

Pfuhl, M., and Pastore, A. (1995) Tertiary structure of an immunoglobulin-like domain from the giant muscle protein titin: a new member of the I-set. *Structure,* **3,** 391–401.

Rief, M., Gautel, M., Oesterhelt, F., Fernandez,, J. M., and Gaub, H. E. (1997) Reversible unfolding of individual titin immunoglobulin domains by AFM. *Science* **276,** 1109–12.

Rief, M., Gautel, M., Schemmel, A., and Gaub, H. E. (1998) The mechanical stability of immunoglobulin and fibronectin III domains in the muscle protein titin measured by atomic force microscopy. *Biophys. J.* **75,** 3008–14.

Schwinger, R. H. G., Bohm, M., Koch, A., Schmidt, U., Morano, I., Eissner, H.-J., Uberfuhr, P., Reichart, B., and Erdmann, E. (1994) The failing human heart is unable to use the Frank-Starling mechanism. *Circ. Res.* **74,** 959–69.

Scott, K. A., Steward, A., Fowler, S. B., and Clarke, J. (2002) Titin; a multidomain protein that behaves as the sum of its parts. *J. Mol. Biol.* **315,** 819–29.

Solaro, R. J., and Rarick, H. M. (1998) Troponin and tropomyosin: proteins that switch on and tune in the activity of cardiac myofilaments. *Circ. Res.* **83,** 471–80.

Soteriou, A., Clarke, A., Martin, S., and Trinick, J. (1993a) Titin folding energy and elasticity. *Proc. R. Soc. Lond. Series B* **254,** 83–6.

Soteriou, A., Gamage, M., and Trinick, J. (1993b) A survey of interactions made by the giant protein titin. *J. Cell Sci.* **104,** 119–23.

Squire, J., Cantino, M., Chew, M., Denny, R., Harford, J., Hudson, L., and Luther, P. (1998) Myosin rod-packing schemes in vertebrate muscle thick filaments. *J. Struct. Biol.* **122,** 128–38.

Stuyvers, B. D., Miura, M., Jin, J.-P., and ter Keurs, H. E. D. J. (1998) Ca^{2+}-dependence of diastolic properties of cardiac sarcomeres: involvement of titin. *Progr. Biophys. Mol. Biol.* **69,** 425–43.

Tatsumi, R., Maeda, K., Hattori, A., and Takahashi, K. (2001) Calcium binding to an elastic portion of connectin/titin filaments. *J. Muscle Res. Cell Motil.* **22,** 149–62.

Trinick, J., Knight, P., and Whiting, A. (1984) Purification and properties of native titin. *J. Mol. Biol.* **180,** 331–56.

Trinick, J. (1994) Titin and nebulin: protein rulers in muscle? *Trends Biochem. Sci.* **10,** 405–9.

Trinick, J. (1996) Cytoskeleton: Titin as a scaffold and spring. *Curr. Biol.* **6,** 258–62.

Trinick, J., and Tskhovrebova, L. (1999) Titin: a molecular control freak. *Trends Cell Biol.* **9,** 377–80.

Trinick, J. A. (1981) End-filaments: a new structural element of vertebrate skeletal muscle thick filaments. *J. Mol. Biol.* **151,** 309–14.

Trombitás, K., and Granzier, H. (1997) Actin removal from cardiac myocytes shows that near the Z-line titin attaches to actin while under tension. *Amer. J. Physiol.* **273,** C662–70.

Trombitás, K., Wu, Y., Labeit, D., Labeit, S., and Granzier, H. (2001) Cardiac titin isoforms are coexpressed in the half-sarcomere and extend independently. *Amer. J. Physiol.* **281,** H1793–9.

Tskhovrebova, L., and Trinick, J. (1997) Direct visualization of extensibility in isolated titin molecules. *J. Mol. Biol.* **265,** 100–6.

Tskhovrebova, L., Trinick, J., Sleep, J. A., and Simmons, R. (1997) Elasticity and unfolding of single molecules of the giant muscle protein titin. *Nature* **387,** 308–12.

Tskhovrebova, L., and Trinick, J. (2000) Extensibility in the titin molecule and its relation to muscle elasticity. In: *Elastic Filaments of the Cell,* H. Granzier and G. H. Pollack (eds.). New York: Kluwer Academic/Plenum Publishers, pp. 163–78.

Tskhovrebova, L., and Trinick, J. (2001) Flexibility and extensibility in the titin molecule: analysis of electron microscope data. *J. Mol. Biol.* **310,** 755–71.

Van der Ven, P. F., Ehler, E., Perriard, J.-C., and Furst, D. O. (1999) Thick filament assembly occurs after the formation of a cytoskeletal scaffold. *J. Muscle Res. Cell Motil.* **20,** 569–79.

Van der Ven, P. F., Bartsch, J. W., Gautel, M., Jockusch, H., and Furst, D. O. (2000) A functional knock-out of titin results in defective myofibril assembly. *J. Cell Sci.* **113,** 1405–14.

Wang, K., Ramirez-Mitchell, R., and Palter, D. (1984) Titin is an extraordinarily long, flexible, and slender myofibrillar protein. *Proc. Natl. Acad. Sci. U.S.A.* **81,** 3685–9.

Wang, K., McCarter, R., Wright, J., Beverly, J., and Ramirez-Mitchell, R. (1991) Regulation of skeletal muscle stiffness and elasticity by titin isoforms – a test of the segmental extension model of resting tension. *Proc. Natl. Acad. Sci. U.S.A.* **88,** 7101–5.

Wang, K. (1996) Titin/connectin and nebulin: Giant protein rulers of muscle structure and function. In: *Advances in Biophysics,* vol. 33, S. Ebashi (ed.). Tokyo, Japan: Elsevier, pp. 123–34.

Whiting, A., Wardale, J., and Trinick, J. (1989) Does titin regulate the length of muscle thick filaments? *J. Mol. Biol.* **205,** 263–8.

Yamasaki, R., Berri, M., Wu, Y., Trombitás, K., McNabb, M., Kellermayer, M. S. Z., Witt, C., Labeit, D., Labeit, S., Greaser, M., and Granzier, H. (2001) Titin-actin interaction in mouse myocardium: Passive tension modulation and its regulation by calcium/S100A1. *Biophys. J.* **81,** 2297–313.

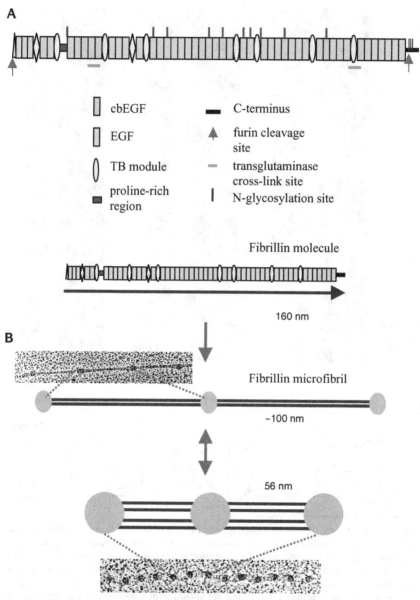

Figure 5.3: Schematic representation of the fibrillin-1 molecule and its assembly into extensible microfibrils. (A) Domain structure of fibrillin, highlighting how the contiguous arrays of calcium-binding epidermal growth factor (cbEGF)-like domains are interspersed with TB modules, and the locations of the proline-rich region, N-glycosylation consensus sequences, and putative transglutaminase cross-link sites. EGF, epidermal growth factor. (B) Diagram highlighting how fibrillin-1 molecules (~160 nm) assemble to form beaded microfibrils with variable periodicity (~56–100 nm).

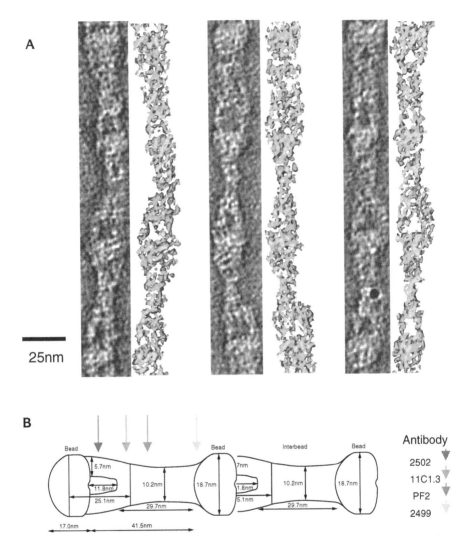

Figure 5.5: Automated electron tomography images of negatively stained untensioned bovine zonular microfibrils. (A) Three Z-slices shown in black and white, compared with the same region of microfibril three-dimensional rendered (green) using AVS Express (Advanced Visual Systems, Inc.). (B) Schematic diagram of two microfibril repeating units with the mean measurements taken from a representative data set. The binding positions of four anti–fibrillin-1 antibodies are shown with coloured arrows. Reproduced with permission from 'The Supramolecular Organization of Fibrillin-Rich Microfibrils', by C. Baldock et al., 2001, *J. Cell Biol.* **152,** 1045–56. © 2001 The Rockefeller University Press.

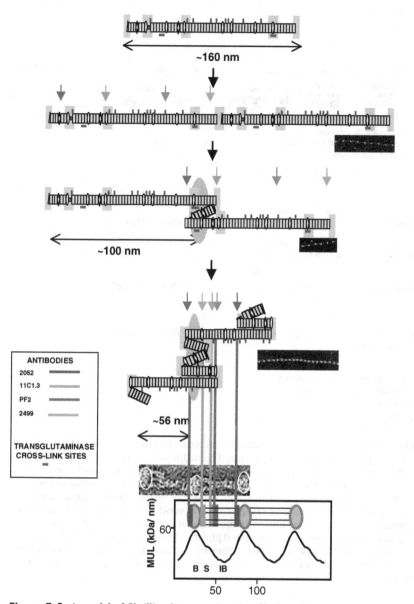

Figure 5.6: A model of fibrillin alignment in microfibrils. Schematic diagram depicting a possible folding arrangement of fibrillin molecules in a beaded microfibril. Two N- and C-terminally processed molecules associate head-to-tail to give ~160 nm periodicity. Subsequent molecular folding events could generate ~100 nm periodicity, then ~56 nm periodicity. The model is supported by automated electron tomography, STEM data, antibody localisations, and microfibril extension studies (Baldock et al., 2001). While the predicted fold sites are shown (orange boxes), it should be noted that the precise packing arrangement of folded segments contributing to the bead remains unresolved. Inset: STEM images of microfibrils with periodicities corresponding to those predicted. Observed antibody binding sites are overlaid on the ~56 nm periodic folding arrangement. The axial mass distribution of ~56 nm microfibrils (shown at the bottom) correlates well with the corresponding predicted molecular folding. B, bead; IB, interbead; MUL, mass per unit length; S, shoulder region.

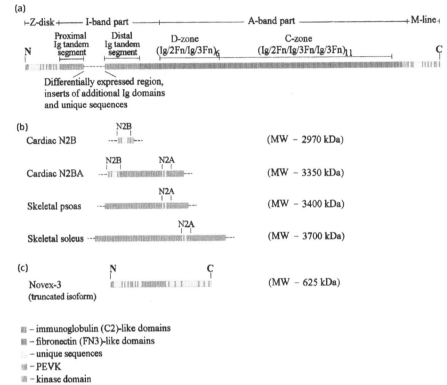

(a)

⊢Z-disk⊣⎯⎯ I-band part ⎯⎯⎯⊢⎯⎯⎯⎯⎯⎯⎯ A-band part ⎯⎯⎯⎯⎯⎯⊢M-line⊣

	Proximal Ig tandem segment	Distal Ig tandem segment	D-zone $(Ig/2Fn/Ig/3Fn)_6$	C-zone $(Ig/2Fn/Ig/3Fn/Ig/3Fn)_{11}$	
N					C

Differentially expressed region, inserts of additional Ig domains and unique sequences

(b)

Cardiac N2B — N2B — (MW – 2970 kDa)

Cardiac N2BA — N2B — N2A — (MW – 3350 kDa)

Skeletal psoas — N2A — (MW – 3400 kDa)

Skeletal soleus — N2A — (MW – 3700 kDa)

(c)

Novex-3 (truncated isoform) — N — C — (MW – 625 kDa)

▦ – immunoglobulin (C2)-like domains
▦ – fibronectin (FN3)-like domains
▦ – unique sequences
▦ – PEVK
▦ – kinase domain

Figure 12.1: Domain structure of the titin molecule and isoform-related differences in the I-band part. (a) Conserved part of titin. This includes the A-band segment and the C- and N-terminal portions of the I-band segment. The central region of the I-band, as well as the Z- and M-line regions, are isoform-specific. (b) Examples of differentially expressed regions of the I-band part of titin molecule. These include the unique sequences and the variable part of the proximal poly-Ig segment. (c) Novex-3, truncated titin isoform. The N-terminal part is similar to the full length molecule and contains the Z-disk region (except Z-repeats) and the constitutively expressed proximal Ig tandem segment. The C-terminal part is isoform specific, and is formed by a giant unique sequence with a few insertions of Ig domains. In muscle, novex-3 is located within and near the Z-disk region. MW, molecular weight.

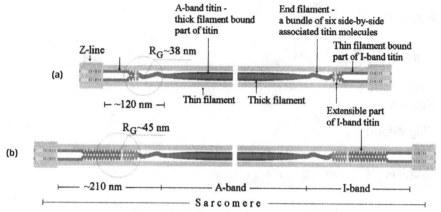

Figure 12.4: Effect of confinement on the configuration of the titin I-band part within (a) psoas, and (b) soleus muscles at slack sarcomere lengths. Scheme is drawn at the scale proportional to the known dimensions and relative positions of major contractile filaments. The I-band lengths in slack sarcomeres are about 120 and 210 nm in psoas and soleus muscles, respectively, and contour lengths of the I-band parts of corresponding titin isoforms are 320 and 440 nm (excluding actin-bound portions). The isolated I-band titin segment in solution will occupy a volume of approximately spherical shape with a radius of 38–45 nm, as can be judged from estimates of radius of gyration. Within the sarcomere the average space per single I-band segment is restricted by a volume of elongated shape with length equal to the I-band and radius 8–10 nm. The latter value is based on a thick filament spacing of 43 nm (Millman, 1998) and the estimate of six titin molecules per half thick filament (Cazorla et al., 2000; Liversage et al., 2001). Thus, there is a discrepancy between the space required and that actually available. This implies that titin strands in the I-band at slack sarcomere length either are arranged in a more ordered way than can be predicted from solution observations or are highly interwired.

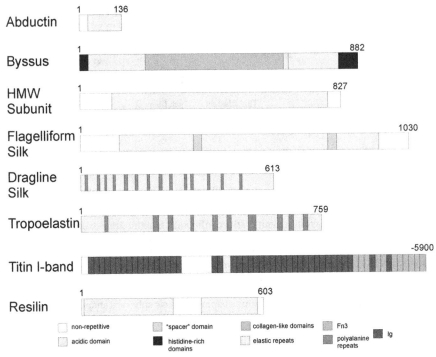

Figure 17.1: Domain structures of the elastomeric proteins: abductin (Cao et al., 1997), byssus (Coyne et al., 1997), HMW subunit (Shewry et al., 1992), flagelliform silk (Hayashi and Lewis, 1998), dragline silk (Guerrette et al., 1996), tropoelastin (Indik et al., 1987), titin I-band (Labeit and Kolmerer, 1995), and resilin (Ardell and Andersen, 2001). I$_g$, immunoglobulin. Redrawn with permission from 'Elastomeric Proteins: Biological Roles, Structures and Mechanisms', by A. S. Tatham and P. R. Shewry, 2000, *Trends Biol. Sci.*, **299**, 567–71. © 2000.

THIRTEEN

Structure and Function of Resilin

Svend Olav Andersen

INTRODUCTION

Biological structures with long-range elasticity are found in most animals, where they serve a number of functions, such as storing kinetic energy, absorbing shocks and vibrations, and acting as antagonists for muscles. Different groups of animals have developed different materials specialised for these purposes; insects and other arthropods have used a rubber-like, exoskeletal protein, resilin, in many places where long-range elasticity is needed.

The protein was discovered by Weis-Fogh during a study of the flight system in locusts, when he realised that energy-saving mechanisms had to be present to allow storage of kinetic energy at the end of wing strokes. A detailed analysis revealed that several elastic elements participate in such energy storage: (a) the thoracic flight muscles themselves possess a passive elastic element, which in the locust accounts for about one-third of the total elasticity of the thorax; (b) the solid, sclerotized cuticle of the thorax, which can store significant amounts of energy, although it cannot be subjected to deformations larger than 2%; and (c) some small, cuticular patches, which possess long-range elasticity, allowing them to suffer considerable deformations and to return rapidly to their equilibrium shape when the deforming forces are released; they account for about one-fourth of the total elasticity of the thorax (Weis-Fogh, 1959, 1961a). Like most other cuticular regions, these patches contain both protein and chitin, and the elastic properties reside in the protein and not in the chitin fraction. Weis-Fogh named the protein resilin, from the latin *resilire*, to spring back, and he selected three samples for detailed characterisation of the elastic properties of resilin: two from the flight system of the desert locust (*Schistocerca gregaria*) and one from dragonflies (*Aeshna grandis* and other *Aeshna* species) (Weis-Fogh, 1960).

The dragonfly sample is a small, hyaline, sausage-shaped region of the tendon connecting the pleuro-subalar muscle to the dorsal thoracic cuticle; both ends of the tendon consist of tough and non-stretchable cuticle, which makes it possible to mount the elastic region of the tendon under the microscope for stretching

259

experiments. In the larger species of dragonflies, the resilin-containing region is about 0.7 mm long, 0.15 mm wide, and contains 5–7 μg of resilin. It consists of almost pure resilin and contains no chitin; in this respect, it differs from most other resilin-containing samples and other types of cuticle.

The two locust samples are (a) the wing-hinge connecting the forewing to the sclerotized cuticle of the solid thoracic wall and (b) the prealar arms which form an elastic connection between the thorax wall and the anterior edge of the movable, dorsal plate between the forewings (the mesotergum). Both samples are reversibly deformed during wing strokes, they are irregularly shaped (making them non-ideal for mechanical measurements), and they contain considerably more resilin than the dragonfly tendon: a single prealar arm contains about 50 μg resilin and 15 μg chitin, and an elastic hinge from the forewing contains about 100 μg resilin and 20 μg chitin. Most of the chemical characterisations of resilin have been performed on locust material, as reasonable amounts of material can more readily be obtained from locusts than from dragonflies.

Nearly all types of arthropod cuticle, both soft and solid, contain non-stretchable chitin filaments embedded in a proteinaceous matrix, and the presence of these filaments has a pronounced influence on the mechanical properties of cuticle. Typically, the filaments run parallel to the cuticular surface, sometimes all in the same direction, but most often in gradually changing directions, resulting in a helicoidal filament pattern; in resilin-containing cuticular regions, the chitin filaments tend to follow the helicoidal pattern, making the locust ligaments difficult to stretch, but easily bent, twisted, and compressed, deformations to which they are exposed during wing movements during flight.

PHYSICAL PROPERTIES OF RESILIN

The dragonfly tendon can be stretched to 3–4 times its resting length before breaking, and it snaps immediately back to its resting position without lasting deformations when the tensile force is released. The quantitative stress–strain relationships at different temperatures and the dependence of birefringence upon tension were reported for dragonfly resilin in 1961 (Weis-Fogh, 1961b, 1961c), and to the best of my knowledge, no such measurements on resilin have been repeated since, although more sophisticated equipment is now available. Resilin in dragonfly elastic tendons was found to have an elastic modulus of 6–7 kg/cm^2, and the elastic modulus of resilin in locust ligaments was determined to be 9 kg/cm^2 (Jensen and Weis-Fogh, 1962); both materials behaved mechanically as true rubbers with very low loss factors. In the electron microscope and by X-ray diffraction studies, dragonfly resilin appears to be unstructured (Elliot et al., 1965), and unstrained resilin is completely isotropic. Birefringence is observed when the sample is stretched, and the relationship between birefringence and degree of stretching is in agreement with the conclusion that resilin behaves as a nearly perfect rubber (Weis-Fogh, 1961c).

According to the theory for rubber-like elasticity, a perfect rubber consists of long, randomly coiled chains, which continuously change form due to thermal agitation, and interactions between chain segments are negligible, apart from a few stable interconnections, cross-links, which prevent the chains from sliding past each other during deformations. Stretching a rubber increases the average distance between interchain links and introduces anisotropy in the material due to a partial alignment of the chains. The increased order of the system corresponds to a decrease in entropy, and when the tensile force is released thermal agitations bring the folding of the chains back to the random distribution in their relaxed state. All the energy used for deformation of a perfect rubber is used for entropy decrease, while no energy is used for changing interactions between chain segments or between chains and solvent.

According to the measurements, resilin comes close to being a perfect rubber when swollen in water (Weis-Fogh, 1961c; Jensen and Weis-Fogh, 1962). I still find the agreement between theory and measurements amazing, since peptide chains will unavoidably interact with each other via electrostatic interactions, hydrogen bonds, and hydrophobic interactions, and they will form hydrogen bonds to the surrounding water molecules. During deformation of resilin, many of these interactions will be interrupted, and true rubber-like behaviour must mean that the amounts of interactions disappearing must energetically balance new interactions formed.

RESILIN CROSS-LINKS

Pieces of resilin from either locust ligaments or dragonfly tendon were shown to be completely insoluble in all solvents which do not attack and degrade peptide chains, but they swell readily in solvents which are effective for dissolving proteins (Weis-Fogh, 1960), such as concentrated solutions of urea, guanidinium chloride, lithium thiocyanate, and sodium dodecyl sulphate; addition of reducing agents, such as mercaptoethanol or dithiothreitol, has no effects on resilin's insolubility. The insolubility combined with the rubber-like elasticity of resilin led Weis-Fogh to suggest that covalent cross-links of an unusual structure had to be present in resilin, and a search for such cross-links soon revealed that resilin contains two strongly fluorescent amino acids with a potential for functioning as cross-links, one of them contains two amino groups and two carboxylic groups and the other contains three of each of these groups (Andersen, 1963). The two amino acids were later (Andersen, 1964) identified as di- and trityrosine (bi- and tertyrosine). Dityrosine can link two peptide chains together, and trityrosine can link three chains together, but it is likely that some of the cross-links involve tyrosine residues from the same peptide chain, resulting in loop formation.

According to quantitative determination of the amounts of the two amino acids, locust resilin contains 0.77 and 0.39 residue percent of dityrosine and trityrosine, respectively, and this corresponds to an average molecular weight of peptide chain segments between cross-links of about 3200 Da, if both amino acids are involved in cross-link formation between separate peptide chains (Andersen, 1966). From the mechanical measurements, Weis-Fogh estimated an average chain weight between cross-links of about 3400 Da in locust resilin and about 5100 Da in dragonfly resilin (Weis-Fogh, 1961c; Jensen and Weis-Fogh, 1962). To the best of my knowledge, the contents of cross-linking amino acids in dragonfly resilin and in other types of resilin have not yet been determined, but the degree of cross-linking will presumably depend upon the functional needs of the various resilin-containing samples.

Di- and trityrosine were originally obtained by peroxidarive oxidation of free tyrosine, and structures where the aromatic rings are linked via *ortho—ortho* bonds were suggested (Gross and Sizer, 1959). The ultraviolet absorbance, fluorescent properties, and complex formation with borate ions of the two fluorescent amino acids obtained from locust resilin were compared with those of authentic *ortho*-bisphenol (2, 2′-dihydroxy-diphenyl), and the results supported the suggestion that the tyrosine residues in di- and trityrosine are linked together via carbon—carbon bonds in *ortho*-position to the phenolic groups (Andersen, 1966). ^{1}H and ^{13}C NMR measurements on purified peroxidation products of tyrosine and tyrosine derivatives have since confirmed the structures (Ushijima et al., 1984; Briza et al., 1986; Huggins et al., 1993; Jacob et al., 1996).

DEFINING RESILIN

Resilin has been reported from a large number of sources, although an unambiguous definition of the term resilin has never been proposed. A preliminary working definition based on the properties of the three test pieces from locusts and dragonflies was suggested by Andersen and Weis-Fogh (1964), and it is still the best definition available. Accordingly, the more a material resembles the locust and dragonfly materials, the more resilin-like it will be, and it is arbitrary, where the borderline between resilin and other structural materials shall be placed. The main criteria used for distinguishing resilin from other types of material are: pure, unstrained resilin is an optically and mechanically isotropic protein, which occurs extracellularly in arthropods, often together with chitin. It is insoluble in all solvents which do not break peptide bonds, and when swollen in water it becomes rubbery, it exhibits long-range deformability, and complete elastic recovery is observed, when the deforming force is released. Resilin is transparent and devoid of colour, and it shows no visible structure in the electron microscope. The histological stains, methylene blue and toluidine blue, stain pure samples of resilin with a sapphire blue colour, and no metachromasia is observed. In ultraviolet light, resilin fluoresces with a bluish colour with maximum emission at 420 nm. At pH values below 7, the excitation maximum is at 285 nm and at higher pH values the maximum shifts to 315 nm; the intensity of fluorescence is stronger in alkaline than in acidic media. Characteristic features of resilin amino acid composition are: a glycine content of about 40%, a proline content of between 7 and 8%, an absence of sulphur-containing amino acids and tryptophan, and presence of the two fluorescent amino acids, di- and trityrosine. Some resilins have amino acid compositions which differ slightly from locust and dragonfly resilins; thus, resilin from abdominal springs in the rhinoceros beetle, *Oryctes rhinoceros*, was found to contain more than 20% of aspartic acid plus asparagine, and its glycine content was only 33% (Andersen, 1971). The mechanical properties of the abdominal springs have not yet been determined, and we do not know whether the high content of aspartic acid has any influence on the elastic properties of the material.

One often encounters cuticular regions which conform to some, but not all of the aforementioned criteria, and it can be a problem to decide whether such regions contain resilin or not. Andersen and Weis-Fogh (1964) introduced tentatively the term 'transitional cuticle' for cuticular regions, which appeared to contain mixtures of resilin and other types of proteins. These regions can be subjected to significant deformations, and recover their original shape, when the deforming force is released, but the recovery may be rather slow; the regions will often stain atypically with histochemical stains, they may show fluorescence and give di- and trityrosine upon acid hydrolysis, but the colour of fluorescence may differ from the typical blue fluorescence of resilin, indicating interference of other materials. Often, there are no well-defined borders between regions of resilin-containing cuticle and the surrounding solid cuticle; it appears that the one type of cuticle changes gradually into the other, suggesting that a zone of transitional cuticle is located between a pure resilin-containing region and the surrounding solid cuticle. In contrast, the borders between the resilin-containing thoracic ligaments in locusts and the solid cuticle are very sharp; it looks as if one group of cells has produced solid cuticle and the neighbouring row of cells has produced resilin, and apparently the materials in the two types of cuticle do not mix. The interface between the two materials is not a region of weakness; if the connection between the wing and the thorax is broken, the fracture occurs across the resilin pad and not at the interface.

OCCURRENCE OF RESILIN

Resilin is widespread in arthropods, but can be difficult to obtain in amounts sufficient for chemical characterisation, as all resilin-containing cuticular regions are minute. Illumination of the interior side of the thorax cuticle of a locust with ultraviolet light demonstrated the presence of many small blue-fluorescent regions, of which the forewing ligaments and the prealar arms were the most prominent, but all the fluorescent regions are likely to contain resilin and contribute to the elasticity of the thoracic structure (Andersen and Weis-Fogh, 1964). By the same method, resilin-like fluorescence was demonstrated in the clypeo-labral springs, the tarsal and areolar pads, and in some regions of the abdominal tergites in locusts (Andersen and Weis-Fogh, 1964), and di- and trityrosine have been obtained from these regions. Fluorescence in ultraviolet light at different pH values has been used to localise the presence of resilin in various regions in the legs of the cockroach, *Periplaneta americana* (Neff et al., 2000).

In many cases, it has been possible to suggest functional roles for the elasticity of the resilin-containing regions. In the flight system of insects, resilin is important as an energy-storing system, contributing to storage of the kinetic energy of the moving wings as elastic energy during the wing strokes. The elastic forces of the ligaments are maximal when the wings are in their extreme upper position, and help the wings to accelerate rapidly during the early phase of the wing downstroke, when the contractile forces of the flight muscles are relatively weak.

Resilin is also present in cuticular structures involved in the jumping of fleas (Neville and Rothschild, 1967) and click beetles (Sannasi, 1969). In both cases, it is essential that the force is delivered by a very rapid movement of some part of the body against a solid substrate. For the rabbit flea, *Spillopsyllus cuniculi*, to jump to a height of 3.5 cm, the necessary force has to be applied for a duration of less than a millisecond, which is not compatible with direct muscle action (Bennet-Clark and Lucey, 1967). By deformation of an elastic resilin-containing ligament, sufficient energy can be stored by prolonged muscular action and released rapidly when needed.

In some cuticular regions, resilin counteracts movements caused by muscles. Resilin is thus involved in the opening mechanism of spiracle 2 in locusts. This spiracle possesses a single closer muscle, but no opener muscle, and resilin elasticity helps in the opening of the spiracle during flight (Miller, 1960).

The extension of the proboscis of butterflies is due to direct muscle contraction combined with an increase in hydrostatic pressure, while the recoiling of the proboscis is caused by decreased pressure in combination with the elasticity of proboscis wall, which contains resilin (Hepburn, 1971).

Haas et al. (2000a) have reported the presence of resilin within some flexible joints of hind wing veins, as well as in some areas of wing membranes in beetles (*Pachnoda marginata* and *Coccinella septempunctata*). When beetles are at rest, the thin hind wings are folded and hidden beneath the thick and solid elytra, and the elasticity of wing resilin assists in the correct folding of the hind wings; furthermore, resilin in wing folds may prevent material damage during repeated folding and unfolding of the wings. Resilin also provides the wings with elasticity to be deformable by aerodynamic forces, and this may result in elastic energy storage in the wing. Elastic, resilin-containing vein-joints have also been descibed from hind wings of earwigs (*Forficula auricularia*) (Haas et al., 2000b) and from damselfly wings (*Enallagma cyathigerum*) (Gorb, 1999).

The presence of resilin has been described in joints connecting segments in legs of the cockroach, *Periplaneta americana* (Neff et al., 2000), and it was suggested that its function is to act as an elastic antagonist to the actions of leg muscles.

Resilin has been reported as part of the sound-producing organs in some insect species. Cicada produce sounds by rapid buckling of a pair of domed tymbals on the first abdominal segment; each tymbal consisting of small regions of sclerotised cuticle separated by and surrounded by strips of resilin. Muscle contraction causes the sclerotised convex tymbal ribs to buckle and become concave, resulting in a pulse of sound. The muscle involved has no muscular antagonist, and re-extension of the muscle is brought about by the elastic strain energy in the buckled tymbal (Young and Bennet-Clark, 1995; Bennet-Clark and Daws, 1999). Resilin-containing sound-producing tymbals have also been described in the moth subfamily Chloephorinae, where males of the species *Pseudoips prasinana* and *Bena bicolorana* possess such tymbals, and it was suggested that the sound clicks they produce are functional in a sexual context (Skals and Surlykke, 1999).

Resilin has been reported in some cuticular structures which are stretchable, but possess no long-range elasticity, such as the abdominal wall of physogastric termite queens (Varman, 1980) and of some ants, where the volume of the abdomen in workers can be increased drastically for the storage of nectar (Varman, 1981). Resilin has also been reported (Sannasi, 1970) in the transparent cuticle covering the compound eyes in the firefly *Photinus pyralis*, a cuticle which is probably not exposed to much deformation. The identification was based upon staining reactions and the presence of the cross-linking amino acids, di- and trityrosine. I have in vain searched for di- and trityrosine in the cornea of the compound eyes in locusts (Andersen, 1971), so resilin is apparently not an obligatory component of the cornea of insects.

Resilin-like material has been reported in cuticular regions of several groups of non-insect arthropods, such as the crayfish *Astacus fluviatilis* (and probably also other crayfishes), which possesses a leg hinge flexed by a flexor muscle, but no extensor muscles are present to return the leg to the stretched position. This movement is caused by the intrinsic elasticity of the leg hinge, due to the presence of a resilin-like material (Andersen and Weis-Fogh, 1964). Resilin-like materials have also been reported from the copepod *Penella elegans*

(Kannupandi, 1976), the scorpion *Palamnaeus swammerdami* (Govindarajan and Rajulu, 1974), and the centipede *Scolopendra morsitans* (Rajulu, 1971). So far, resilin appears to be restricted to arthropods.

Resilin was the first protein in which the natural occurrence of di- and trityrosine was established, and the presence of these amino acids has occasionally been used as sole indicator for the presence of resilin, but they have also been reported from several unrelated biological systems, such as some silks and keratins (Raven et al., 1971), bovine dental enamel (Booij and ten Bosch, 1982), the fertilisation membrane in sea urchins (Foerder and Shapiro, 1977; Hall, 1978), the wall of yeast ascospores (Briza et al., 1986), the cuticle of nematodes (Fujimoto, 1975; Lassandro et al., 1994), and the egg chorion of fruit flies (Petri et al., 1976), mosquitoes (Li et al., 1996), mealworms (Kawasaki et al., 1975), and dragonflies (Kawasaki et al., 1974). Di- and trityrosine, together with mono- and dibromodityrosine and monobromotrityrosine, have been isolated from the hard, calcified exoskeleton of the crab, *Cancer pagurus* (Welinder et al., 1976). In many of these cases, the role of di- and trityrosine is presumably to add extra strength to extracellular materials by cross-linking their proteins, but the proteins need not have long-range elasticity and they appear to be unrelated to resilin. When free radicals are produced due to exposure of tissues to oxidative stress, tyrosine residues in various proteins in the mammalian body can be transformed into di- and trityrosine (Huggins et al., 1993), indicating that the occurrence of these amino acids need not always have a physiological function.

In summary: resilin appears to be present in arthropods in places where long-range elasticity is needed for either energy storage, for damping of vibrations, for restoring structures to their resting position without involving muscular movements, and for allowing frequently repeated rapid deformations without risking fatigue of the material.

A RESILIN GENE

Resilin is readily degraded by proteolytic enzymes (Weis-Fogh, 1960), and in 1993, Lombardi and Kaplan published amino acid sequences for a number of peptides obtained by tryptic digestion of cockroach (*Periplaneta americana*) resilin, and later Ardell and Andersen (2001) obtained sequences for tryptic peptides from locust (*Schistocerca gregaria*) resilin. The cockroach and locust resilin peptides resembled each other, but only little information on the structure of resilin could be deduced directly from the peptide sequences. Searching the complete genome of the fruit fly, *Drosophila melanogaster*, when it became available in 2000, revealed that fruit flies possess two genes coding for proteins, which on tryptic digestion would give peptides homologous to three of the peptides from locust resilin and three of the peptides from cockroach resilin.

MFKLLGLTLLMAMVVLGRPEPPVNS<u>YLPPSDSYGAPGQSG</u>P<u>GGRPSDSYG</u>　50

<u>APGGGNGGRPSDSYGAPGQGQGQGQGQGGY</u>AGKPSDTYGAPGGGNGNGGR　100

<u>PSSSYGAPGGGNGGRPSDTYGAPGGGNGGRPSDTYGAPGGGGNGNGGRPS</u>　150

<u>SSYGAPGQGQGNGNGGRSSSSYGAPGGGNGGRPSDTYGAPGGGNGGRPSD</u>　200

<u>TYGAPGGGNNGGRPSSSYGAPGGGNGGRPSDTYGAPGGGNGNGSGGRPSS</u>　250

<u>SYGAPGQGQGGFGGRPSDSYGAPGQNQKPSDSYGAPGSGNGNGGRPSSSY</u>　300

<u>GAPGSGPGGRPSDSYGPPASGSGAGGAGGSGPGGADYDNDE</u><u>PAKYEFNYQ</u>　350

<u>VEDAPSGLSFGHSEMRDGDFTTGQYNVLLPDGRKQIVEYEADQQGYRPQI</u>　400

<u>RY</u>EGDANDGSG<u>PSGPGGPGGQNLGA</u>D<u>GYSSGRPGNGNGNGNGGYSGGRPG</u>　450

<u>GQDLGPSGYSGGRPGGQDLGAGGYSNGKPGGQDLGPGGYSGGRPGGQDLG</u>　500

RD<u>GYSGGRPGGQDLGASGYSNGRPGGNGNGGS</u>DGGRVIIGGRVIGGQDGG　550

DQ<u>GYSGGRPGGQDLGRDGYSSGRPGGRPGGN</u>GQDSQDGQ<u>GYSSGRPGQGG</u>　600

<u>RN</u>GFGPGGQNGDNDGSGYRY　620

Figure 13.1: Predicted amino acid sequence of *Drosophila melanogaster* gene product CG15920, a putative fruit fly resilin. The signal peptide (residues 1–17) is indicated by italics, the conserved central region (residues 342–403) is indicated by double underlining, and the repeat sequences in the N- and C-terminal regions are indicated by single underlining. The numbers on the right refer to amino acid residue number in the preprotein.

A single locust peptide and several cockroach peptides did not have significant similarity to tryptic peptides from any *Drosophila* proteins (Ardell and Andersen, 2001).

The sequence of the predicted translation product of one of the genes with similarity to resilin peptides (gene product CG15920) is shown in Figure 13.1. The seventeen amino acid residues constituting the N-terminus of the gene product corresponds to a signal peptide (Nielsen et al., 1997), indicating that the protein is secreted into the extracellular space. The calculated amino acid composition of the gene product minus the signal peptide corresponds closely to the amino acid analyses published for locust and dragonfly resilins, and it was concluded that gene product CG15920 is likely to be a *Drosophila* resilin precursor (Ardell and Andersen, 2001).

The protein is 620 amino acid residues long, including the signal peptide of 17 residues. The relative mass of the protein after removal of the signal peptide is 56771, and its calculated isoelectric point is 5.0, which is close to the pH value at which dragonfly resilin has minimum swelling in water (Andersen and Weis-Fogh, 1964). The protein consists of three domains: a central 62-residue domain (residues 342–403 in the unprocessed gene product) which contains the sequences homologous to the locust and cockroach tryptic resilin peptides, a 324 residue long N-terminal region (residues 18–341), and a 217 residue long C-terminal region (residues 404–620). The N-terminal region is dominated by 18 slightly varying copies of a 15-residue motif: GGRPSDSYGAPGGGN while the C-terminal region is dominated by 11 variants of a 13-residue motif: GYSG GRPGGQDLG. Tripartite structures have repeatedly been observed in structural proteins from soft and solid types of arthropod cuticle, as well as in proteins from insect egg chorions, and the central 62-residue domain of gene product CG15920 shows significant sequence similarity to the central regions in cuticular proteins, which contain variants of the Rebers-Riddiford consensus sequence (Rebers and Riddiford, 1988), such as the so-called RR-2 consensus sequence in proteins from solid cuticles and the RR-1 consensus sequence in proteins from soft cuticles (Andersen, 1998). No pronounced similarity has been found between the terminal regions in fruit fly resilin and sequence regions in other structural proteins.

The deduced amino acid sequence of the other fruit fly protein with similarity to resilin peptides (gene product CG9036) is shown in Figure 13.2. The protein does not contain a typical signal peptide sequence, and its amino acid composition is quite different from that of resilin. It has a tripartite structure, and it contains a 62-residue sequence region with homology to the corresponding region in gene product CG15920, but from the lack of a signal peptide it was concluded that this protein may not be a resilin protein (Ardell and Andersen, 2001), although its high content of proline and the occurrence of regions dominated by glycine and asparagine residues could indicate that the protein has a structural function.

```
MTAVRDGVFNQGLSTQNENCNVNATLNFHSQAFTSIALLVCLAAWTHAEP    50

PVPQNQYLPPNQSPQAPSNNYLPPTQGYQSPSSNYLPPQRAGGNGGAPSN   100

SYGAPIAPPQGQYGAPALTGAIFKGGNGNGNGGYGGGNGNGNGYGQRDEE   150

QYGPAKYEFKYDVQDYESGNDFGHMESRDGDLAVGRYYVLLPDGRKQIVE   200

YEADQNGYRPTIRYEQVGNGNGNGNGNGRNGGGYDSNAQQGKFNGY      246
```

Figure 13.2: Predicted amino acid sequence of *Drosophila melanogaster* gene product CG9036. The conserved central region (residues 154–215) is indicated by double underlining, and the sequence regions rich in glycine and asparagine are indicated by single underlining. The numbers on the right refer to amino acid residue numbers in the protein.

THE MOLECULAR BASIS FOR RESILIN ELASTICITY

Convincing evidence has been published supporting the suggestion that the RR-2 sequence in proteins from solid cuticle and the corresponding RR-1 sequence from soft cuticle proteins are involved in binding the cuticular proteins to the chitin filament system (Iconomidou et al., 1999 , 2001; Rebers and Willis, 2001), and the N- and C-terminal regions of these proteins, as well as the cuticular proteins devoid of RR sequences, have been suggested to occupy the space between the filaments (Andersen, 1998). The protein chains of resilin are probably organised in a similar way, with the 62-residue central region anchoring the chains to the chitin filaments, while the long terminal regions occupy the interfilament space and constitute the major part of the cuticular matrix. The repeated 15-residue motif in the N-terminal region is rich in glycine with 2 prolines and a single tyrosine, while the 13-residue motif in the C-terminal region is also rich in glycine with one tyracine, but contains only one proline residue. The tyrosines are probably involved in forming di- and trityrosine cross-links, which are characteristic for resilin, and the glycine and proline residues may influence the folding and flexibility of the peptide chains. Proteins possessing long-range elasticity are often dominated by shorter or longer repeat sequences rich in glycine and proline (Tatham and Shewry, 2000), and presence of glycines and prolines tend to introduce beta-turns in peptide chains. Elastin from vertebrates contains several copies of the repeat sequence VPGVG (Sandberg et al., 1985), and the highly stretchable material secreted from the flagelliform silk glands in orb web spiders contains the repeat GPGGX (Hayashi and Lewis, 1998). The elastin molecule has been predicted to fold into easily deformable beta-spirals (Urry, 1988; Urry et al., 1995), and it is likely that the N- and C-terminal repeats in resilin also have a tendency to fold into spiral structures. Since the repeats in resilin vary somewhat in sequence, its spiral turns will be slightly irregular, which will diminish the tendency to form local crystalline regions in an easily deformed material. As long as elastin is kept above its transition temperature, its beta-spirals are stabilised by hydrophobic interactions between the turns of the spirals, while below the transition temperature the hydrophobic interactions are weakened due to pronounced hydration of the hydrophobic amino acid residues, and the material becomes softer. The peptide chain in resilin is much less hydrophobic than that of elastin, and hydrophobic interactions will be of minor importance in resilin. Swollen resilin contains 50–60% water (Weis-Fogh, 1960), so the peptide chains will be surrounded by water, and no transition temperature has been observed for the elasticity of resilin. Presumably, the peptide groups in resilin will form hydrogen bonds to water more willingly than to each other, and a slightly irregular chain folding will prevent cooperativity in formation of interchain hydrogen bonds. This model for the structure of resilin is only a suggestion, but a lack of cooperative interactions between chain segments can explain the observation (Weis-Fogh, 1961c) that the elasticity of resilin is dominated by entropic contributions from changes in chain conformations.

BIOSYNTHESIS OF RESILIN

The resilin-containing cuticle in adult locusts is produced during a period lasting from about 3 days before to 2–3 weeks after ecdysis, and cross-linking of the secreted resilin seems to occur as soon as it is deposited (Neville, 1963b). During cross-linking, protein-bound tyrosine residues are oxidatively linked to each other, resulting in the formation of di- and trityrosine. If a young adult locust is injected with radioactively labelled tyrosine, a narrow layer of radioactivity can be observed afterwards in the resilin-containing ligaments, and two consecutive injections of radioactive tyrosine given at an interval of 6 hr result in two clearly separated layers, indicating that the protein is immobilised rapidly after it is deposited extracellularly (Kristensen, 1966, 1968). Weis-Fogh noticed the presence of a narrow non-fluorescent zone between the apical, resilin-secreting cell surface and the strongly fluorescent resilin of the locust ligaments, and he concluded that cross-linking occurs as soon as the secreted proresilin attaches to the surface of the fully cross-linked resilin (Weis-Fogh, 1970). If so, we should expect the enzyme responsible for oxidising the tyrosine residues to be situated either on the cell surface, probably anchored to the apical cell membrane with its active site on the extracellular side, or to be part of the already cross-linked resilin.

It is likely that the oxidative coupling of tyrosine residues proceeds via a free radical mechanism. Di- and trityrosine can be formed from free tyrosine or protein-bound tyrosine residues by metal-catalysed oxidation or radiolysis (Huggins et al., 1993) or by oxidation with hydrogen peroxide and a peroxidase (Gross and Sizer, 1959). Peroxidases are involved in di- and trityrosine formation in several biological systems, such as hardening of egg chorions of some insects (Mindrinos et al., 1980; Li et al., 1996; Han et al., 2000b) and hardening of the fertilisation membrane in sea urchins (Foerder and Shapiro, 1977; Nomura and Suzuki, 1995), and a specific peroxidase is presumably responsible for formation of di- and trityrosine in resilin, although the evidence for the presence of such an enzyme is incomplete. Coles (1966) studied the problem, but the methods then available were not sensitive enough to conclusively demonstrate the presence or absence of peroxidase involvement in resilin cross-linking. Peroxidase activity has been demonstrated in insect solid cuticle, where it may play a role in cuticular hardening, but not necessarily by formation of di- and trityrosine (Locke, 1969).

If the tyrosine-oxidising enzyme in resilin is a peroxidase, the problem is: How is the necessary hydrogen peroxide provided? The peroxidase involved in cross-linking *Aedes aegypti* egg chorion is able to catalyse NADH oxidation with concomitant reduction of molecular oxygen to hydrogen peroxide, which then can be used by the same peroxidase to produce dityrosine. A malate/NAD$^+$ oxidoreductase is present in the chorion, and by oxidising malate, this enzyme can supply the NADH necessary for the production of hydrogen peroxide (Han et al., 2000a). It is possible that a similar system of coupled enzyme reactions provides the hydrogen peroxide needed for resilin cross-linking, but I do not know whether anyone has looked into this possibility.

RESILIN COMPARED WITH OTHER CUTICULAR PROTEINS

The cuticle of arthropods is a continuous extracellular layer, which protectively covers the complete animal and is a secreted product of the underlaying layer of epidermal cells. The properties of the various regions of the cuticle are functionally optimised and are, to a large extent, determined by the proteins present in the cuticular matrix, but it is not yet possible to account for regional differences in cuticular properties on the basis of the structures of the proteins. Both soft, pliable cuticles and stiff, solid cuticles contain their own, characteristic proteins, which differ from the resilin protein in several respects, although some similarities are present. It may be of interest to compare the various types of cuticular proteins to see whether some of the differences in protein structure can be correlated to differences in mechanical properties.

Soft as well as solid cuticles contain complex mixtures of proteins, while resilin-containing cuticles are dominated by a few or just one single, major protein. Proteins from soft cuticles are hydrophilic like resilin, but their content of glycine is significantly lower, and most of the proteins from solid cuticles are rather hydrophobic. As mentioned previously, many of the proteins from soft as well as solid cuticles have a tripartite structure resembling that in fruit fly resilin, and the central region in resilin shows some sequence similarity to the central regions in the other cuticular proteins. It has been demonstrated that the central regions in the other cuticular proteins are involved in linking the proteins to the cuticular chitin filaments (Rebers and Willis, 2001), and the central region in resilin has probably the same function. But while the N- and C-terminal regions in resilin are dominated by very regular sequence repeats, no such structures and motifs have been reported from terminal regions in soft cuticle proteins, and the terminal regions in proteins from solid cuticles tend to be dominated by short alanine-proline-valine-rich motifs, such as Ala-Ala-Pro-Val (Andersen et al., 1995), or by long glycine-rich stretches (Bouhin et al., 1992; Charles et al., 1992). Studies of the interactions between cuticular proteins and the hydrophobicity probe, 8-anilinonaphthalene sulphonate (ANS), indicate that both soft and solid cuticle proteins at physiological pH values tend to fold into rather open structures, readily accessible for the probe. ANS is only weakly fluorescent in polar solvents, such as water, but is strongly fluorescent when present in nonpolar surroundings. The cuticular proteins in solution can only shield the probe from the quenching effects of water, when pH in the solution is close to their isoelectric points, where the electrostatic repulsions between chain segments are minimal. Near pH 7, the folding of most of the cuticular proteins will result in structures so open that water has access to the polypeptide backbone, and while segments of the peptide chains may stick together due to hydrophobic interactions, these interactions will not be very stable and can easily be broken (Andersen, 2001 and unpublished data).

The alanine/proline-rich regions in proteins from solid cuticles have been suggested to fold in irregular, elastin-like beta-spirals, resulting in an easily deformable structure (Andersen et al., 1995). These proteins have a pronounced tendency to aggregate, and like elastin-related peptides they tend to be more soluble at low than at high temperatures (Hillerton and Vincent, 1985; Andersen, unpublished). Warming cold solutions of the proteins to room temperature results in formation of two fluid phases, of which the heavy, protein-rich phase precipitates as a coacervate. The temperature for coacervation depends upon protein concentration, pH, and ionic strength of the solution. The proteins are probably present as a coacervate in the newly deposited, not-yet sclerotised cuticle when an insect emerges from its old cuticle in the process of moulting, and after emergence, when it has been expanded to its proper size and shape, the cuticle is partly dehydrated, and the proteins are stabilised and cross-linked by incorporation of oxidation products of N-acetyldopamine and N-beta-alanyldopamine (Andersen et al., 1996). In the sclerotised cuticle, the proteins will be in water-poor surroundings, and, as they are linked to each other, they cannot move much, but presumably their peptide chains will have retained their spiral structure and be able to suffer limited deformations. Such a model could explain the combination of stiffness and elasticity described for solid cuticles by Jensen and Weis-Fogh (1962).

Measurements of their mechanical properties indicate that soft cuticles are viscous or viscoelastic materials where the chitin filaments act as fillers (Hillerton, 1984). Covalent cross-links connecting the proteins to each other are not present in soft cuticles, and little is known of the noncovalent interactions between the proteins, although such interactions must be of importance for the mechanical properties of the material.

SOME PROBLEMS FOR THE FUTURE

Among the resilin problems to be studied in more detail, I can suggest: Is resilin composed of a single protein or is the material a mixture of different, but related gene products? Gene product CG15920 is the only gene product in *Drosophila* which corresponds to locust resilin in both amino acid composition and sequence of tryptic peptides and is probably the main component in fruit fly resilin, but, other, minor protein components may be present in the resilin-containing ligaments and tendons. The elastic tendon in dragonflies is in vivo stretched by rapid unidirectional tensile forces, so all regions in the tendon will at any moment be exposed to the same force, and there will probably not be any physiological reasons for local differences in elasticity. The problem is more complex for structures where the various local regions are exposed to different forces and deformation during bending and twisting, and differences in elasticity could be of functional value here. When locust ligaments (wing hinges and prealar arms) are studied by fluorescence microscopy, it is evident that there are marked differences in fluorescence from the resilin deposited during the last 3 days before ecdysis and that deposited after ecdysis, indicating that the pre- and postecdysial resilin layers are cross-linked to different degrees, but do the two parts of the ligaments contain the same proteins and do they have the same mechanical properties? The pre- and postecdysial layers of locust solid cuticle contain different mixtures of proteins, indicating that the epidermal cells at ecdysis switch from secreting one set of proteins to producing a completely different set. Could a similar switch operate in case of the resilin-producing regions of the epidermis, so that the pre- and postecdysial layers of resilin differ in elasticity? I have reported (Andersen, 1966) that locust resilin ligaments, containing both pre- and postecdysial material, yielded at least two different N-terminal amino acids after dinitrophenylation, indicating the presence of different proteins. The question has not been studied since, but it would be interesting to know whether the two parts of the resilin-containing ligaments contain different proteins or whether they are identical in protein content.

In the so-called transitional cuticles, which possess long-range elasticity, and where no sharp borders are present between the resilin-containing elastic parts and the surrounding solid cuticle, it appears likely that a gradient of various cuticular proteins exists, going from the central, highly elastic region to the more peripheral and less elastic regions. A decreasing gradient in resilin-characteristic fluorescence can be observed microscopically, but we have no evidence for the local distribution of the proteins characteristic for solid cuticle. Impregnation of resilin with smaller, less elastic proteins would be a way of gradually modifying the mechanical properties of cuticular regions. The problem deserves to be better investigated, for instance by use of antibodies specific for individual proteins.

Another problem worth studying is whether external conditions, such as different temperatures, will influence the degree of cross-linking of resilin. Slight differences in fluorescence have been described for resilin deposited during day and night conditions (Neville, 1963a,1963b), and the differences are probably due to small differences in di- and trityrosine content. But how wide are the variations in cross-link density that can be obtained within the temperature range at which locusts can be grown?

As mentioned previously, the enzyme (or enzymes) responsible for resilin cross-linking should be studied in much more detail, and the possibilities for interfering with the cross-linking process in vivo explored. A single publication exists, which indicates that the hormone bursicon in blowflies may be involved in resilin deposition and/or cross-linking (Sabaratnam, 1973), and this observation needs to be followed up by a more detailed study.

The various problems outlined here are all concerned with aspects of the biochemistry of resilin, but the recent identification of a gene for resilin in fruit-flies should make it possible to start investigating a number of other problems. It could thus be interesting to synthesise various model polypeptides related to the two types of repeat sequences in resilin to study their physical and chemical behaviour, to produce mutant versions of resilin, such as resilin without the central, chitin-binding region, for studies of the elasticity of resilin modified in controlled ways, and to investigate in detail the cellular and hormonal control of resilin gene expression. It is my hope that such a fascinating material as resilin will be a source for enjoyable research for many years to come for biochemists, biophysicists, and cell biologists.

REFERENCES

Andersen, S. O. (1963) Characterization of a new type of cross-linkage in resilin, a rubber-like protein. *Biochim. Biophys. Acta* **69,** 249–62.

Andersen, S. O. (1964) The crosslinks in resilin identified as dityrosine and trityrosine. *Biochim. Biophys. Acta* **93,** 213–15.

Andersen, S. O. (1966) Covalent crosslinks in a structural protein, resilin. *Acta Physiol. Scand.* **66**(Suppl. 263), 1–81.

Andersen, S. O. (1971) Resilin. In: *Comprehensive Biochemistry*, vol. 26C, M. Florkin and E. H. Stotz (eds.). Amsterdam: Elsevier, pp. 633–57.

Andersen, S. O. (1998) Amino acid sequence studies on endocuticular proteins from the desert locust, *Schistocerca gregaria. Insect Biochem. Mol. Biol.* **28,** 421–34.

Andersen, S. O. (2001) Matrix proteins from insect pliable cuticles: Are they flexible and easily deformed? *Insect Biochem. Mol. Biol.* **31,** 445–52.

Andersen, S. O., and Weis-Fogh, T. (1964) Resilin. A rubberlike protein in arthropod cuticle. *Adv. Insect Physiol.* **2,** 1–65.

Andersen, S. O., Højrup, P., and Roepstorff, P. (1995) Insect cuticular proteins. *Insect Biochem. Mol. Biol.* **25,** 153–76.

Andersen, S. O., Peter, M. G., and Roepstorff, P. (1996) Cuticular sclerotization in insects. *Comp. Biochem. Physiol.* **113B,** 689–705.

Ardell, D. H., and Andersen, S. O. (2001) Tentative identification of a resilin gene in *Drosophila melanogaster. Insect Biochem. Mol. Biol.* **31,** 965–70.

Bennet-Clark, H. C., and Lucey, E. C. A. (1967) The jump of the flea: A study of the energetics and a model of the mechanism. *J. Exp. Biol.* **47,** 59–76.

Bennet-Clark, H. C., and Daws, A. G. (1999) Transduction of mechanical energy into sound energy in the cicada *Cyclochila australasiae. J. Exp. Biol.* **202,** 1803–17.

Booij, M., and ten Bosch, J. J. (1982) A fluorescent compound in bovine dental enamel matrix compared with synthetic dityrosine. *Arch. Oral Biol.* **27,** 417–21.

Bouhin, H., Charles, J.-P., Quennedey, B., and Delachambre, J. (1992) Developmental profiles of epidermal mRNAs during the pupal-adult molt of *Tenebrio molitor* and isolation of a cDNA clone encoding an adult cuticular protein: effects of a juvenile hormone analogue. *Dev. Biol.* **149,** 112–22.

Briza, P., Winkler, G., Kalchhauser, H., and Breitenbach, M. (1986) Dityrosine is a prominent component of the yeast ascospore wall. A proof of its structure. *J. Biol. Chem.* **261,** 4288–94.

Charles, J.-P., Bouhin, H., Quennedey, B., Courrent, A., and Delachambre, J. (1992) cDNA cloning and deduced amino acid sequence of a major, glycine-rich cuticular protein from the coleopteran *Tenebrio molitor*. Temporal and spatial distribution of the transcript during metamorphosis. *Eur. J. Biochem.* **206,** 813–19.

Coles, G. C. (1966) Studies on resilin biosynthesis. *J. Insect Physiol.* **12,** 679–91.

Elliot, G. F., Huxley, A. F., and Weis-Fogh, T. (1965) On the structure of resilin. *J. Mol. Biol.* **13,** 791–95.

Foerder, C. A., and Shapiro, B. M. (1977) Release of ovoperoxidase from sea urchin eggs hardens the fertilization membrane with tyrosine crosslinks. *Proc. Natl. Acad. Sci. USA* **74,** 4214–18.

Fujimoto, D. (1975) Occurrence of dityrosine in cuticlin, a structural protein from *Ascaris* cuticle. *Comp. Biochem. Physiol.* **51B,** 205–57.

Gorb, S. N. (1999) Serial elastic elements in the damselfly wing: Mobile vein joints contain resilin. *Naturwissenschaften* **86,** 552–5.

Govindarajan, S., and Rajulu, G. S. (1974) Presence of resilin in a scorpion *Palamnaeus swammerdami* and its role in the food-capturing and sound-producing mechanisms. *Experientia* **30,** 908–9.

Gross, A. J., and Sizer, I. W. (1959) The oxidation of tyramine, tyrosine, and related compounds by peroxidase. *J. Biol. Chem.* **234,** 1611–14.

Haas, F., Gorb, S., and Blickhan, R. (2000a). The function of resilin in beetle wings. *Proc. R. Soc. Lond. B* **267,** 1375–81.

Haas, F., Gorb, S., and Wootton, R. J. (2000b) Elastic joints in dermapteran hind wings: Materials and wing folding. *Arthropod Struct. Dev.* **29,** 137–46.

Hall, H. G. (1978) Hardening of the sea urchin fertilization envelope by peroxidase-catalyzed phenolic coupling of tyrosines. *Cell* **15,** 343–55.

Han, Q., Li, G., and Li, J. (2000a) Chorion peroxidase-mediated $NADH/O_2$ oxidoreduction cooperated by chorion malate dehydrogenase-catalyzed NADH production: a feasible pathway leading to H_2O_2 formation during chorion hardening in *Aedes aegypti* mosquitoes. *Biochim. Biophys. Acta* **1523,** 246–53.

Han, Q., Li, G., and Li, J. (2000b) Purification and characterization of chorion peroxidase from *Aedes aegypti* eggs. *Arch. Biochem. Biophys.* **378,** 107–15.

Hayashi, C. Y., and Lewis, R. V. (1998) Evidence from flagelliform silk cDNA for the structural basis of elasticity and modular nature of spider silks. *J. Mol. Biol.* **275,** 773–84.

Hepburn, H. R. (1971) Proboscis extension and recoil in Lepidoptera. *J. Insect Physiol.* **17,** 637–56.

Hillerton, J. E. (1984) Cuticle: Mechanical properties. In: *Biology of the Integument, Vol. 1.*

Invertebrates, J. Bereiter-Hahn, A. G. Matoltsy, and K. S. Richards (eds.). Berlin, Germany: Springer-Verlag. pp. 626–37.

Hillerton, J. E., and Vincent, J. F. V. (1985) In vitro aggregation of proteins from insect cuticle. *Entomol. Gener.* **11**, 1–9.

Huggins, T. G., Wells-Knecht, M. C., Detorie, N. A., Baynes, J. W., and Thorpe, S. R. (1993) Formation of O-tyrosine and dityrosine in proteins during radiolytic and metal-catalyzed oxidation. *J. Biol. Chem.* **268**, 12341–47.

Iconomidou, V. A., Willis, J. H., and Hamodrakas, S. J. (1999) Is beta-pleated sheet the molecular conformation which dictates formation of helicoidal cuticle? *Insect Biochem. Mol. Biol.* **29**, 285–92.

Iconomidou, V. A., Chryssikos, G. D., Gionis, V., Willis, J. H., and Hamodrakas, S. J. (2001) "Soft"-cuticle protein secondary structure as revealed by FT-Raman, ATR FT-IR and CD spectroscopy. *Insect Biochem. Mol. Biol.* **31**, 877–85.

Jacob, J. S., Cistola, D. P., Hsu, F. F., Muzaffar, S., Mueller, D. M., Hazen, S. L., and Heinecke, J. W. (1996) Human phagocytes employ the myeloperoxidase-hydrogen peroxide system to synthesize dityrosine, trityrosine, pulcherosine, and isodityrosine by a tyrosyl radical-dependent pathway. *J. Biol. Chem.* **271**, 19950–56.

Jensen, M., and Weis-Fogh, T. (1962) Biology and physics of locust flight. V. Strength and elasticity of locust cuticle. *Phil. Trans. R. Soc. Lond. Ser. B* **245**, 137–69.

Kannupandi, T. (1976) Occurrence of resilin and its significance in the cuticle of *Pennella elegans*, a copepod parasite. *Acta Histochem.* **56**, 73–9.

Kawasaki, H., Sato, H., and Suzuki, M. (1974) Structural proteins in the egg envelopes of dragonflies, *Sympetrum infuscatum* and *S. frequens. Insect Biochem.* **4**, 99–111.

Kawasaki, H., Sato, H., and Suzuki, M. (1975) Structural proteins in the egg envelopes of the mealworm beetle, *Tenebrio molitor. Insect Biochem.* **5**, 25–34.

Kristensen, B. I. (1966) Incorporation of tyrosine into rubber-like cuticle of locusts studied by autoradiography. *J. Insect Physiol.* **12**, 173–77.

Kristensen, B. I. (1968) Time course of incorporation of tyrosine into rubber-like cuticle of locusts. *J. Insect Physiol.* **14**, 1135–40.

Lassandro, F., Sebastiano, M., Zei, F., and Bazzicalupo, P. (1994) The role of dityrosine formation in the crosslinking of CUT-2, the product of a second cuticlin gene of *Caenorhabditis elegans. Mol. Biochem. Parasitol.* **65**, 147–59.

Li, J., Hodgeman, B. A., and Christensen, B. M. (1996) Involvement of peroxidase in chorion hardening in *Aedes aegypti. Insect Biochem. Mol. Biol.* **26**, 309–17.

Locke, M. (1969) The localization of a peroxidase associated with hard cuticle formation in an insect, *Calpodes ethlius* Stoll, Lepidoptera, Hesperiidae. *Tissue Cell* **1**, 555–74.

Lombardi, E. C., and Kaplan, D. L. (1993) Preliminary characterization of resilin isolated from the cockroach, *Periplaneta americana. Mat. Res. Soc. Symp. Proc.* **292**, 3–7.

Miller, P. L. (1960) Respiration in the desert locust. II. The control of the spiracles. *J. Exp. Biol.* **37**, 237–63.

Mindrinos, M. N., Petri, W. H., Galanopoulos, V. K., Lombard, M. F., and Margaritis, L. H. (1980) Crosslinking of the *Drosophila* chorion involves a peroxidase. *Wilhelm Roux's Arch.* **189**, 187–96.

Neff, D., Frazier, S. F., Quimby, L., Wang, R.-T., and Zill, S. (2000) Identification of resilin in the leg of cockroach, *Periplaneta americana*: Confirmation by a simple method using pH dependence of UV fluorescence. *Arthropod Struct. Devel.* **29**, 75–83.

Neville, A. C. (1963a) Daily growth layers in locust rubber-like cuticle influenced by an external rhythm. *J. Insect Physiol.* **9,** 177–86.

Neville, A. C. (1963b) Growth and deposition of resilin and chitin in locust rubber-like cuticle. *J. Insect Physiol.* **9,** 265–78.

Neville, A. C., and Rothschild, M. (1967) Fleas—Insects which fly with their legs. *Proc. R. Ent. Soc. Lond.* (*C*) **32,** 9–10.

Nielsen, H., Engelbrecht, J., Brunak, S., and von Heijne, G. (1997) Identification of prokaryotic and eukaryotic signal peptides and prediction of their cleavage sites. *Protein Eng.* **10,** 1–6.

Nomura, K., and Suzuki, N. (1995) Sea urchin ovoperoxidase: Solubilization and isolation from the fertilization envelope, some structural and functional properties, and degradation by hatching enzyme. *Arch. Biochem. Biophys.* **319,** 525–34.

Petri, W. H., Wyman, A. R., and Kafatos, F. C. (1976) Specific protein synthesis in cellular differentiation. III. The egg shell proteins in *Drosophila melanogaster* and their program of synthesis. *Dev. Biol.* **49,** 185–99.

Rajulu, G. S. (1971) Presence of resilin in the cuticle of the centipede, *Scolopendra morsitans* L. *Indian J. Exp. Biol.* **9,** 122–23.

Raven, D. J., Earland, C., and Little, M. (1971) Occurrence of dityrosine in Tussah silk fibroin and keratin. *Biochim. Biophys. Acta* **251,** 96–99.

Rebers, J. E., and Riddiford, L. M. (1988) Structure and expression of a *Manduca sexta* larval cuticle gene homologous to *Drosophila* cuticle genes. *J. Mol. Biol.* **203,** 521–31.

Rebers, J. E., and Willis, J. H. (2001) A conserved domain in arthropod cuticular proteins binds chitin. *Insect Biochem. Mol. Biol.* **31,** 1083–93.

Sabaratnam, M. (1973) The effect of digging on development of adult characters in blowflies *Calliphora erythrocephala. J. Insect Physiol.* **19,** 2143–54.

Sandberg, L. B., Leslie, J. G., Leach, C. T., Alvarez, V. L., Torres, A. R., and Smith, D. W. (1985) Elastin covalent structure as determined by solid phase amino acid sequencing. *Pathol. Biol.* **33,** 266–74.

Sannasi, A. (1969) Resilin in the cuticle of click beetles. *J. Georgia Entomol. Soc.* **4,** 31–32.

Sannasi, A. (1970) Resilin in the lens-cuticle of the firefly, *Photinus pyralis* (L.). *Experientia* **26,** 154.

Skals, N., and Surlykke, A. (1999) Sound production by abdominal tymbal organs in two moth species: The green silver-line and the scarce silver-line (Noctuoidea: Nolidae: Chloephorinae). *J. Exp. Biol.* **202,** 2937–49.

Tatham, A. S., and Shewry, P. R. (2000) Elastomeric proteins: Biological roles, structures and mechanisms. *Trends Biochem. Sci.* **25,** 567–71.

Urry, D. W. (1988) Entropic elastic processes in protein mechanisms. I. Elastic structure due to an inverse temperature transition and elasticity due to internal chain dynamics. *J. Protein Chem.* **7,** 1–34.

Urry, D., Luan, C.-H., and Peng, S. Q. (1995) Molecular biophysics of elastin. Structure, function and pathology. *Ciba Found. Symp.* **192,** 4–30.

Ushijima, Y., Nakano, M., and Goto, T. (1984) Production and identification of bityrosine in horseradish peroxidase-H_2O_2-tyrosine system. *Biochem. Biophys. Res. Comm.* **125,** 916–18.

Varman, A. R. (1980) Resilin in the cuticle of physogastric queen termites. *Experientia* **36,** 564.

Varman, A. R. (1981) Resilin in the abdominal cuticle of workers of the honey-ants, *Myrmecocystus mexicanus. J. Georgia Entomol. Soc.* **16,** 11–13.

Weis-Fogh, T. (1959) Elasticity in arthropod locomotion: A neglected subject, illustrated by the wing system of insects. *Proc. 15th Int. Congr. Zool. 1958* 393–95.

Weis-Fogh, T. (1960) A rubber-like protein in insect cuticle. *J. Exp. Biol.* **37,** 889–907.

Weis-Fogh, T. (1961a) Power in flapping flight. In: *The Cell and the Organism,* J. A. Ramsay and V. B. Wigglesworth (eds.). Cambridge, UK: Cambridge University Press, pp. 283–300.

Weis-Fogh, T. (1961b) Thermodynamic properties of resilin, a rubber-like protein. *J. Mol. Biol.* **3,** 520–31.

Weis-Fogh, T. (1961c) Molecular interpretation of the elasticity of resilin, a rubber-like protein. *J. Mol. Biol.* **3,** 648–67.

Weis-Fogh, T. (1970) Structure and formation of insect cuticle. *Symp. R. Ent. Soc. Lond.* **5,** 165–85.

Welinder, B. S., Roepstorff, P., and Andersen, S. O. (1976) The crustacean cuticle—IV. Isolation and identification of cross-links from *Cancer pagurus* cuticle. *Comp. Biochem. Physiol.* **53B,** 529–33.

Young, D., and Bennet-Clark, H. C. (1995) The role of the tymbal in cicada sound production. *J. Exp. Biol.* **198,** 1001–19.

FOURTEEN

Gluten, the Elastomeric Protein of Wheat Seeds

Peter R. Shewry, Nigel G. Halford, Peter S. Belton,
and Arthur S. Tatham

INTRODUCTION

Wheat is one of the three most important crops in the world, together with maize
and rice. Approximately 600 million tonnes are harvested annually with culti-
vation extending over a vast geographical area, from Scandinavia to Argentina,
including higher elevations in the tropics. Although the ability to give high yields
under a range of conditions has contributed to the success of wheat, the most
important factor has been the unique properties of wheat dough which allow it
to be processed into a range of foodstuffs, notably bread, other baked products,
and pastas. These properties are usually described as viscoelasticity, with the
balance between the extensibility and elasticity determining the end use quality. For example, highly elastic ('strong') doughs are required for breadmaking,
but more extensible doughs for making cakes and biscuits.

The viscoelastic properties of dough are determined by the grain proteins,
and in particular by the storage proteins which form a network in the dough
called gluten (Schofield, 1994). Consequently, the gluten proteins have been
widely studied over a period exceeding 250 years to determine their structures
and properties and to provide a basis for manipulating and improving end use
quality (Shewry et al., 1995).

THE ORIGIN OF THE WHEAT GLUTEN NETWORK

Gluten can be readily prepared by gently washing dough under a stream of
running water. This removes the bulk of the soluble and particulate matter
to leave a proteinaceous mass which retains its cohesiveness on stretching
[Figure 14.1(A)]. Gluten comprises some 75% protein on a dry weight basis, with
most of the remainder being starch and lipids. Furthermore, the vast majority
of the proteins are of a single type called prolamins.

Prolamins are a group of proteins which were initially defined based on their
solubility in alcohol-water mixtures (Osborne, 1924), typically 60–70% (v/v)
ethanol. This definition has since been extended to include related proteins

(A)

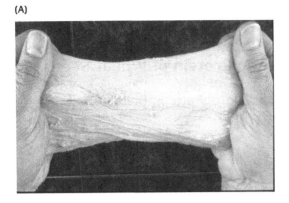

(B)

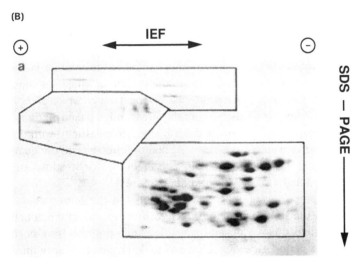

Figure 14.1: (A) Sheet of gluten being stretched to demonstrate its cohesive properties. (B) Two-dimensional analysis [isoelectric focusing (IEF) followed by SDS-PAGE] of wheat gluten proteins shows multiple components.

which are not soluble in alcohol-water mixtures in the native state owing to their presence in polymers stabilised by interchain disulphide bonds. In wheat, these groups of monomeric and polymeric prolamins are called gliadins and glutenins, respectively, and together form gluten (Shewry et al., 1986).

Wheat prolamins are the major storage proteins present in the starchy endosperm cells of the grain, where they are synthesised and deposited via the secretory system. Thus, the individual polypeptides are synthesised on ribosomes on the rough endoplasmic reticulum and pass via the usual translocation machinery into the lumen, with the loss of an N-terminal signal peptide.

Once within the lumen, it is probable that protein folding and disulphide bond formation occur with no further post-translational modifications taking place (i.e., no glycosylation or proteolysis as may occur with other types of seed storage protein).

The subsequent fate of the proteins may also vary with the protein type and with the age and stage of development of the tissue. Some of the proteins appear to be transported via the Golgi apparatus into the vacuole where they form protein deposits (Shewry, 1999). However, others appear to accumulate directly within the lumen of the endoplasmic reticulum (ER) to form a second population of protein bodies. Galili (1997) has proposed that the latter subsequently become engulfed by vesicles and 'internalised' into vacuoles, but this has not so far been confirmed by other workers. What is known is that, during the later stages of grain maturation, the starchy endosperm cells become disrupted and die, and the protein bodies fuse to form a continuous matrix which surrounds the starch granules and engulfs other organelles and membranes. The overall pathway is summarised in Figure 14.2.

Galili (1997) has also suggested that the gliadins are preferentially transported to the vacuole via the Golgi apparatus, which is consistent with the lack of a classical ER retention sequence. We have also shown that a γ-gliadin is rapidly degraded in leaves and seeds of transgenic tobacco (presumably in the vacuole) unless a C-terminal ER retention sequence (HDEL or KDEL) is added (Napier et al., 1997). However, the glutenins also lack an obvious ER retention sequence but are, nevertheless, proposed to be preferentially retained in the ER (Galili, 1997). In this case, retention could be determined by their rapid assembly into high M_r polymers which precipitate and accumulate directly within the ER lumen. It is also possible to envisage how the relative rates of trafficking via the ER and Golgi routes could vary with the level of protein synthesis and age of the tissue.

As a result of the formation of a protein matrix, individual cells of wheat flour contain networks of gluten proteins which are brought together during dough mixing. The precise changes which occur during mixing are still not completely understood, but an increase in dough stiffness occurs which is generally considered to result from 'optimisation' of protein:protein interactions within the gluten network. In molecular terms, this 'optimisation' may include some exchange of disulphide bonds as mixing in air, oxygen and nitrogen result in different effects on the sulphydryl and disulphide contents of dough (Tsen and Bushuk 1963; Mecham and Knapp, 1966).

Of course, the natural fate of the wheat grain is not to provide flour for humankind but to germinate to produce a new plant. The biological role of the gluten protein is, therefore, to provide a store of carbon, nitrogen, and sulphur to support seed germination and seedling growth. The gluten proteins have no

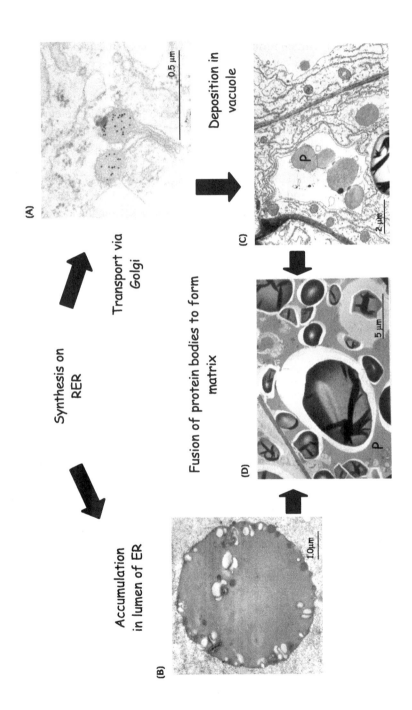

(A)

0.5 μm

Transport via *Golgi*

Synthesis on RER

Deposition in vacuole

(C)

P

2 μm

Fusion of protein bodies to form matrix

(D)

P

5 μm

Accumulation in lumen of ER

(B)

1.0μm

other known biological role, and their viscoelastic properties appear to be a purely fortuitous consequence of their sequences and interactions.

WHEAT GLUTEN PROTEINS

Wheat gluten is a highly complex mixture of proteins with at least 50 individual components being separated by two-dimensional isoelectric focusing/sodium dodecyl sulphate-polyacrylamide gel electrophoresis (SDS-PAGE) of reduced total fractions [Figure 14.1(B)]. Furthermore, there is great variation in the patterns of components present in different genotypes. This high level of polymorphism initially limited attempts to isolate and characterise individual components, but we now know details of the structures and sequences of all of the major gluten protein types (Shewry et al., 1999).

Gluten proteins are classically divided into two fractions: the gliadins, which are readily soluble in alcohol/water mixtures [e.g., 60–70% (v/v) ethanol or 50% (v/v) propan-1-ol] and the glutenins, which are only soluble if a reducing agent (2-mercaptoethanol or dithiothreitol) is included. It is now known that this difference is determined by the fact that the gliadins are monomeric proteins, whereas the glutenins comprise subunits assembled in polymers which are stabilised by interchain disulphide bonds. The gliadins can be further separated in α-, β-, γ-, and ω-gliadins by electrophoresis at low pH and the reduced glutenin subunits into high molecular weight (HMW) and low molecular weight groups by SDS-PAGE (Figure 14.3).

However, despite this apparent diversity, comparisons of amino acid sequences indicate that all gluten proteins are related and can be classified into three broad groups called sulphur-rich (S-rich), sulphur-poor (S-poor), and HMW prolamins, as summarised in Figure 14.3.

The S-poor prolamins comprise the ω-gliadins and a minor group of low molecular weight glutenin subunits (D-type), while the HMW prolamins

---

Figure 14.2: Summary of the pathway of synthesis and deposition of gluten proteins in developing grain of wheat. RER, rough endoplasmic reticulum. (A) Immunolabelling of protein in Golgi vesicles in cells at 14 days after anthesis (cv. Highbury) using a γ-gliadin polyclonal antibody. From 'The Synthesis, Processing and Deposition of Gluten Proteins in the Developing Wheat Grain', by P. R. Shewry, 1999, *Cereal Foods World*, **44**, 587–89. © 1999 with permission. (B) Protein accumulation within the ER of wheat (50 mg endosperm fresh weight). From 'Cereal Storage Proteins and Their Effects on Technological Properties', by B. J. Miflin et al., 1983, in: *Seed Proteins*, J. Daussant et al. (eds.), London, UK: Academic Press, pp. 255–319. © 1983 with permission. (C) Protein deposition (P) within the vacuole of a cell at 13 days after anthesis (cv. Maris Freeman). (D) Proteinaceous matrix (P) formed from disruption and coalescence of protein bodies in an endosperm cell at 46 days after anthesis (cv. Maris Freeman). (C) and (D) From 'Biotechnology of Breadmaking: Unravelling and Manipulating the Multi-Protein Gluten Complex', by P. R. Shewry et al., 1995, *Bio/Technology*, **13**, 1185–90. © 1995 with permission.

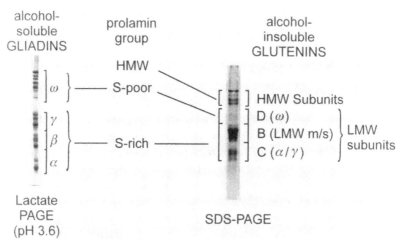

Figure 14.3: Summary of the classification and nomenclature of wheat gluten proteins separated by electrophoresis at low pH (left) and SDS-PAGE (right). The D group of low molecular weight (LMW) subunits are only minor components and are not clearly resolved. From 'The Prolamins of the Triticeae', by P. R. Shewry et al., 1999, in: *Seed Proteins*, P. R. Shewry and R. Carey (eds.), Dordrecht, The Netherlands: Kluwer Academic Publishers, pp. 11–33. © 1999 with permission.

comprise only the HMW subunits of glutenin. In contrast, the S-rich prolamins comprise major groups of gliadins (α-type gliadins comprising α- and β-gliadins and γ-type gliadins) and glutenin subunits (B- and C-types).

Despite variation in their molecular masses, amino acid compositions, and properties, all gluten proteins are rich in the amino acids glutamine and proline. This results from the presence of extensive repeated sequences, which are located in the N-terminal regions (domains) of the S-rich prolamins, but in the centre of the HMW and S-poor proteins (Figure 14.4). Indeed, in the ω-gliadin shown in Figure 14.4, the repeated sequences account for all except 23 of the 261 amino acids. The repeated sequences present in the S-rich prolamins are based on one or two short motifs, and are clearly related to those present in the S-poor ω-gliadins. In contrast, the repeated sequences present in the HMW subunits are based on two or three motifs which are not related to those in S-rich or S-poor prolamins.

Comparisons of the sequences of the non-repetitive domains of the S-rich and HMW prolamins also show regions of clear homology, indicating a common evolutionary origin (Kreis et al., 1985; Shewry and Tatham, 1999).

It is probable that all of the proteins discussed previously contribute to aspects of gluten protein functionality. However, work over the past 20 years has focused largely on one group of proteins, the HMW subunits, which appear to play a major role in determining gluten elasticity.

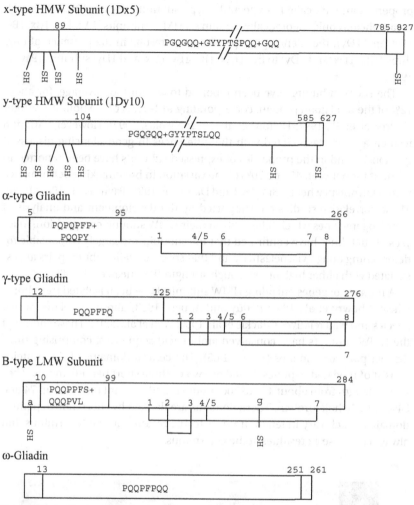

Figure 14.4: Summary of the domain structures of 'typical' wheat gluten proteins. The consensus repeat motifs in the repetitive domains are shown using standard single letter abbreviations for amino acids (G, glycine; F, phenylalanine; L, leucine; P, proline; Q, glutamine; S, serine; T, threonine; V, valine; Y, tyrosine). Based on sequences reported by Kasarda et al. (1984), Bartels et al. (1988), Anderson et al. (1989), Colot et al. (1989), and Hsia and Anderson (2001). LMW, low molecular weight; SH, sulphydryl.

THE HMW GLUTENIN SUBUNITS

Bread wheat is a hexaploid species with three genomes (called A, B, and D) derived from related wild grass species. Single loci encoding HMW subunits are present on the long arms of the group 1 chromosomes (1A, 1B, and 1D), with each locus comprising two genes encoding subunits which differ in their properties and are called x-type and y-type subunits (Payne, 1987). Although bread wheats could theoretically contain six HMW subunits (1Ax, 1Ay, 1Bx, 1By, 1Dx, and 1Dy), the silencing of specific genes results in the presence of only three (1Bx, 1Dx, and 1Dy) to five (1Ax, 1Bx, 1By, 1Dx, and 1Dy) subunits (Payne, 1987).

The HMW subunits have been reported to account, on average, for about 12% of the total grain protein, corresponding to 1–1.7% of the flour dry weight (Seilmeier et al., 1991; Halford et al., 1992; Nicolas, 1997). However, variation in their amount (associated with the differences in gene silencing discussed previously) and in the properties of expressed subunits have been reported to account for between 45 and 70% of the variation in breadmaking performance within European wheats (Branlard and Dardevet, 1985; Payne et al., 1987, 1988). These correlative studies are supported by the development and analysis of near-isogenic lines which differ only in their HMW subunit composition. Analyses of such lines have confirmed that the subunits are largely responsible for determining dough viscoelasticity and that specific allelic subunit pairs are associated with either high or low dough strength (Popineau et al., 1994).

A number of genes encoding HMW subunits have been isolated from bread wheat (Shewry et al., 1992; Reddy and Appels, 1993) and from related wheat species and wild relatives (Mackie et al., 1996; Wan et al., 2001). These show that the HMW subunits have conserved amino acid sequences, comprising three distinct parts or domains (Figure 14.5). The central domains of the proteins consist of repeated peptides, based on two or three short peptide motifs. They vary in length from about 420 to 700 residues, and account for between 74 and 84% of the whole protein. These domains are flanked by short, non-repetitive domains which vary in length from 81 to 104 residues at the N-terminus, but always comprise 42 residues at the C-terminus.

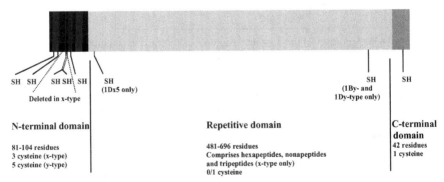

Figure 14.5: Schematic summary of the sequences of x-type and y-type HMW subunits. SH, sulphydryl.

SEQUENCES OF THE REPETITIVE DOMAINS

The x-type and y-type subunits have essentially similar repeat structures, comprising mainly nonapeptide and hexapeptide motifs. Also, whereas tandem blocks of hexapeptides may be present, the nonapeptides are always interspersed with hexapeptides. Consequently, it is convenient to consider them as forming a 15 amino acid motif. The x-type subunits also differ from the y-type in having additional tripeptide motifs, which also only occur in tandem with hexapeptides, forming a second nonapeptide motif. Figure 14.6 shows the sequences of the repetitive domains of typical x-type and y-type subunits (1Dx5 and 1Dy10, respectively) arranged to show their repeated block structure. The repeat motifs are rich in glutamine, proline, and glycine, which together account for more than 70% of the total amino acid residues. No major differences are apparent between the homeoallelic proteins of bread wheat (A, B, and D) or related genomes present in other wheat species or wild relatives (A, C, and D), so combined data for five x-type and seven y-type subunits are presented in Table 14.1.

Comparison of the patterns of amino acid substitutions shows that some positions of the motifs appear to be more highly conserved than others. In particular, glutamine tends to be more highly conserved at specific positions than other consensus amino acids, at positions 3, 5, and 6 of the hexapeptide, 8 and 9 of the nonapeptide, and 2 and 3 of the tripeptide. This may relate to the role of glutamine residues in stabilising the structures and interactions of the subunits. Similarly, serine is conserved at position 6 of the nonapeptides. In contrast, positions 1 and 4 of the hexapeptide are poorly conserved, as is position 7 of the nonapeptide.

Some differences are also observed between the x-type and y-type subunits. Thus, replacement of Pro with Ser at position 1 of the hexapeptides is more common in x-type subunits as is replacement of Gln with Pro at position 6. However, the latter only occurs in hexapeptides within a 15 residue (6 + 9) motif, rather than in the tandemly arranged hexapeptides. Similarly, replacement of Tyr with His at position 2 and Thr with Ala at position 5 of the nonapeptides are more common in y-type subunits, and these two substitutions usually occur together, giving the two consensus motifs GYYPTSLQQ and GHYPASLQQ, compared with GYYPTSPQQ for x-type subunits.

There is no evidence that amino acid substitution lead to replacement with similar amino acid residues (i.e., conservative substitutions). Instead, analysis of codons indicates that most replacements are due to single nucleotide changes, with substitutions resulting from double nucleotide changes occurring more rarely. For example, proline (CCA) occurs at position 1 in 55% of x-type hexapeptides with single nucleotide changes leading to the occurrence of leucine (CTA, 12%) and serine (TCA, 30%) and two nucleotide changes to isoleucine (ATA, 3%).

1Dx5

```
RYYPSVTCPQQ
VSYYPGQASPQR
PGQGQQ
PGQGQQGYYPTSPQQ
PGQWQQ
PEQGQPRYYPTSPQQ
SGQLQQ
PAQGQQ
PGQGQQGQQ
PGQGQPGYYPTSSQLQ
PGQLQQ
PAQGQQGQQ
PGQAQQGQQ
PGQGQQ
PGQGQQGQQ
PGQGQQ
PGQGQQGQQ
LGQGQQGYYPTSLQQ
SGQGQPGYYPTSLQQ
LGQGQSGYYPTSPQQ
PGQGQQ
PGQLQQ
PAQGQQ
PGQGQQGQQ
PGQGQQGQQ
PGQGQQ
PGQGQPGYYPTSPQQ
SGQGQPGYYPTSSQQ
PTQSQQ
PGQGQQGQQ
VGQGQQAQQ
PGQGQQ
PGQGQPGYYPTSPQQ
SGQGQPGYYLTSPQQ
SGQGQQ
PGQLQQ
SAQGQKGQQ
PGQGQQ
PGQGQQGQQ
PGQGQQGQQ
PGQGQPGYYPTSPQQ
SGQGQQ
PGQWQQ
PGQGQPGYYPTSPLQ
PGQGQPGYDPTSPQQ
PGQGQQ
PGQLQQ
PAQGQQGQQ
LAQGQQGQQ
PAQVQQGQR
PAQGQQGQQ
PGQGQQGQQ
LGQGQQGQQ
PGQGQQGQQ
PAQGQQGQQ
PGQGQQGQQ
PGQGQQGQQ
PGQGQQ
PGQGQPWYYPTSPQE
SGQGQQ
PGQWQQ
PGQGQPGYYLTSPLQ
LGQGQQGYYPTSLQQ
PGQGQQ
PGQWQQ
SGQGQHWYYPTSPQL
SGQGQR
PGQWLQ
PGQGQQGYYPTSPQQ
PGQGQQ
LGQWLQ
PGQGQQGYYPTSLQQ
TGQGQQ
SGQGQQGYY
```

1Dy10

```
GYYPGVTSPRQ
GSYYPGQASPQQ
PGQGQQ
PGKWQE
PGQGQQWYYPTSLQQ
PGQGQQ
IGKGQQGYYPTSLQQ
PGQGQQGYYPTSLQH
TGQRQQ
PVQGQQ
PEQGQQ
PGQWQQGYYPTSPQQ
LGQGQQ
PRQWQQ
SGQGQQGHYPTSLQQ
PGQGQQGHYLASQQQ
PGQGQQGHYPASQQQ
PGQGQQGHYPASQQQ
PGQGQQGHYPASQQE
PGQGQQGQIPASQQQ
PGQGQQGHYPASLQQ
PGQGQQGHYPTSLQQ
LGQGQQIGQ
PGQKQQ
PGQGQQ
TGQGQQ
PEQEQQ
PGQGQQGYYPTSLQQ
PGQGQQ
QGQGQQGYYPTSLQQ
PGQGQQGHYPASLQQ
PGQGQ
PGQRQQ
PGQGQH
PEQGKQ
PGQGQQGYYPTSPQQ
PGQGQQ
LGQGQQGYYPTSPQQ
PGQGQQ
PGQGQQGHCPTSPQQ
SGQAQQ
PGQGQQ
IGQVQQ
PGQGQQGYYPTSVQQ
PGQGQQ
SGQGQQ
SGQGHQ
PGQGQQ
SGQEQQGYD
```

Figure 14.6: Amino acid sequences of the repetitive domains of typical x-type (1Dx5) and y-type (1Dy10) HMW subunits arranged to show their repeat unit structures.

Table 14.1. Frequency of occurrence of different amino acid residues

A	Hexapeptides (%)						Tripeptides (%)		
	1	2	3	4	5	6	1	2	3
	Pro 62	Gly 84	Gln 99	Gly 75	Gln 94	Gln 80	Gly 89	Gln 99	Gln 99
	Ser 26	Ala 7	Other 1	Trp 9	Leu 3	Pro 15	Asp 5	Arg 1	Arg 1
	Leu 10	Glu 4		Leu 7	Other 3	Ser 2	Ala 2		
	Ile 1	Arg 3		Glu 4		Arg 1	Arg 2		
	Other 2	Thr 2		Arg 2		Leu 1	His 1		
				Ala 1		Other 2			
				Other 1					

Nonapeptides (%)								
1	2	3	4	5	6	7	8	9
Gly 84	Tyr 98	Tyr 97	Pro 90	Thr 96	Ser 100	Pro 70	Gln 88	Gln 94
Arg 6	His 2	Asp 2	Leu 8	Ile 4		Ser 13	Leu 8	Leu 4
Glu 3		Phe 1	Ser 2			Leu 11	Trp 3	Glu 2
Trp 3						Ala 2	Arg 1	
Val 2						Glu 2		
Ala 1						— 2		
Lys 1						Arg 1		

B	Hexapeptides (%)					
	1	2	3	4	5	6
	Pro 65	Gly 92	Gln 96	Gly 76	Gln 94	Gln 94
	Ser 12	Glu 6	Lys 4	Glu 7	His 2	Glu 2
	Leu 10	Lys 2		Trp 7	Lys 2	His 2
	Ile 7	Other 1		Arg 4	Other 1	— 2
	Thr 4			Ala 2		
	Gln 1			Val 2		
	Other 1			Lys 1		
				Other 2		

Nonapeptides (%)								
1	2	3	4	5	6	7	8	9
Gly 96	Tyr 54	Tyr 85	Pro 91	Thr 60	Ser 97	Leu 54	Gln 97	Gln 90
Trp 2	His 41	Cys 4	Leu 5	Ala 37	Tyr 2	Pro 21	His 3	His 7
Arg 1	Gln 5	Asp 2	Arg 2	Ser 2	Phe 1	Gln 19		Glu 2
Tyr 1		His 2	Ser 1	Ile 1		Val 4		Stop 1
		Ile 2	Thr 1			Gly 2		
		Phe 2				Ser 2		
		Arg 1				Ala 1		
		Asn 1						
		Glu 1						

Note: (A) The 347 hexapeptide, 103 nonapeptide, and 81 tripeptide repeat motifs of x-type HMW subunits 1Ax1, 1Bx7, and 1Dx5 (*T. aestivum*); 1Ax (*T. timopheevi*); and 1Dx (*A. cylindrica*). (B) The 339 hexapeptide and 123 nonapeptide repeat motifs of y-type HMW subunits 1Ay (not expressed), 1Bx7 and 1Dx5 (*T. aestivum*); 1Ay (*T. timopheevi*); 1Cy and 1Dy (*A. cylindrica*); and 1Dy (*T. tauschii*). (Percentages may not add up to 100 because of rounding off. Residues present at less than 1% are either included as 'Other' if together they add up to 1% or are not shown.)

Failure to detect any appreciable differences between the consensus motifs and degree of conservation of the repetitive sequences present in the HMW subunits of cultivated and wild species (Wan et al., 2001) indicates that selection by plant breeders for dough strength, which has been carried out systematically for the last century and perhaps unconsciously over the 10,000 year-life of bread wheat, has had little or no impact on the sequences (and hence structure) of the subunits. However, it is possible that the differences in degree of conservation within the motifs, and the precise amino acid residues which are present as substitutions at different positions within the motifs, may relate to their role in determining the structure adopted by the domain.

Cysteine residues occur only rarely in the repetitive sequences, with single cysteine residues present towards the C-terminal end of the repetitive domains (at position −73) of 1By and 1Dy subunits only. In addition, subunit 1Dx5 differs from all other subunits whose sequences are known in that a single additional cysteine is present at position 8 relative to the N-terminal end of the domain.

STRUCTURE OF THE HMW SUBUNIT REPETITIVE DOMAIN

Although several workers have attempted to determine the structure adopted by the HMW subunit repeats by X-ray crystallography of whole subunits or repetitive peptides, the crystals produced have failed to give clear diffraction patterns. Similarly, analysis of synthetic peptides based on the repetitive sequence motifs has not so far led to the determination of three-dimensional structures. Consequently, our current view of HMW subunit structure comes from a range of indirect studies.

Early hydrodynamic studies of subunit 1Bx20 purified from pasta wheat indicated that it had an extended rod-shaped conformation in solution, the dimensions ranging from about 500×17.5 Å to 620×15 Å depending on the solvent (Field et al., 1987). Detailed spectroscopic studies of whole subunits, of recombinant repetitive peptides, and of linear and circular synthetic peptides (Tatham et al., 1985; Field et al., 1987; van Dijk et al., 1997a, 1997b; Gilbert et al., 2000) have also been reported. The results are consistent with the repetitive sequences forming β-reverse turns which may be in equilibrium with poly-L-proline II structure, the latter predominating at low temperature (Gilbert et al., 2000). It has also been proposed that the β-turns are organised to give a regular spiral structure (termed a β-spiral) similar to that demonstrated for a synthetic polypentapeptide based on a repeat motif of elastin (Urry, 1988). Molecular modelling can be used to generate such spiral structures (Figure 14.7) whose dimensions (diameter, pitch, and length) are consistent with those determined by viscometric analysis and revealed by scanning tunnelling microscopy of purified proteins in the hydrated solid state (Miles et al., 1991). However, Kasarda

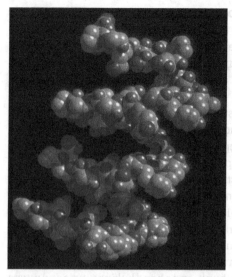

Figure 14.7: Molecular model developed for a β-spiral structure based on the amino acid sequence of a repetitive domain of a HMW subunit. Only the backbone structure is shown. (Unpublished results of D. J. Osguthorpe, O. Parchment, P. R. Shewry, and A. S. Tatham.)

and coworkers (1994) have proposed that an alternative type of spiral structure is formed, based on γ-turns rather than β-turns.

SEQUENCES AND STRUCTURES OF THE NON-REPETITIVE DOMAINS

The N-terminal domains vary in length, being 81–89 residues in the x-type subunits and 104 residues in the y-type. This difference results from a deletion in the x-type subunits, compared with the y-type, which involves the loss of two cysteine residues. Consequently, the N-termini of the x-type subunits usually contain three cysteine residues and those of the y-type subunits five. Structure prediction and molecular modelling studies indicate that this domain is 'globular', with one or more α-helices (Tatham et al., 1984, 1985; van Dijk et al., 1998; Köhler et al., 1997).

The C-terminal domains of all of the subunits comprise 42 residues with single cysteine residues at position-13 with respect to the C-terminus. Structure prediction indicates that this domain may be α-helical (Tatham et al., 1984) and nuclear magnetic resonance spectroscopy of a synthetic peptide dissolved in 40% (v/v) aqueous trifluoroethanol (a 'structure-inducing' solvent) allowed a low-resolution structure containing two α-helices to be determined (Bekkers et al., 1996).

HMW SUBUNIT STRUCTURE AND GLUTEN ELASTICITY

The HMW subunits are only present in glutenin polymers, particularly in HMW polymers, the amounts of which are positively correlated with dough strength (Field et al., 1983). This provides support for the genetic evidence (discussed previously) that the HMW subunits are the major determinants of dough and gluten elasticity.

Two features of HMW subunit structure may be relevant to their role in glutenin elastomers: the number and distribution of disulphide bonds and the properties and interactions of the repetitive domains.

Direct sequence analysis of disulphide-linked peptides released by enzymic digestion of glutenin or gluten fractions has revealed a number of inter- and intrachain disulphide bonds involving HMW subunits (Köhler et al., 1991, 1993, 1994; Tao et al., 1992; Keck et al., 1995). These are summarised diagramatically in Figure 14.8 and include one interchain disulphide bond within the N-terminal domain of an x-type subunit, two parallel disulphide bonds between the N-termini of y-type subunits, an interchain bond between a y-type subunit and a low molecular mass glutenin subunit, and a bond linking y-type and x-type subunits in a 'head-to-tail' fashion. The latter is consistent with the results obtained by partial reduction of glutenin, which leads to the release of dimers comprising x-type + y-type subunits (Lawrence and Payne, 1983; Tao et al., 1992). Such dimers have therefore been proposed to form the 'building blocks' of glutenin (Graveland et al., 1985). However, our knowledge of the detailed

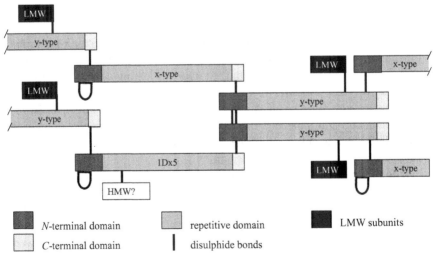

Figure 14.8: Schematic model of the structure of HMW subunit polymers, based on mapped disulphide bonds (Köhler et al., 1991, 1993, 1994, Tao et al., 1992; Keck et al., 1995). LMW, low molecular weight.

disulphide structure of glutenin is not sufficiently complete to allow us to relate disulphide distribution to biomechanical properties.

Although it is now widely accepted that disulphide-linked glutenin chains provide an 'elastic backbone' to gluten, evidence from spectroscopic studies (using nuclear magnetic resonance and Fourier-Transform Intra-Rod spectroscopy) of HMW subunits and of model peptides based on the repeat motif suggests that non-covalent hydrogen bonding between glutenin subunits and polymers may also be important (Belton et al., 1994, 1995, 1998; Wellner et al., 1996; Gilbert et al., 2000). These studies have shown that the dry proteins are disordered with little regular structure, but that their mobility increases and β-sheet structures form on hydration. Further changes occur if hydration continues, with a further increase in protein mobility and the formation of turn-like structures at the expense of β-sheet.

These observations led to the development of a "loop and train" model (Belton, 1999) which is summarised in Figure 14.9. This proposes that the low hydration state has many protein:protein interactions, via hydrogen bonding of glutamine residues in the β-spiral structures. As the hydration level increases, the system is plasticised allowing the orientation of the β-turns in adjacent β-spirals to form structures resembling 'interchain' β-sheet. Further hydration leads to the breaking of some of the interchain hydrogen bonds in favour of hydrogen bonds between glutamine and water, leading to the formation of loop regions. However, it does not result in the complete replacement of interchain hydrogen bonds, and hence solution of the protein, as the number of glutamine residues is high and the statistical likelihood of all the interchain bonds breaking simultaneously is therefore low. The result is an equilibrium between hydrated 'loop' regions and hydrogen bonded 'chain' regions, with the ratio between these being dependent on the hydration state.

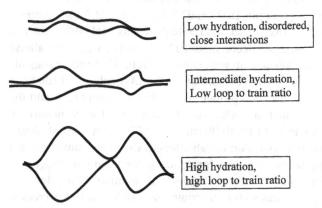

Low hydration, disordered, close interactions

Intermediate hydration, Low loop to train ratio

High hydration, high loop to train ratio

Figure 14.9: Model for the effect of hydration on the loop to train ratio of HMW subunits.

The equilibrium between 'loops' and 'trains' may also contribute to the elasticity of gluten, as extension of the dough will result in stretching of the 'loops' and 'unzipping' of the 'trains'. The resulting formation of extended chains may be a mechanism by which elastic energy is stored in the dough, providing an explanation for the increased resistance to extension which occurs during dough mixing. The formation of interchain hydrogen bonds between glutamine residues may also account for the observations that the esterification of glutamine residues results in decreased resistance to extension while mixing in the presence of deuterium oxide rather than water results in increased resistance (Beckwith et al., 1963; Mita and Matsumoto, 1981; Bushuk, 1998).

MANIPULATION OF HMW SUBUNIT COMPOSITION
IN TRANSGENIC WHEAT

The major aim of determining the structures of the HMW subunits and their role in gluten and dough elasticity is to facilitate the improvement of the end use properties of wheat. Substantial improvement in the processing performance of wheat has already been achieved by a combination of classical plant breeding and optimisation of the agronomic and processing conditions. However, it is unlikely that these approaches will be sufficient in the long term, and genetic engineering therefore provides an important additional approach. We are, therefore, using genetic engineering of wheat to further develop the role of the HMW subunits in determining processing properties and to define strategies for the production of improved germ plasm for incorporation into plant breeding programmes.

Most of our work so far has focused on transformation of two 'model' lines of wheat with two different HMW subunit genes. The model lines form part of a near isogenic series which have been produced by crossing lines differing in their expression of HMW subunit genes. Thus, line L88-31 expresses only two HMW subunit genes (encoding subunits 1Bx17 and 1By18), while L88-6 also expresses genes encoding subunits 1Ax1, 1Dx5, and 1Dy10 (Lawrence et al., 1988). The two genes used for transformation encode subunits 1Ax1 and 1Dx5, the latter always occurring as part of an 'allelic pair' with subunit 1Dy10. The three transgenic lines which have been studied in detail, express the 1Ax1 subunit in L88-31 at a level of about 5.7% of the total protein (cf. 0% in the control line) and the 1Dx5 subunit in L88-31 and L88-6 at 8.7% (cf. 0%) and 17% (cf. 4.2%) of the total protein, respectively [Figure 14.10(A)] (Barro et al., 1997; Popineau et al., 2001).

The effects of the transgenes on dough strength were determined using a Mixograph. This measures the energy input during the mixing of dough and is routinely used for quality testing in a number of countries. When dough is mixed, the resistance increases up to a certain level, after which it decreases. The increase in resistance may result from limited exchange of disulphide bonds

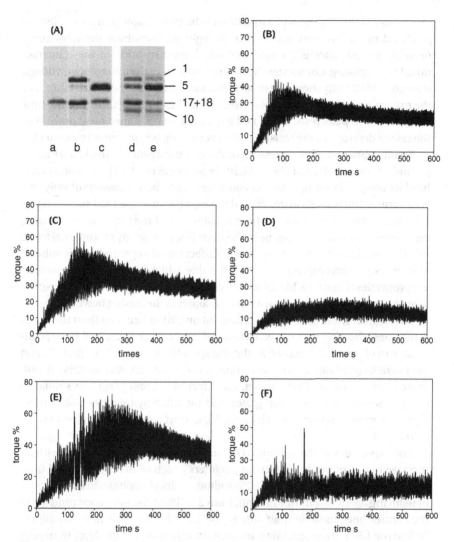

Figure 14.10: Analysis of the mixing properties of transgenic wheats expressing additional HMW subunits using the 2 g mixograph. (A) SDS-PAGE of the HMW subunits from (a) control line L88-31 (subunits 1Bx17 + 1By18); (b) L88-31 expressing the 1Ax1 transgene; (c) L88-31 expressing the 1Dx5 transgene; (d) control line L88-6 (subunits 1Ax1, 1Dx5 + 1Dy10, and 1Bx17 + 1By18); (e) L88-6 expressing the 1Dx5 transgene. (B–F) are mixographs of (B) L88-31; (C) L88-31 expressing the 1Ax1 transgene; (D) L88-31 expressing the 1Dx5 transgene; (E) L88-6; and (F) L88-6 expressing the 1Dx5 transgene. The resistance is given as torque% and the mixing time in seconds (s). From Prolamin Aggregation, Gluten Viscoelasticity, and Mixing Properties of Transgenic Wheat Line Expressing 1Ax and 1Dx High Molecular Weight Glutenin Subunit Transgenes', by Y. Popineau et al., 2001, *J. Agric. Food Chem.*, **49,** 395–401. © 2001 with permission.

(as discussed previously) and formation of the most stable patterns of hydrogen bonding (i.e., to form extensive 'train' regions). In contrast, the subsequent decrease in resistance is thought to result from disruption of these interactions by overmixing. Consequently, beneficial effects of the transgenes on dough strength and stability should be observed as increases in the peak resistance (i.e., the maximum resistance which is observed) and the mixing time (i.e., the time taken to mix to peak resistance), and a decrease in resistance breakdown (i.e., the rate of decrease in the resistance on overmixing beyond peak resistance).

Results obtained with expression of the two transgenes in the L88-31 background are summarised in Figure 14.10 (Popineau et al., 2001). The control line has low dough strength, which is consistent with the expression of only two endogenous HMW subunit genes, and the expression of the 1Ax1 transgene results in substantial increases in peak resistance and mixing time. In contrast, expression of the 1Dx5 transgene in the same line was clearly detrimental to the mixing properties. An even more extreme effect was observed when the subunit 1Dx5 transgene was expressed in the L88-6 line, which had much stronger mixing properties (Figure 14.10). In fact, both lines expressing the 1Dx5 transgene failed to absorb water and form a normal dough in the mixing bowl.

Rheological studies were also carried out on gluten fractions from the transgenic lines, showing that the expression of subunit 1Dx5 resulted in large increases in elasticity (measured as the storage and loss moduli, G' and G'', and the viscoelastic plateau, $Gn°$) while only a small increase was associated with expression of subunit 1Ax1. In fact, the effect of subunit 1Dx5 was similar to that previously observed when gluten was modified by treatment with transglutaminase to introduce interchain lysyl-glutamyl cross-links (Popineau et al., 2001).

The expression of the subunit 1Dx5 transgene was also associated with an increase in the amounts of glutenin polymers which were only extracted from flour by sonication with detergent (2% sodium dodecyl sulphate) in the presence of reducing agent (1% dithiothreitol), from 2 to 3% of the total flour proteins in the control lines to more than 18% in the L88-31 transgenic line and almost 30% in the L88-6 transgenic line. In contrast, expression of the 1Ax1 transgene was associated with a modest increase in the amount of polymers extracted by sonication in the absence of reducing agent, but had no effect on the amount of insoluble polymers.

These results suggest that the proteins encoded by the two transgenes had fundamentally different effects on the structure of the glutenin polymers in the two lines, with the 1Dx5 protein leading to the formation of highly cross-linked polymers which resulted in high gluten strength, unusual hydration behaviour, and failure to form a homogeneous network during mixing. In contrast, the expression of subunit 1Ax1 resulted in similar effects on gluten composition and properties to those observed when comparing near-isogenic lines differing in HMW subunit composition.

As discussed previously, the 1Dx5 subunit protein differs from other characterised subunits in the presence of an additional cysteine residue within the repetitive domain, and this may be responsible for the formation of highly cross-linked polymers in the transgenic lines. However, it must also be borne in mind that subunit 1Dx5 is always found together with 1Dy10 and that dimers of these subunits are released by partial reduction of glutenin. Consequently, a precise molar balance of these two subunits may be required to give a 'normal' glutenin polymer structure.

These results demonstrate, therefore, that transformation of bread wheat with different HMW subunit genes may have fundamentally different effects on gluten structure and properties which may relate to the expression levels, structures and interactions of the individual proteins.

ACKNOWLEDGEMENTS

Long Ashton Research Station and The Institute of Food Research receive grant-aided support from the Biotechnology and Biological Sciences Research Council of the United Kingdom. We thank Dr. Don Bechtel (USDA, Manhattan, KS, USA) and Dr. Mary Parker (The Institute of Food Research, Norwich, UK) for providing parts of Figure 14.2.

REFERENCES

Anderson, O. D., Greene, F. C., Yip, R. E., Halford, N. G., Shewry, P. R., and Malpica-Romero J.-M. (1989) Nucleotide sequences of the two high-molecular-weight glutenin genes from the D-genome of a hexaploid bread wheat, *Triticum aestivum* L. cv Cheyenne. *Nucleic Acids Res.* **17**, 461–62.

Barro, F., Rooke, L., Békés, F., Gras, P., Tatham, A. S., Fido, R. J., Lazzeri, P., Shewry, P. R., and Barcelo, P. (1997) Transformation of wheat with HMW subunit genes results in improved functional properties. *Nature Biotech.* **15**, 1295–99.

Bartels, D., Altosaar, I., Harberd, N. P., Barker, R. F., and Thompson, R. D. (1986) Molecular analysis of γ-gliadin gene families at the complex *Gli-1* locus of bread wheat (*T. aestivum* L.). *Theoret. Appl. Genet.* **72**, 845–53.

Beckwith, A. C., Wall, J. S., and Dimmler, R. J. (1963) Amide Groups as Interaction Sites in Wheat Gluten Proteins: Effects of Amide-Ester Conversion. *Arch. Biochim. Biophys.* **103**, 319–30

Bekkers, A. C., van Dijk, A., de Boef, E., van Swieten, E., Robillard, G., and Hamer, R. J. (1996) HMW glutenins:structure-function relationships step by step. In: *Gluten 96—Proceedings of the 6th International Wheat Gluten Workshop*, Sydney, Australia, September 1996, pp. 190–94.

Belton P. S., Gil, A. M., and Tatham, A. S. (1994) ^{1}H NMR relaxation times studies of the hydration of the barley protein C-hordein. *J. Chem. Soc. Faraday Trans.* **90**, 1099.

Belton, P. S. (1999) On the elasticity of wheat gluten. *J. Cereal Sci.* **29**, 103–7.

Belton, P. S., Colquhoun, I. J., Field J. M., Grant, A., Shewry, P. R., Tatham, A. S., and Wellner, N. (1995) FTIR and NMR studies on the hydration of a high M_r subunit of glutenin. *Int. J. Biol. Macromol.* **17,** 74.

Belton, P. S., Gil, A. M., Grant, A., Alberti, E., and Tatham, A. S. (1998) Proton and carbon NMR measurements of the effects of hydration on the wheat protein omega gliadin. *Spectrochim. Acta A* **54,** 955–66.

Branlard, G., and Dardevet, D. (1985) Diversity of grain protein and bread wheat quality. II. Correlation between high molecular weight subunits of glutenin and flour quality characteristics. *J. Cereal Sci.* **3,** 345–54.

Bushuk, W. (1998) Interactions in wheat doughs. In: *Interactions, the Keys to Cereal Quality,* R. J. Hamer and R. C. Hoseney (eds.). St. Paul, MN: American Association of Cereal Chemists, pp. 1–14.

Colot, V., Bartels, D., Thompson, R., and Flavell R. (1989). Molecular characterization of an active wheat LMW glutenin gene and its relation to other wheat and barley prolamin genes. *Mol. Gen. Genet.* **216,** 81–90.

Field, J. M., Shewry, P. R., and Miflin, B. J. (1983) Solubilization and characterization of wheat gluten proteins; correlations between the amount of aggregated proteins and baking quality. *J. Sci. Food Agric.* **34,** 370–77.

Field, J. M., Tatham, A. S., and Shewry, P. R. (1987) The structure of a high M_r subunit of durum wheat (*T. durum*) gluten. *Biochem. J.* **247,** 215–21.

Galili, G. (1997) The prolamin storage proteins of wheat and its relatives. In *Cellular and Molecular Biology of Plant Seed Development*, B. A. Larkins and I. K. Vasil (eds.). Dordrecht, The Netherlands: Kluwer Academic Publishers, pp. 221–56.

Gilbert, S. M., Wellner, N., Belton, P. S., Greenfield, J. A., Siligardi, G., Shewry, P. R., and Tatham, A. S. (2000) Expression and characterisation of a highly repetitive peptide derived from a wheat seed storage protein. *Biochim. Biophys. Acta* **1479,** 135–46.

Graveland, A., Bosveld, P., Lichtendonk, W. J., Marseille, J. P., Moonen, J. H. E., and Scheepstra, A. (1985) A model for the molecular structure of the glutenins from wheat flour. *J. Cereal Sci.* **3,** 1–16.

Halford, N. G., Field, J. M., Blair, H., Urwin, P., Moore, K., Robert, L., Thompson, R., Flavell, R. B., Tatham, A. S., and Shewry, P. R. (1992) Analysis of HMW glutenin subunits encoded by chromosome 1A of bread wheat (*Triticum aestivum* L.) indicates quantitative effects on grain quality. *Theoret. Appl. Genet.* **83,** 373–78.

Kasarda, D. D., King, G., and Kumosinski, T. F. (1994) Comparison of spiral structures in wheat high-molecular-weight glutenin subunits and elastin by molecular modeling. In: *Molecular Modeling: From Virtual Tools to Real Problems*, T. F. Kumosinski and M. N. Liebman (eds.). Washington, DC: American Chemical Society, ACS Symposium Series No. 576, pp. 210–20.

Kasarda, D. D., Okita, T. W., Bernardin, J. E., Baecker, P. A., Nimmo, C. C., Lew, E. J.-L., Dietler, M. D., and Greene, F. (1984) Nucleic acid (cDNA) and amino acid sequences of α-type gliadins from wheat (*Triticum aestivum*). *Proc. Natl. Acad. Sci. U.S.A.* **81,** 4712–16.

Keck, B., Köhler, P., and Wieser, H. (1995) Disulphide bonds in wheat gluten: Cystine peptides derived from gluten proteins following peptic and thermolytic digestion. *Z. Lebensm. Unters. Forsch.* **200,** 432–39.

Köhler, P., Belitz, H.-D., and Wieser, H. (1991) Disulphide bonds in wheat gluten: Isolation of a cysteine peptide from glutenin. *Z. Lebensm. Unters. Forsch.* **192,** 234–39.

Köhler, P., Belitz, H.-D., and Wieser, H. (1993) Disulphide bonds in wheat gluten: Further cysteine peptides from high molecular weight (HMW) and low molecular weight (LMW) subunits of glutenin and from γ-gliadins. *Z. Lebensm. Unters. Forsch.* **196,** 239–47.

Köhler, P., Keck, B., Müller, S., and Wieser, H. (1994) Disulphide bonds in wheat gluten. In: *Wheat Kernal Proteins, Molecular and Functional Aspects.* Proc. Int. Meeting S. Martino at Cimino, Viterbo, Italy, 28–30 September 1994, pp. 45–54.

Köhler, P., Keck-Gassenmeier, B., Wieser, H., and Kasarda, D. D. (1997) Molecular modelling of the *N*-terminal regions of high molecular weight glutenin subunits 7 and 5 in relation to intramolecular disulphide bond formation. *Cereal Chem.* **74,** 154–58.

Kreis, M., Forde, B. G., Rahman, S., Miflin, B. J., and Shewry, P. R. (1985) Molecular evolution of the seed storage proteins of barley, rye and wheat. *J. Mol. Biol.* **183,** 499–502.

Lawrence, G. J., and Payne, P. I. (1983) Detection by gel electrophoresis of oligomers formed by the association of high-molecular-weight glutenin protein subunits of wheat endosperm. *J. Exp. Bot.* **34,** 254–67.

Lawrence, G. J., MacRitchie, F., and Wrigley, C. W. (1988) 'Dough and baking quality of wheat lines deficient in glutenin subunits controlled by the *Glu-A1*, *Glu-B1* and *Glu-D1* loci'. *J. Cereal Sci.* **7,** 109–12.

Mackie, A. M., Sharp, P. J., and Lagudah, E. S. (1996) The nucleotide and derived amino acid sequence of a HMW glutenin gene from *Triticum tauschii* and comparison with those from the D genome of bread wheat. *J. Cereal Sci.* **24,** 73–78.

Mecham, D. K., and Knapp, C. (1966) The sulphydryl contents of doughs mixing under nitrogen. *Cereal Chem.* **43,** 226.

Miflin, B. J., Field, J. M., and Shewry, P. R. (1983) Cereal storage proteins and their effects on technological properties. In: *Seed Proteins*, J. Daussant, J. Mosse, and J. Vaughan (eds.). London, UK: Academic Press, pp. 255–319.

Miles, M. J., Carr, H. J., McMaster, T., I'Anson, K. J., Belton, P. W., Morris, V. J., Field, J. M., Shewry, P. R., and Tatham, A. S. (1991) Scanning tunnelling microscopy of a wheat gluten protein reveals details of an unusual supersecondary structure. *J. Mol. Biol.* **88,** 68–71.

Mita, T., and Matsumoto, H. (1981) Flow properties of aqueous gluten and gluten methyl ester dispersions. *Cereal Chem.* **58,** 57–61.

Napier, J. A., Richard, G., Turner, M. F. P., and Shewry, P. R. (1997) Trafficking of wheat gluten proteins in transgenic tobacco plants: γ-Gliadin does not contain an endoplasmic reticulum-retention signal. *Planta* **203,** 488–94.

Nicolas, Y. (1997) Les prolamines de blé: Extraction exhaustive et développement de dosages chromatographiques en phase inverse et de dosages immunochimiques à l'aide d'anticorps anti-peptide. Thesis, University of Nantes, France.

Osborne, T. B. (1924) *The Vegetable Proteins*, 2nd ed. London, UK: Longmans, Green and Co., pp. 1–154.

Payne, P. I. (1987) Genetics of wheat storage proteins and the effect of allelic variation on breadmaking quality. *Ann. Rev. Plant Physiol.* **38,** 141–53.

Payne, P. I., Nightingale, M. A., Krattiger, A. F., and Holt, L. M. (1987) The relationship between HMW glutenin subunit composition and the breadmaking quality of British grown wheat varieties. *J. Sci. Food Agric.* **40,** 51–65.

Popineau, Y., Cornec, M., Lefebvre, J., and Marchylo, B. (1994) Influence of high M_r glutenin subunits on glutenin polymers and rheological properties of gluten and gluten subfractions of near-isogenic lines of wheat Sicco. *J. Cereal Sci.* **19,** 231–41.

Popineau, Y., Deshayes, G., Lefebvre, J., Fido, R., Tatham, A. S., and Shewry, P. R. (2001) Prolamin aggregation, gluten viscoelasticity, and mixing properties of transgenic wheat lines expressing 1Ax and 1Dx high molecular weight glutenin subunit transgenes. *J. Agric. Food Chem.* **49,** 395–401.

Reddy, P., and Appels, R. (1993) Analysis of a genomic DNA segment carrying the wheat high-molecular-weight (HMW) glutenin Bx17 subunit and its use as an RFLP marker. *Theoret. Appl. Genet.* **85,** 616–24.

Schofield, J. D. (1994) Wheat proteins: Structure and functionality in milling and bread-making. In: *Wheat Production, Properties and Quality,* W. Bushuk and V. F. Rasper (eds.). London, UK: Chapman and Hall, pp. 72–106.

Seilmeier, W., Belitz, H. D., and Wieser, H. (1991) Separation and quantitative determination of high-molecular-weight subunits of glutenin from different wheat varieties and genetic variants of the variety Sicco. *Z. Lebensm. Unters. Forsch.* **192,** 124–29.

Shewry, P. R. (1999) The synthesis, processing and deposition of gluten proteins in the developing wheat grain. *Cereal Foods World* **44,** 587–89.

Shewry, P. R., and Tatham, A. S. (1999) The characteristics, structures and evolutionary relationships of prolamins. In: *Seed Proteins,* P. R. Shewry and R. Casey (eds.). Dordrecht, The Netherlands, Kluwer Academic Publishers, pp. 11–33.

Shewry, P. R., Halford, N. G., and Tatham, A. S. (1992) The high molecular weight subunits of wheat glutenin. *J. Cereal Sci.* **15,** 105–20.

Shewry, P. R., Tatham, A. S., and Halford, N. G. (1999) The prolamins of the Triticeae. In: *Seed Proteins,* P. R. Shewry and R. Casey (eds.). Dordrecht, The Netherlands, Kluwer Academic Publishers, pp. 33–78.

Shewry, P. R., Tatham, A. S., Barro, F., Barcelo, P., and Lazzeri, P. (1995) Biotechnology of breadmaking: Unravelling and manipulating the multi-protein gluten complex. *Bio/Technology* **13,** 1185–90.

Shewry, P. R., Tatham, A. S., Forde, J., Kreis, M., and Miflin, B. J. (1986) The classification and nomenclature of wheat gluten proteins: A reassessment. *J. Cereal Sci.* **4,** 97–106.

Tao, H. P., Adalsteins, A. E., and Kasarda, D. D. (1992) Inter-molecular disulfide bonds link specific high-molecular-weight glutenin subunits in wheat endosperm. *Biochim. Biophys. Acta* **1159,** 13–21.

Tatham, A. S., Miflin, B. J., and Shewry, P. R. (1985) The β-turn conformation in wheat gluten proteins: Relationship to gluten elasticity. *Cereal Chem.* **62,** 405–12.

Tatham, A. S., Shewry, P. R., and Miflin, B. J. (1984) Wheat gluten elasticity: A similar molecular basis to elastin? *FEBS Lett.* **177,** 205–8.

Tsen, C. C., and Bushuk, W. (1963) Changes in sulphydryl and disulphide contents of doughs during mixing under various conditions. *Cereal Chem.* **40,** 399.

Urry, D. W. (1988) Entropic elastic processes in protein mechanisms. I. Elastic structure due to an inverse temperature transition and elasticity due to internal chain dynamics. *J. Protein Chem.* **7,** 1–34.

van Dijk, A. A., de Boef, E., Bekkers, A., Wijk, L. L., van Swieten, E., Hamer, R. J., and Robillard, G. T. (1997a) Structure characterization of the central repetitive domain of high molecular weight gluten proteins. II. Characterization in solution and in the dry state. *Protein Sci.* **6,** 649–56.

van Dijk, A. A., van Swieten, E., Kruize, I. T., and Robillard, G. T. (1998) Physical characterisation of the *N*-terminal domain of high-molecular-weight gluten subunit Dx5 from wheat. *J. Cereal Sci.* **28,** 115–26.

van Dijk, A. A., van Wijk, L. L., van Vliet, A., Haris, P., van Swieten, E., Tesser, G. I., and Robillard, G. T. (1997b) Structure characterization of the central repetitive domain of high molecular weight gluten proteins. I. Model studies using cyclic and linear peptides. *Protein Sci.* **6,** 637–48.

Wan, Y. F., Wang, D., Shewry, P. R., and Halford, N. G. (2002) Isolation and characterization of five novel high molecular weight subunit genes from *Triticum timopheevi* and *Aegilops cylindrica. Theor. Appl. Genet.* **104,** 828–39.

Wellner, N., Belton, P. S., and Tatham, A. S. (1996) Fourier transform IR spectroscopic study of hydration induced structure changes in the solid state of omega gliadins. *Biochem. J.* **319,** 741–47.

FIFTEEN

Biological Liquid Crystal Elastomers

David P. Knight and Fritz Vollrath

INTRODUCTION

Liquid crystals have been defined as a remarkable, paradoxical state of matter intermediate between true liquids and true crystals because their molecules or regions within the molecules have a mobility similar to that of true liquids, yet show orientational and sometimes positional order similar to that of solid crystals (Collings, 1990). While most types of liquid crystal are able to flow, liquid crystal elastomers (LCEs), also described as solid liquid crystals are not. LCEs have only recently been suggested to be a novel state of matter (Warner and Terentjev, 1996), although the theoretical existence of this state was first postulated more than 25 years ago by de Gennes (1969, 1975) and the term 'solid liquid crystal' was originally used in a now outmoded way to describe the more heavily cross-linked nematic resins (Bouligand et al., 1974). However, it was not until the late 1980s and early 1990s that LCEs were first synthesised by Finkelman's group at Freiburg University (Barclay and Ober, 1993; Schneider et al., 2000).

Both classical rubbers and LCEs are weakly cross-linked polymers showing reversible extensibility and energy storage (Legge et al., 1991). A major difference between these two classes of materials is that the polymer chains in classical rubbers are highly disordered at rest, while those of LCEs show orientational order earning them the term 'anisotropic elastomers' (Texeira, 1999). LCEs can be defined as lightly cross-linked main chain or side chain (Northolt and Sikkema, 1991) liquid crystalline polymers showing elastomeric properties. Their mesogenic (liquid crystal forming) segments can exist in a number of different liquid crystalline phases (mesophases), including nematic, smectic, twisted nematic (cholesteric) (Meier and Finkelman, 1993), hexagonal (Schneider et al., 2000), and lamellar (Fischer et al., 1995). These mesophases represent different energy minima that depend on different balances between weak attractive and repulsive forces acting on the mesogenic units (Collings, 1990; de Gennes 1974). Mesogenic groups must be chiral (displaying a single-handedness), relatively stiff, and rod-shaped or prolate- or oblate-spheroids (de Gennes, 1974). Lamellar

LCEs show an alternation of hydrophobic and hydrophilic segments in alternating layers (lamellae) and are formed from amphiphilic AB block copolymers (Fischer et al., 1995; King et al., 1999; Fraaije et al., 2000; Hamley, 2000). The formation of lamellae is driven by the segregation of the block copolymer's hydrophobic A-blocks from the more hydrophilic B-blocks. Amphiphilic AB-block copolymers also show a remarkable ability to self-assemble into nanofibrils and other hierarchical textures (Chan et al., 1992; Putthanarat et al., 2000; Qu et al., 2000).

The mesogenic units in LCEs as in other liquid crystals show a statistically defined orientation distribution about a single direction known as the liquid crystal director (de Gennes, 1974). The director field is coherent throughout a volume known as a domain. Monodomain LCEs are constructed from a single domain, and are more common and better understood than the polydomain ones (with many domains each with their own director field). Cross-links at nodes hold together the polymer chains of LCEs into a solid meshwork, but the mesogenic units retain at least some degree of liquid-like mobility (Texeira, 1999). This mobility depends on three factors: the flexibility of spacer links (Mitchell et al., 1992) which join the mesogenic segments to the cross-linked nodes of the meshwork; the relatively small number of cross-linked regions in the network (Warner & Terentjev, 1996); and as in other liquid crystals (Collings, 1990), on weak repulsive forces balancing the attractive ones between the mesogenic units. The mesogenic domains are free to change in orientation and in position relative to one another (Mao et al., 1998).

The elasticity of LCEs is thought to depend on a mechanical coupling through cross-links between the applied strain and the orientation of the mesogenic units (Legge et al., 1991), though the detailed workings of this coupling are still far from clear (Uchida, 2000). When a deforming force is applied to the mesogenic units, there is usually a change in their statistical orientation relative to the original director and a consequential change in shape of the polymer molecule as a whole accompanied by an increase in free energy. The increase in free energy can be used to restore the chains to their preferred, energy-minimised state. This means that the classical Gaussian theory of rubber elasticity (Treloar, 1975) can be applied with suitable extension, to LCEs (Warner et al., 1988; Mao et al., 1998). Under certain circumstances, however, the whole distribution of the mesogenic units (director) can be rotated by a deforming force in such a way that the shape of the molecules continues to reflect the shape of the sample of material as a whole. This rotation therefore occurs without distortion of the polymer chains and consequently without a change in free energy giving them what is known as soft elasticity, one of several exotic properties found in LCEs. Soft elasticity results in wholly or partly flattened regions in the material's stress/strain curve not seen in isotropic rubbers (Texeira and Warner, 1999; Uchida, 2000).

Monodomain LCEs can be prepared in two ways: (1) by cross-linking a liquid crystalline monomer in its liquid crystalline state; (2) by lightly cross-linking a liquid crystalline monomer in its isotropic (disordered) state, then applying sufficient mechanical stress to order the molecules and finally cross-linking them again (Warner and Terentjev, 1996).

Having described LCEs from a material science perspective, we aim to show that fibrils of transversely banded collagens and the nanofibrils of spider dragline silks can be regarded as lamellar liquid crystal elastomers (LLCEs). To do this, it is necessary to satisfy the following criteria taken from the aforementioned description of the structure, formation and properties of LCEs: (1) the fibrils must have the structure of an LLCE – that is, they must be lightly cross-linked networks in which some segments exhibit liquid-like mobility; and there is an alternation of hydrophobic and hydrophilic lamellae. (2) Initial assembly of the fibrils is from mesogenic, amphiphilic AB-block copolymeric 'monomers'; is thermodynamically reversible; and depends on the existence of the balance between weak attractive and repulsive interactions, such as hydrophobic/hydrophilic effects and coulombic forces. (3) The fibrils show reversible extensibility.

We then consider the predictive value of classifying collagens and spider dragline silks as LLCEs. Finally, we suggest that other amphiphilic block copolymeric fibrous proteins forming transversely banded fibrous fibrils may be LLCEs.

EVIDENCE THAT FIBRILLAR COLLAGENS ARE LLCEs

Liquid Crystalline Structure

The first evidence for the liquid crystallinity of a fibrillar collagen with a classical 68 nm axial banding pattern came from the study of elastoidin, which has tapered giant fibrils up to 2 mm in diameter. These give resilience to the fins of cartilaginous fish (Woodhead-Galloway and Knight, 1977), and the term 'elastoidin' derives from their rubber-like properties when wet, completely and rapidly straightening after being bent double. Elastoidin shrinkage temperature and insolubility are both high (Woodhead-Galloway and Knight, 1977), indicating that the molecules are cross-linked into a network, a necessary prerequisite for an LCE. The low-angle X-ray scattering of wet elastoidin showed only diffuse scattering on the equator, indicating that the transverse arrangement of the molecules was completely liquid-like despite the highly ordered longitudinally staggered arrangement of molecules closely similar to that of conventional mammalian collagen fibrils (Woodhead-Galloway and Knight, 1977) found for example in tendons.

Findings on elastoidin led to a re-examination of the low-angle equatorial X-ray diffraction pattern of rat tail tendon fibrils. These demonstrated a combination of sharp Bragg reflections and diffuse scatter, suggesting that the side-to-side arrangement in some regions of the axially repetitive structure of the fibril were crystalline while other regions were liquid-like (Woodhead-Galloway and Knight, 1977). A comparison of the X-ray diffraction of wet and dry elastoidin gave further evidence for liquid crystallinity as follows. The molecules were axially orientated when wet, but tilted sharply on drying, and a similar tilting occurs in other collagen fibrils (Hukins et al., 1976) as would be expected in a smectic A to smectic C liquid crystal phase transition. Taken together, these observations led to the suggestion that elastoidin and other transversely banded collagen fibrils were liquid crystalline (Hukins and Woodhead-Galloway, 1977, 1978). The diffuse equatorial scattering of elastoidin has been recently reinvestigated and shown to involve two components, one of which is Gaussian. The latter component changed as the water content was altered (Lees, 1998) and is thought to result from side-to-side movements of collagen segments produced by thermal agitation, suggesting the mobility required for liquid crystallinity. In rat tail tendon fibrils, a high proportion (80%) of the equatorial scattering comes from liquid-like segments while only 20% comes from ordered segments (Hulmes et al., 1995), as might be expected in an LCE.

Further support for the liquid crystallinity of collagen fibrils came from improved models for rat tail tendon collagen fibrils. A crimped model in which highly mobile molecular segments of collagen in the gap regions are linked by highly ordered straight, tilted segments in the overlap region explained the sharp reflections and diffuse scatter of the low-angle equatorial X-ray diffraction pattern (Parry, 1988). The gap region has a much higher water content, lower packing fraction, and lower content of imino- and aromatic amino acids thought to contribute to rigidity of the collagen helix (Fraser et al., 1987). Thus, evidence from X-ray diffraction, together with evidence for formation of networks by systematic rather than random intermolecular covalent cross-linking (Fraser, 1998; Knott and Bailey, 1998), suggests a strong resemblance between collagen fibrils and LCEs.

The mobility of different transverse zones of the conventional collagen fibril is incompletely understood. The 4:1 ratio of mobile to immobile segments (Hulmes et al., 1995) suggests that the mobile segments are not limited solely to the gap region. Evidence from molecular modelling suggests that the regions of the collagen molecule rich in hydrophilic residues are less perfectly triple helical than those rich in hydrophobic amino acids (Brown et al., 1997), and this may contribute to differences in flexibility. In addition, we have suggested that an anomalous region of the rat tail tendon fibril may be particularly mobile and serve as dashpot close to the principal intermolecular cross-link (Hu et al., 1997).

In addition to work on conventionally banded collagen fibrils, there is evidence that the unconventionally banded fibrils of dogfish egg case (EC) collagen also have the structure of an LCE. These fibrils show an axial periodicity of 40 nm and are formed from a much shorter collagen molecule than that of rat tail

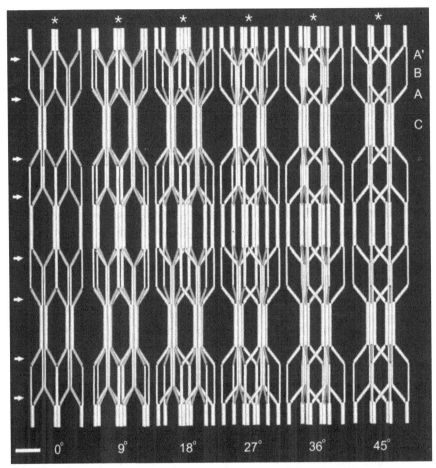

Figure 15.1: A stereo gallery showing four unit cells to illustrate the open network molecular construction of dogfish EC collagen fibrils. The structural similarity to an LLCE is immediately obvious. The figure can be viewed as a series of stereo images of the unit cell at different rotations. Arrows point to the A or A′ bands. The axial repeat period is 40 nm. From 'Collagen Packing in the Dogfish Egg Case Wall', by C. Knupp et al., 1998, *J. Struct. Biol.*, **122**, 101–10. ©1998 with permission.

tendon collagen (Knight et al., 1996). Like elastoidin, dogfish EC collagen is highly insoluble and has a very high thermal shrinkage temperature, indicating the existence of a cross-linked meshwork. Figure 15.1 shows the regularly kinked, meshwork structure of these fibrils (Knupp et al., 1998; Knupp and Squire, 1998). The meshwork is thought to be held together in part by covalent bonds between the ends of the molecules (Knight and Vollrath, 2001a), which are thought to lie in narrow bands that strongly exclude negative stain (Knight and Hunt, 1974). The fibrils have been described as a cross-linked lamellar liquid crystal (Knight et al., 1993). The structure of assembly states in other unconventional collagens, such as the hexagonal columns seen in type VIII collagen (Ricard-Blum et al., 2000), suggests that many other collagens could be liquid crystalline and represent other LC phases.

The hydrophobic residues of collagens tend to be grouped into several transverse bands within the axial period of the fibril (Chapman et al., 1990; Gathercole et al., 1993; Ortolani et al., 1994) as in LLCEs.

Liquid Crystalline Assembly

Martin et al. (2000) have shown that type I procollagen, the precursor for mammalian type I collagen fibrils, forms nematic and precholesteric liquid crystals in vitro and have argued that this accounts for the assembly of crimped fibrils and other assemblies in vivo. We have also presented evidence from chemically fixed material that dogfish EC collagen fibrils are assembled from a liquid crystalline precursor, which undergoes a number of liquid crystal phase transitions during storage and secretion (Knight et al., 1993). The length-to-width ratio (approximately 20) of dogfish EC collagen molecules is highly favourable for nematic liquid crystallisation (Knight et al., 1996).

Sequence data obtained for many other members of the collagen family clearly demonstrates that they are block copolymers with an amphiphilic nature arising from the tendency of hydrophobic and hydrophilic amino acids to be segregated into separate regions in the molecule (Ricard-Blum et al., 2000; Knupp and Squire, 2001). This molecular structure is prerequisite for the formation of LLCEs.

Weak forces – including hydrophobic, hydrophilic, and coulombic interactions – are important, in establishing the stagger which gives rise to the axially banded structure of fibrillar collagens (Hulmes et al., 1973; Leikin et al., 1995; Wallace, 1985) as would be expected for lyotropic liquid crystals (Collings, 1993). An analysis of the assembly of highly ordered sheets of type VI collagen, closely analogous to those of dogfish EC collagen, clearly demonstrates that type VI collagen is an amphiphilic AB-block copolymer and that hydrophobic and coulombic interactions are important in its assembly (Knupp and Squire, 2001).

The earliest stages in the assembly of type I collagen fibrils are thermodynamically reversible as would be expected for a liquid crystal phase transition (see brief review in Knight et al., 1993).

Elastomeric Properties

Thus, the structure of transversely banded collagens, the mobility of some of their molecular segments, and their method of assembly strongly suggest that they are LLCEs, but do they show elastomeric properties? A short answer to this question is complicated by the fact that tissues or structures containing fibrillar collagen are by definition fibrous composites. This means that the bulk mechanical properties depend at least to some degree on the interfibrillar matrix and the arrangement of fibrils within the material. Inferences about the mechanical properties of the fibrils themselves cannot therefore be derived directly from the bulk properties. However, the practically uniaxial orientation of collagen fibrils in slightly stretched tendons and the small concentration of matrix in this tissue simplifies the problem, enabling the conclusion that tendon collagen fibrils must be quite stiff (Vincent, 1982). Subtler approaches are required to examine the tensile properties of the fibrils themselves.

The effect of strain on conventional collagen fibrils has been investigated by low-angle synchrotron X-ray diffraction (Misof et al., 1997), demonstrating a completely reversible extension of the axial period during cyclical loading in the toe region of the stress/strain curve and an increase in lateral molecular order and reduction of mobility with strain. The extension of the axial period is thought to arise from the straightening out of thermally activated molecular kinks. Such tensile behaviour is compatible with that of an LLCE.

Reversible extension of the axial period has also been demonstrated in dogfish EC collagen (Hepworth et al., 1994). This material is much tougher than tendon and can be extended by up to 39% before rupture. Low-angle X-ray diffraction and transmission electron microscopy (TEM) indicated that the axial period increased reversibly by 8% when the material was subjected to a global strain of 24%. As in conventional collagen fibrils, this reversible extension could be accounted for by a straightening out of the molecular kinks.

The mobility and rigidity of the collagen molecules under progressive strain has also been investigated by solid-state proton NMR in mammalian arterial wall after selective removal of elastin. This approach indicated that a reduction in mobility of the chains accounts for the slope of the stress/strain curve of the bulk material at high strain (Vinee et al., 1993) as would be expected in an elastomer. These findings suggested a model for the elasticity of collagen closely similar to that proposed by Misof et al. (1997).

Finally, the effect of temperature on increasing the stiffness of collagen fibrils in suspension has been successfully modelled using the assumption that the fibrils behave mechanically like lyotropic liquid crystal polymers (Rosenblatt et al., 1993).

EVIDENCE THAT ORB WEB SPIDER DRAGLINE SILKS ARE LLCEs

Liquid Crystalline Structure

There is consensus that the dragline silk of orb web spiders contains β-crystallites largely composed from the polyalanine segments (Jelinski, 1998; Riekel et al., 1999) separated by the more mobile and less ordered glycine-rich segments

(Gosline et al., 1984, 1994; Termonia, 1994). X-ray diffraction indicated that the β-crystallites were approximately axially oriented with respect to the fibre (see reviews and further evidence in Riekel et al., 1999; Grubb and Ji, 1999). These observations support a model (Termonia, 1994) for silk as a classical entropic rubber in which isolated β-crystallites with an axial orientation are set in a tangle of disordered segments. It is, however, unclear what orientated the isolated crystallites in Termonia's model and the model did not fit all the experimental data (Termonia, 1994; Kümmerlen et al., 1996; Simmons et al., 1996).

Evidence has accumulated for an improved model for spider dragline silk in which the major component for orb web dragline silk is an orientated nanofibril composed of transverse lamellae which show a regular alternation of axially orientated β-crystallites and less crystalline glycine-rich domains (Figure 15.2). Four lines of evidence strongly support such a model: (1) ^{13}C two-dimensional nuclear magnetic resonance (NMR) spin-diffusion measurements on *Nephila* dragline silk indicated that both polyalanine β-crystallites and 3(1) helices derived from the glycine-rich segments are orientated fairly precisely in line with the thread (Kümmerlen et al., 1996; van Beek et al., 1999). (2) A study of amide I shifts obtained by Raman microscopy from single spider dragline threads (Sirichaisit et al., 1999; Yeh and Young, 1999) showed that both β-crystallites and glycine-rich segments experience uniform stress during extension of the thread. This strongly suggests that both components lie within lamellar micro- or nanofibrils in the thread. (3) TEM (see Figures 15.3 and 15.4), scanning electron microscopy, small-angle X-ray diffraction, and atomic force microscopy gave strong evidence for the existence of nanofibrils with a diameter of approximately 5 nm in spider dragline silk (for review, see Putthanarat et al., 2000). Furthermore, nanofibrils of this diameter can be assembled from dilute aqueous solutions of spider silk dope (Chen et al., 2001) or a polypeptide multi-block co-polymer modelled on spider dragline silk (Qu et al., 2000). (4) Meridional small-angle X-ray scattering revealed two orders of a lamellar spacing of 8.4 nm (Yang et al., 1997). Calculations show that an extended glycine-rich segment with an average of 30 residues together with a polyalanine segment with an average of 8 residues (figures derived from the spidroin I sequence published by Xu and Lewis, 1990) can be accommodated into this lamellar spacing with slight twisting or bending of the glycine-rich segments (C. Riekel, 2000 personal communication).

These observations strongly indicate that the spidroin nanofibril has the structure of an LLCE. For such a structure, variations in the detailed configuration of the glycine-rich domains are likely to have only minor impact on the mechanical properties provided that these segments remain mobile. By contrast, variations in the extent of β-crystallization are likely to have much larger effects. Thus, it is hardly surprising that a comparative study of dragline silks

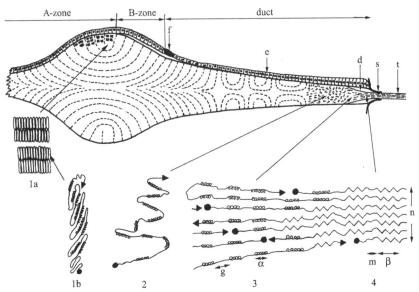

Figure 15.2: Proposed stages in lamellar liquid crystalline assembly of the nanofibrils of the major ampulate (dragline) silk thread in orb web spiders (see text) showing: a diagrammatic optical section of the gland and duct (upper part of illustration); nematic discotic units (1a); rod-shaped molecules of spidroin (1b); partly unfolded molecules (2); early stage of formation of the solid fibre (3); and fully formed nanofibril (4) (see text). The lumen of the gland has been drawn much wider in proportion to length, with only a small number of bi-layer discs (top left) and the epithelium (e), only on one side of the duct and gland. Dotted lines represent the molecular director field. This lies at right angles to the slow axis of polarisation as a result of assembly of the compactly wound, rod-shaped molecules of spidroin (1b) into bi-layered discs of the nematic discotic phase (1a). These are present as an escaped nematic texture in the gland proper and first half of the duct (upper half of figure). Funnel (f), draw-down taper (d), spigot (s), thread (t), glycine-rich segments (g), polyalanine segments (α), mobile segments (m), β-crystallites (β), and nanofibril (n). Based on Knight and Vollrath (1999, 2001a, 2001b); Vollrath and Knight (2001); and Chen et al., (2001).

suggests that the glycine-rich segments are highly variable, but the polyalanine ones giving rise to the β-crystallites are highly conserved (Gatesy et al., 2001).

The mobility of the glycine-rich segments necessary for liquid crystsalline elasticity has been demonstrated by ^{13}C and proton NMR (Jelinski, 1998; Yang et al., 2000).

Liquid Crystalline Assembly

A number of dragline silks from the most primitive spiders to the advanced orb web spinners have now been partially sequenced (Gatesy et al., 2001). This shows that their principal fibrous proteins, spidroins, like the analogous silkworm fibroins, are largely constructed from highly repetitive AB-block

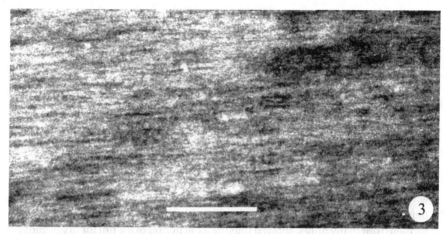

Figure 15.3: Transmission electron micrograph of dragline silk from the orb web spider, *Nephila edulis* dispersed with a Silversen homogeniser at 4°C and negatively stained in aqueous 2% ammonium molybdate. A fibril dispersed from a thread appears to be largely constructed from axially orientated nanofibrils, with a diameter of about 5 nm lying approximately parallel. Scale bar = 100 nm.

copolymers showing a regular alternation of hydrophobic and less hydrophobic blocks. The hydrophobic blocks consist of 4–10 alanine residues linked together, sometimes interrupted by a single serine residue. The less hydrophobic segments are longer and more variable in length and composition, and contain a high concentration of glycine residues.

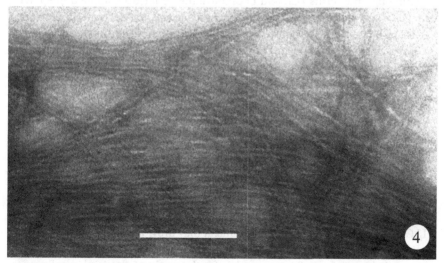

Figure 15.4: As Figure 15.1, but showing an area of nanofibrils dispersed to a greater extent. Scale bar = 100 nm.

Evidence that coulombic interactions play an important part in the assembly of spider dragline silk has been presented elsewhere (Chen et al., 2001).

Strong evidence that the precursor (spinning dope) for orb web spider dragline silk is liquid crystalline has been reviewed elsewhere (Vollrath and Knight, 2001).

Elastomeric Properties

The bulk tensile properties of dragline silk from orb web spiders are those of an elastomer when wet (Gosline, 1984, 1994; Shao et al., 1999). The stress/strain curve is non-linear (Gosline et al., 1984, 1986; Madsen et al., 1999) as it is in LCEs (Uchida, 2000). The tensile behaviour of the nanofibril is less well characterised than the bulk properties of the thread. Two groups have attempted to determine X-ray moduli for spider dragline silk. Yang et al. (1997), using a loop formed from many fibres, showed that the axial period of 8.4 nm for dry *Nephila* silk stretched reversibly by 4% when the fibre was stretched by 10%. Although this demonstrates short-range elasticity, it is not clear whether the greater bulk extension compared with the extension of the nanofibril arises from individual fibres or nanofibrils slipping past each other when strained. Similar objections together with uncertainties about the packing fraction apply to the work of Becker et al. (1994), who used a similar method to determine the X-ray modulus. Future work with single fibres and a synchrotron X-ray source may provide detailed information about the tensile behaviour of the nanofibril, though it seems likely that the nanofibril will have a considerably higher modulus than the bulk material.

Although the terms LCE and LLCE have not so far been applied to spider dragline silk, the materials have been modelled as an elastomeric network (Gosline et al., 1984; Termonia, 1994), and many of these authors' conclusions are applicable to a composite containing lamellar liquid crystalline nanofibrils as well as the amorphous matrix model they have used.

DISCUSSION

Thus, we have presented strong evidence that the fibrous collagens and the dragline silk of orb web spiders are lamellar liquid crystal elastomers. Of the two routes by which LCEs can be assembled (see Introduction), Nature appears to have mainly used the cross-linking of a monodomain liquid crystal rather than the more complex processing of an initially isotropic phase (see Introduction).

The suggested classification of the fibrous collagens and dragline silks as LLCEs has considerable predictive value. It may provide: (a) a testable hypothesis for the assembly of spider silk nanofibrils (see Figure 15.2) that fits the presently known facts; (b) new models for the origin of elasticity in collagens and silks; (c) new approaches for the production of biomimetic materials; and (d) the prediction that exotic properties of LCEs may be found in biological elastomers.

Assembly of Spider Silk

A hypothesis for the assembly of spider dragline silk as an LLCE is illustrated in Figure 15.2. Spidroin is thought to be stored for use as compactly folded molecules (Hijirida et al., 1996). In these, the polyalanine segments are thought to assume an α-helical conformation while the glycine-rich ones show a combination of β-turns and 'random coil' conformation (Hijirida et al., 1996). The flexible glycine-rich segments may allow the stiffer α-helical segments to take up a nematic configuration within the compactly coiled molecule as has been suggested by Pitard et al. (1997) for other proteins (1b in Figure 15.2). The molecule may be rod-shaped and would be chiral as are all proteins. The rod may be markedly amphiphilic, with one end more hydrophilic than the other. These characteristics would enable the molecules to assemble into nematic discotic units (1a in Figure 15.2) in the very highly concentrated dope in the first and second limbs of the duct (Vollrath and Knight, 2001). The secretion of potassium and hydrogen ions into the lumen of the third limb of the duct (Chen et al., 2001; Knight and Vollrath, 2001a, 2001b) is thought to encourage the molecules to partly refold by disrupting their water shell and altering coulombic forces and intramolecular hydrogen bonding (see 2 in Figure 15.2). They would then extend somewhat, align, and be brought closer together in the extensional flow field of the draw-down taper found in the distal part of the duct (see 3 in Figure 15.2). This would align the molecules along the duct, but with no preference for direction back or forwards, thus contributing to an antiparallel arrangement (see 3 in Figure 15.2). The retention of some hairpin loops arising from incomplete extension of the molecules would help to provide an antiparallel arrangement. Incomplete rather than full extension is also likely on the grounds that it would reduce the force required to form the silk thread. Hydrophobic effects operating on the repetitive amphiphilic AB-block copolymeric structure would enable staggered, antiparallel chains to assemble in a lamellar structure. In addition, coulombic forces arising from charged amino acids are thought to play a role in the formation of the lamellar nanofibrils (Chen et al., 2001) as in collagens. As the hydrophobic polyalanine segments of spidroin align and are drawn closer together by extensional flow, they would find themselves in an increasingly hydrophobic environment, and this could trigger their conversion from an α-helical or random coil to a β-pleated structure resulting in the formation of numerous interchain hydrogen bonds (see 4 in Figure 15.2). The latter would act as multifunctional cross-links at nodes between the more mobile glycine-rich segments (see 4 in Figure 15.2). Thus, the assembly of the thread can be seen as a liquid crystalline phase transition involving separation into polymer-rich and solvent-rich phases (Knight et al., 2000).

Tensile Properties

It may be possible to apply LCE theory to silks and collagens to understand their tensile properties further and, in particular, to predict the influence of variation in the length of mobile segments and cross-link density in different silks. Although extensible molecules lying between and tightly bonded to rather stiff nanofibrils constructed from spidroin I may be important for the extensibility of spider silk, this could be incorporated into an LCE model. An extended Gaussian theory may be applicable to biological LCEs, but this mechanism may not be solely responsible for their elastomeric force as energy storage in helical springs, such as β-turns or spirals and other entropic effects, may also be involved (Tatham and Shewry, 2000). The complexity of mechanical behaviour of biological LCE may be further increased because liquid crystals are highly tunable (Collings, 1990). Thus, changes in the content of water, ions, and small molecules may account for some of the variation in the properties of spider dragline silk from day-to-day, individual-to-individual, and species-to-species observed by Madsen et al. (1999). It may also help to account for the variation in the structure and mechanical behaviour of fibrillar collagens in health and disease (Royce and Steinman, 1993) and with collagen type (Ricard-Blum et al., 2000).

Production of Biomimetic Materials

It may be possible to use LCE theory to design polymers that mimic the properties of fibrous collagens, dragline silks, and other biological elastomers. Three routes might be used for this: genetic engineering of spidroin analogues; direct use of natural amphiphilic block copolymeric proteins from agricultural products, such as wheat germ (Tatham and Shewry, 2000); and conventional synthesis of block copolymers. The ability of amphiphilic block copolymers to form lamellar liquid crystals, sometimes with nanofibrillar textures (Bauer et al., 1995; Morishima, 2000; Roschinski and Kulicke, 2000), is particularly important in this connection and provides a possible route for the production of nanocomposite materials.

Exotic Properties

LCEs have a range of exotic properties (Teixeira, 1999), some of which may have been exploited by living organisms. These may include transparency (Anderson and Ström, 1991) as in the collagenous cornea; piezoelectricity (Meier and Finkelmann, 1993), which may provide a feedback loop relating mechanical loading to increased synthetic activity in connective tissues; and anomalous sensitivity of shape change to electrical fields and mechanical strain (Verwey

and Warner, 1994), which may help to define collagen orientations in morphogenesis and tissue remodelling.

Finally, we note that other nano- or microfibrils, including those of elastin
(Kim et al., 1999), fibrillin (Keene et al., 1991), fibrin (Mosesson et al, 1995),
Bombyx mori silk (Putthanarat et al., 2000), and ciliary rootlets (Anstrom, 1992),
may also be LLCEs. The transversely banded costas of protists (Brugerolle, 1999)
may be of particular interest for the development of biomimetic microactua-
tors as they show a remarkable phase transition, reversibly shortening on the
addition of low concentrations of calcium ions.

ACKNOWLEDGEMENTS

We thank David Porter for encouraging us to think of collagen and spider silk
as LCEs; the Danish Science Research Council (SNF), the Science Faculty of
Aarhus University, the Carlsberg Foundation, the Danish Academy of Science,
the Biotechnology and Biological Science Research Council (BBSRC) and Engineering and Physical Science Research Council (EPSRC), and the European
Science Foundation (ESF) for funding; Terry Chen, Peter Fratzl, David Kaplan,
Fred Keeley, Arthur Tatham, Carlo Knupp, Eugene Terentjev, and John Gosline
for helpful discussion; and Academic Press for permission to reproduce
Figure 15.1. D.P.K. thanks the Biological Imaging Centre, Southampton University, for technical assistance.

REFERENCES

Anderson, D. M., and Ström, P. (1991) Polymerized lyotropic liquid crystals as contact lens
materials. *Physica A* **176,** 151–67.
Anstrom, J. A. (1992) Organization of the ciliary basal apparatus in embryonic cells of the
sea urchin *Lytechinus pictus. Cell Tissue Res.* **269,** 305–13.
Barclay, G. G., and Ober C. K. (1993) Liquid-crystalline and rigid-rod networks. *Prog. Polym.
Sci.* **18,** 899–945.
Bauer, H., Böse, N., and Stern, P. (1995) Viscoplastic flow and shear thickening in concen-
trated diblock copolymer solutions. *Colloid Polym. Sci.* **273,** 480–89.
Becker, M. A., Mahoney, D. V., Lenhert, P. G., Eby, R. K., Kaplan, D., and Adams, W. W. (1994)
X-ray moduli of silk fibres from *Nephila clavipes* and *Bombyx mori. ACS Symp Ser.* **544,**
185–95.
Bouligand, Y., Cladis, P. E., Liebert, L., and Strzelecki, L. (1974) *Mol. Cryst. Liq. Cryst.* **25,**
253.
Brown, E. M., King, G., and Chen, J. M. (1997) Model of the helical portion of a type I
collagen microfibril. *J. Am. Leather Chem. Assn.* **92,** 1–7.
Brugerolle, G. (1999) Fine structure of *Pseudotrypanosoma giganteum* of *Porotermes,* a
trichomonad with a contractile costa. *Eur. J. Protistol.* **35,** 121–28.
Chan, Y. N. C., Craig, G. S. W., Schrock, R. R., and Cohen, R. E. (1992) Synthesis of palladium
and platinum nanoclusters within microphase-separated diblock copolymers. *Chem.
Mat.* **4,** 885–94.
Chapman, J. A., Tzaphlidou, M., Meek, K. M., and Kadler, K. E. (1990) The collagen

fibril – A model system for studying the staining and fixation of a protein. *Electron Microsc. Rev.* **3**, 143–82.

Chen, X., Knight, D. P., and Vollrath, F. (2001) Rheological characterization of *Nephila* spidroin solution. *Biomacromolecules* **3**, 644–48

Collings, P. J. (1990) *Liquid Crystals: Nature's Delicate Phase of Matter,* 1st ed. Princeton, NJ: Princeton University Press.

de Gennes, P. G. (1969) Possibilités offertes par la réticulation de polymères en présence d'un cristal liquide. *Phys. Lett.* **28A**, 725.

de Gennes, P. G. (1974) *The Physics of Liquid Crystals,* 1st ed. Oxford; UK: Clarendon Press.

de Gennes, P. G. (1975) Réflexions sur un type de polymères nématiques. *C. R. Acad. Sci.* **B281**, 101–3.

Fischer, P., Schmidt, C., and Finkelmann, H. (1995) Amphiphilic liquid crystalline networks. Phase-behaviour and alignment by mechanical fields. *Macromolecules* **16**, 435–47.

Fraaije, J. G. E. M., Zvelindovsky, A. V., Sevink, G. J. A., and Maurits, N. M. (2000) Modulated self-organization in complex amphiphilic systems. *Mol. Simul.* **25**, 131–44.

Fraser, R. D. B., MacRae, T. P., and Miller, A. (1987) Molecular packing in type I collagen fibrils. *J. Mol. Biol.* **193**, 115–26.

Fraser, W. D. (1998) The collagen crosslinks pyridinoline and deoxypyridinoline: A review of their biochemistry, physiology, measurement, and clinical applications. *Clin. Ligand Assay,* **21**, 102–10.

Gatesy, J., Hayashi, C., Motrluk D., Woods, J., and Lewis, R. (2001) Extreme diversity, conservation and convergence of spider silk fibroin sequences. *Science* **291**, 2603–5.

Gathercole, L. J., Miles, M. J., McMaster, T. J., and Holmes, D. F. (1993) Scanning probe microscopy of collagen-I and Pn-collagen-I assemblies and the relevance to scanning-tunneling-microscopy contrast generation in proteins. *J. Chem. Soc.-Faraday Trans.* **89**, 2589–94.

Gosline, J., Denny, M., and DeMont, M. (1984) Spider silk as rubber. *Nature* (Lond.) **309**, 551–52.

Gosline, J. M., DeMont, M. E., and Denny, M. W. (1986) The structure and properties of spider silk, *Endeavour, N.S.* **10**, 31–43.

Gosline, J. M., Pollack, C. C., Guerette, P. A., Cheng, A., DeMont, M. E., and Denny, M. W. (1994) Elastomeric network models for the frame and viscid silks from the orb web of the spider *Araneus diadematus.* In: *Silk Polymers. Materials Science and Biotechnology,* D. Kaplan, W. W. Adams, B. Farmer, and C. Viney (eds.). Washington; DC: American Chemical Society, pp. 328–41.

Grubb, D. T., and Ji, G. (1999) Molecular chain orientation in supercontracted and re-extended spider silk. *Int. J. Biol. Macromol.* **24**, 203–10.

Hamley, I. W. (2000) The effect of shear on ordered block copolymer solutions. *Curr. Opin. Colloid Interface Sci.* **5**, 342–50.

Hepworth, D. G., Gathercole, L. J., Knight, D. P., Feng, D., and Vincent, J. F. V. (1994) Correlation of ultrastructure and tensile properties of a collagenous composite material, the egg capsule of the dogfish, *Scyliorhinus spp.,* a sophisticated collagenous material. *J. Struct. Biol.* **112**, 231–40.

Hijirida, D. H., Do, K. G., Michal, C., Wong, S., Zax, D., and Jelinski, L. W. (1996) C13 NMR of *Nephila clavipes* major ampullate silk gland. *Biophys. J.* **71**, 3442–47.

Hu, X. W., Knight, D. P., and Chapman, J. A. (1997) The effect of non-polar liquids and non-ionic detergents on the ultrastructure and assembly of rat tail tendon collagen fibrils *in vitro. Biochim. Biophys. Acta-General Subjects,***1334,** 327–37.

Hukins, D. W. L., Woodhead-Galloway, J., and Knight, D. P. (1976) Molecular tilting in dried elastoidin and its implications for the structures of other collagen fibrils. *Biochem. Biophys. Res. Commun.***73,** 1049–55.

Hukins, D. W. L., and Woodhead-Galloway, J. (1977) Collagen fibrils as examples of smectic A biological fibres. *Mol. Cryst. Liq. Cryst.* **41,** 33–39.

Hukins, D. W. L., and Woodhead-Galloway, J. (1978) Liquid crystalline model for the organization of molecules in collagen fibrils. *Biochem. Soc. Trans.* **6,** 238–39.

Hulmes, D. J. S., Miller, A., Parry, D. A. D., Piez, K. A., and Woodhead-Galloway, J. (1973) Analysis of the primary structure of collagen for the origins of molecular packing. *J. Mol. Biol.* **79,** 137–48.

Hulmes, D. J. S., Wess, T. J., Prockop, D. J., and Fratzl, P. (1995) Radial packing, order, and disorder in collagen fibrils. *Biophys. J.* **68,** 1661–70.

Jelinski, L. W. (1998) Establishing the relationship between structure and mechanical function in silks. *Curr. Opin. Solid State Mater. Sci.* **3,** 237–35.

Keene, D. R., Maddox, B. K., Kuo, H. J., Sakai, L. Y., and Glanville, R. W. (1991) Extraction of extendable beaded structures and their identification as fibrillin-containing extracellular matrix microfibrils. *J. Histochem. Cytochem.* **39,** 441–49.

Kim, J. S., McHugh, S. K., and Swager, T. M. (1999) Nanoscale fibrils and grids; aggregated structures from rigid-rod conjugated polymers. *Macromolecules* **32,** 1500–7.

King, M. R., White, S. A., Smith, S. D., and Spontak, R. J. (1999) Mesogel networks via selective swelling of lamellar triblock copolymers. *Langmuir* **5,** 7886–89.

Knight, D. P., and Hunt, S. (1974) Fibril structure of collagen in egg capsule of dogfish. *Nature* (Lond.) **249,** 379–80.

Knight, D. P., Feng, D., Stewart, M., and King, E. (1993) Changes in macromolecular organisation in collagen assemblies during secretion in the nidamental gland and formation of the egg capsule wall in the dogfish. *Scyliorhinus canicula. Phil. Trans. R. Soc. B* **341,** 419–36.

Knight, D. P., Feng, D., and Stewart, M. (1996) Structure and function of the selachian egg case. *Biol. Rev.* **76,** 81–111.

Knight, D. P., and Vollrath, F. (1999) Liquid crystals and flow elongation in a spider's silk production line. *Proc. R. Soc. B* **266,** 519–23.

Knight, D. P., Knight, M. M., and Vollrath, F. (2000) Beta transition and stress-induced phase separation in the spinning of spider dragline silk. *Int. J. Biol. Macromol.* **27,** 205–10.

Knight, D. P., and Vollrath, F. (2001a) A comparison of liquid crystal spinning of selachian egg case ply sheet and orb web spider dragline filaments. *Macromolecules* **2,** 323–34.

Knight, D. P., and Vollrath, F. (2001b) Changes in element composition in the secretory pathway for dragline silk in a *Nephila* spider. *Naturwissenschaften* **88,** 179–82.

Knott, L., and Bailey, A. J. (1998) Collagen cross-links in mineralizing tissues: A review of their chemistry, function, and clinical relevance. *Bone* **22,** 181–87.

Knupp, C., and Squire, J. (1998) X-ray diffraction analysis of the 3D organization of collagen fibrils in the wall of the dogfish egg case. *Proc. R. Soc. Lond. Ser. B-Biol. Sci.* **265,** 2177–86.

Knupp, C., Chew, M., and Squire, J. (1998) Collagen packing in the dogfish egg case wall. *J. Struct. Biol.* **122,** 101–10.

Knupp, C., and Squire, J. (2001) A new twist in the collagen story-the type VI segmented supercoil. *EMBO J.,* **20,** 372–76.

Kümmerlen, J., van Beek, J., Vollrath, F., and Meier, B. (1996) Local structure in spider dragline silk investigated by two-dimensional spin-diffusion nuclear magnetic resonance. *Macromolecules* **29,** 2920– 28.

Lees, S. (1998) Interpreting the equatiorial diffraction pattern of collagenous tissues in the light of molecular motion. *Biophys. J.* **75,** 1058–61.

Legge, C. H., Davis, F. J., and Mitchell, G. R. (1991) Memory effects in liquid-crystal elastomers. *J. Phys. II France* **1,** 1253–61.

Leikin, S., Rau, D. C., and Parsegian, V. A. (1995) Temperature-favored assembly of collagen is driven by hydrophilic not hydrophobic interactions. *Nature Struct. Biol.* **2,** 205–10.

Mao, Y., Warner, M., Terentjev, E. M., and Ball, R. C. (1998) Finite extensibilty effects in nematic elastomers. *J. Chem. Phys.* **108,** 8743–48.

Madsen, B., Shao, Z., and Vollrath, F. (1999) Variability in the mechanical properties of spider silks on three levels: Interspecific, intraspecific and intraindividual. *Int. J. Biol. Macromol.* **24,** 301–6.

Martin, R., Farjanel, J., Eichenberger, D., Colige, A., Kessler, E., Hulmes, D. J. S., and Giraud-Guille, M.-M. (2000) Liquid crystalline ordering of procollagen as a determinant of three-dimensional extracellular matrix architecture. *J. Mol. Biol.* **301,** 11–17.

Meier, W., and Finkelmann, H. (1993) Piezoelectricity of cholesteric elastomers. 1. Influence of the helicoidal pitch on the piezoelectric coefficient. *Macromolecules* **26,** 1811–17.

Misof, K., Rapp, G., and Fratzl, P. (1997) A new molecular model for collagen elasticity based on synchtrotron x-ray scattering evidence. *Biophys. J.* **72,** 1376–81.

Mitchell, G. R., Coulter, M., Davis, F. J., and Guo W. (1992) The effect of the spacer length on the nature of coupling in side-chain liquid crystal polymers and elastomers. *J. Phys. II* **2,** 1121–32.

Morishima, Y. (2000) Self-assembling amphiphilic polyelectrolytes and their nanostructure. *Chin. J. Polym. Sci.* **18,** 323–24.

Mosesson, M. W., Siebenlist, K. R., and Hanfield, J. F. (1995) The covalent structure of factor XIIIA cross-linked fibrinogen fibrils. *J. Struct. Biol.* **115,** 88–101.

Northolt, M. G., and Sikkema, D. J. (1991) Lyotropic main chain liquid-crystal polymers. *Adv. Polym. Sci.* **98,** 115–17.

Ortolani, F. Raspanti, M., and Marchini, M. (1994) Correlations between amino acid hydrophobicity scales and stain exclusion capacity of type I native collagen fibrils. *J. Electron Microsc.* **43,** 32–38.

Parry, D. A. D. (1988) The molecular and fibrillar structure of collagen and its relationship to the mechanical properties of connective tissue. *Biophys. Chem.* **29,** 195–209.

Pitard, E., Garel, T., and Orland, H. (1997) Protein folding, anisotropic collapse and blue phases. *J. Phys. I,* **17,** 1201–10.

Putthanarat, S., Stribeck, N., Fossey, S. A., Eby, R. K., and Adams, W. W. (2000) Investigation of the nanofibrils of silk fibers. *Polymer* **41,** 7735–47.

Qu, Y., Payne S. C., Apkarian, R. P., and Conticello, V. P. (2000) Self-assembly of a polypeptide multi-block copolymer modelled on dragline silk protein. *J. Am. Chem. Soc.* **122,** 5014–15.

Ricard-Blum, S., Dublet, B., and van der Rest, M. (2000) *Unconventional Collagens*. New York: Oxford University Press.

Riekel, C., Bränden, C., Craig, C., Ferrero, C., Heidelbach, G., and Müller, M. (1999) Aspects of x-ray diffraction on single spider fibers. *Int. J. Biol. Macromol.* **24**, 179–86.

Roschinski, C., and Kulicke, W.-M. (2000) Rheological characterization of aqueous hydroxypropyl cellulose solutions close to phase separation. *Macromol. Chem. Phys.* **201**, 2031–40.

Rosenblatt, J., Devereux, B., and Wallace, D. G. (1993) Dynamic rheological studies of hydrophobic interactions in injectable collagen biomaterials. *J. Appl. Polym. Sci.* **50**, 953–63.

Royce, P. M., and Stenimann, B. (eds.). (1993) *Connective Tissue and Its Heritable Disorders*. New York: Wiley-Liss.

Schneider, A., Muller, S., and Finkelmann, H. (2000) Lyotropic mesomorphism of AB block copolymers in nematic solvents. *Macromol. Chem. Phys.* **201**, 184–91.

Shao, Z., Young, R. J., and Vollrath, F. (1999) The effects of solvents on spider silk studied by mechanical testing and single-fibre Raman spectroscopy. *Int. J. Biol. Macromol.* **24**, 295–300.

Simmons, A., Michal, C., and Jelinski, L. (1996) Molecular orientation and two-component nature of the crystalline fraction of spider dragline silk. *Science* **271**, 84–87.

Sirichaisit, S., Young, R. J., and Vollrath, F. (1999) Molecular deformation in spider dragline silk subjected to stress. *Polymers* **41**, 1223–27.

Tatham, A. S., and Shewry, P. R. (2000) Elastomeric proteins: Biological roles, structures and mechanisms. *Trends Biol. Sci.* **25**, 567–71.

Termonia, Y. (1994) Molecular modeling of spider silk elasticity. *Macromolecules* **27**, 7378–81.

Texeira, P. I. C. (1999) Dynamics of reorientation of a constrained nematic elastomer. *Eur. Phys. J. B* **9**, 471–77.

Texeira, P. I. C., and Warner, M. (1999) Dynamics of soft and semisoft nematic elastomers. *Phys. Rev. E* **60**, 603–9.

Treloar, L. R. G. (1975) *The Physics of Rubber Elasticity*. Oxford, UK: Clarendon Press.

Uchida, N. (2000) Soft and nonsoft structural transitions in disordered nematic networks. *Phys. Rev. E.* **62**, 5119–34.

van Beek, J. D., Kümmerlen, D., Vollrath, F., and Meier, B. H. (1999) Supercontracted spider dragline silk: A solid-sate NMR study of the local structure. *Int. J. Biol. Macromol.* **24**, 173–78.

Verwey, G. C., and Warner, M. (1995) Soft rubber elasticity. *Macromolecules* **28**, 4303–6.

Vincent, J. F. V. (1982) *Structural Materials*. London, UK: Macmillan.

Vinee, P., Meurer, B., Constantinesco, A., Kohlberger, B., Hauenstein, K. H., Lauberberger, J., and Petkov, S. (1993) Characterization of human aortic collagen's elasticity by nuclear magnetic resonance. *Magn. Res. Imaging* **11**, 395–99.

Vollrath, F., and Knight, D. P. (2001) Liquid crystalline spinning in Nature. *Nature* (Lond.) **410**, 541–48.

Wallace, D. (1985) The role of hydrophobic bonding in collagen fibril formation – A quantitative model. *Biopolymers* **24**, 1705–20.

Warner, M., Gelling, K. P., and Vilgis, T. A. (1988) Theory of nematic networks. *J. Chem. Phys.* **88**, 4008–13.

Warner, M., and Terentjev, E. M. (1996) Nematic elatomers—A new state of matter? *Prog. Polym. Sci.* **21,** 853–91.

Woodhead-Galloway, J., and Knight, D. P. (1977) Some observations on the fine structure of elastoidin. *Proc. R. Soc. Lond. B* **195,** 355–64.

Xu, M., and Lewis R. V. (1990) Structure of a protein superfiber: Spider dragline silk. *Proc. Natl. Acad. Sci. U.S.A.* **87,** 7120–24.

Yang, Z., Grubb, D. T., and Jelinski, L. W. (1997) Small-angle x-ray scattering of spider dragline silk. *Macromolecules* **30,** 8254–61.

Yang, Z. T., Liivak, O., Seidel, A., LaVerde, G., Zax, D. B., and Jelinski, L. W. (2000) Super-contraction and backbone dynamics in spider silk: C-13 and H-2 NMR studies. *J. Am. Chem. Soc.* **122,** 9019–25.

Yeh, W.-Y., and Young, R. J. Molecular deformation processes in aromatic high modulus polymer fibres. *Polymer* **40,** 857–70.

SIXTEEN

Restraining Cross-Links in Elastomeric Proteins

Allen J. Bailey

INTRODUCTION

We have seen in the preceding chapters that elastomeric proteins must firstly contain independent monomeric chains that are flexible and generally conformationally free, and secondly be cross-linked at specific points to form a network to ensure elastic recoil. The elastic properties are modulated by the length and properties of elastic domains, which are in turn delineated by the extent and location of cross-linking. These cross-links tend to form in small 'crystalline' domains within a generally amorphous "rubber-like" structure. The ability to control the specific location and nature of the intermolecular cross-links in biological macromolecules has evolved and been widely exploited by organisms throughout the animal kingdom. It has allowed tissues to evolve optimal characteristics of strength and elasticity for protection against the stresses of the environment, since malfunctioning can at worst be lethal. An understanding of these cross-linking mechanisms can provide another dimension to our knowledge of the properties of elastic proteins, and at the same time an insight into the changes in physical properties during growth and maturation. In some instances, the location of the cross-links may also provide data on the alignment of the molecules in the supramacromolecular assemblies involved in elastomeric proteins.

COVALENT CROSS-LINKS

Peroxidase-Induced Di-tyrosine Cross-links

Resilin

The first intermolecular cross-links to be identified in an elastomeric protein were isolated from resilin by Andersen (1964) as di- and tri-tyrosine (Figure 16.1 (see chapter 13)) (Table 16.1). The stable di-tyrosine cross-links are formed by peroxidase oxidation of tyrosine to connect the resilin molecules in a

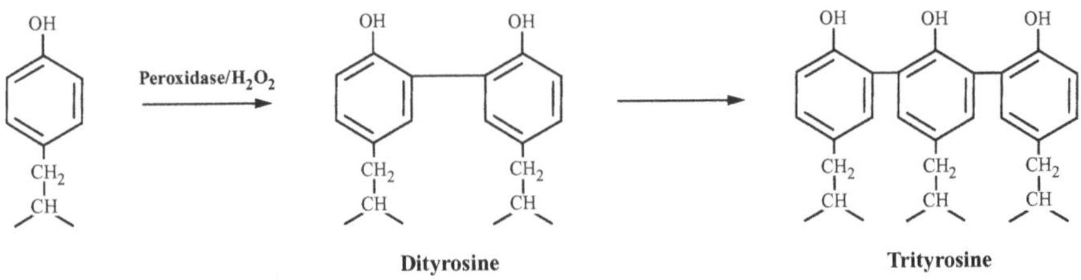

Figure 16.1: Di- and tri-tyrosine cross-links present in resilin and formed by peroxidation to build up a cross-linked network in resilin. These tyrosine cross-links are also believed to be present in abductin.

three-dimensional network. Only preliminary amino acid sequences are available. Based on the *Drosophila* gene for resilin, the peptide chains between the cross-links have been shown to be hydrophobic involving a random distribution of Gly.Ala.Pro. The sequence of the cross-link domain appears to involve the N- and C-flanking regions which differ from the central region in being dominated by small residues and tyrosine, the latter being regularly spaced, often 14–16 residues apart (Ardell and Andersen, 2001). The specificity of the enzyme in determining which specific tyrosine residues are cross-linked is not yet clear. Possibly, the regions rich in glycine and alanine allow close approach of the tyrosines.

Our knowledge of the role of peroxidase in the conversion of tyrosine to dityrosine is unfortunately primarily based on studies demonstrating the ability of this enzyme to oxidise tyrosine in vitro. Only in the case of the membrane of the egg capsule of the sea urchin has the presence of these cross-links been

Table 16.1. Cross-link types in elastomeric proteins

Protein	Bond Type	Cross-link
Resilin	Covalent	Di-tyrosine
Abductin	Covalent	Di-tyrosine ?
Elastin	Covalent	Lysyl aldehyde/desmosines
Collagen	Covalent	Lysyl aldehyde/pyridinolines
Lamprin	Covalent	Pyridinoline
Fibrillin	Covalent	Glutamyl-lysine
	Covalent	Disulphide
Gluten	Covalent	Disulphide
	Non-covalent	Hydrogen bond
Byssus	Coordinate	Zinc-ion histidine
	Covalent	DOPA-lysine
Silks	Non-covalent	Hydrophobic bonds
	Non-covalent	Hydrogen bonds
Spectrin	Unknown	Unknown
Titin	Unknown	None ? single chain

definitely associated with the presence of ovoperoxidase activity in vivo (Foerder and Shapiro, 1977).

In some tissues, di-tyrosine may be associated with diether linked tri-tyrosines (e.g., the collagenous cuticle of *Ascaris lumbricoides*), but neither their mechanism of formation nor their functional role has been elucidated (Fujimoto et al., 1981).

Di-tyrosine has also been reported to be present in collagenous tissues, for example, in cataractous human lens capsules (type IV collagen), collagen, and elastin. However, the presence of di-tyrosine in these latter tissues may be due to ultraviolet (UV) irradiation (Waykole and Heidemann, 1976) and is not therefore a functional cross-link restraining the elastin from over-extension. Our own studies on UV-irradiated collagen failed to detect the presence of di-tyrosine (Miles et al., 2000).

Abductin
The inner hinge ligament of bivalve molluscs is a highly insoluble elastic protein, and preliminary sequence studies indicate that tyrosines are located at the N- and C-terminii of two subunits. However, the role of these sites in the formation of di-tyrosine cross-links remains to be confimed.

Lysyl Aldehyde-Derived Cross-links
The ε-amino groups of specific peptide lysines in the connective tissue proteins collagen and elastin undergo oxidative deamination by lysyl oxidase to form lysyl aldehydes, which then react further to form a complex series of different cross-links, depending on the nature and age of the tissue (Bailey, 2001). Inhibition of lysyl oxidase by lathyritic agents such as β-aminopropionitrile, which binds irreversibly to the lysyl oxidase and inhibits its action, results in very fragile tissues which rupture readily under light loading. Lysyl oxidase is a copper-containing enzyme and, consequently, copper deficiency in the organism reduces the activity of the enzyme and similarly leads to fragile tissues. Recent studies have revealed at least five lysyl oxidases (Maki and Kivirikko, 2001), all of which are located on different chromosomes, but the contribution of a particular gene product to cross-linking is unknown. They may be related to particular tissues, although the enzymes appear to have overlapping tissue distribution.

Elastin
Elastin is present in tissues where elastic recoil is an essential part of its function (e.g., the major arteries, lungs, and ligaments). However, prevention of over-extension, which would destroy effective recoil, is also essential, and this is achieved by the formation of cross-links at specific points in the peptide chains, thus forming a network within the fibre.

Newly synthesised tropoelastin spontaneously coacervates to an insoluble aggregate following secretion from the cell. Lysyl oxidase acts on the aggregate, but not on the tropoelastin itself. Clearly, there is a conformational change in the elastin molecule on coacervation to form fibrils. About 85% of the lysines in elastin are oxidised to lysine aldehyde such that the fibrillar elastin only contains five lysine residues. These residual lysines are in the alanine-rich α-helical domains of the elastin fibre. During coacervation, these α-helical domains are presumably brought into apposition, thus permitting the formation of interchain cross-links. The cross-links therefore occur in short, α-helical domains while the remainder of the peptide chain is comprised mainly of hydrophobic residues and randomly distributed glycines and prolines resulting in a random organisation of free independent peptide chains. However, some order within the chains appears to be present through the formation of β-spirals (see Chapter 4 in this volume).

The initial reactions involve the formation of divalent lysinonorleucine from lysine aldehyde and lysine, lysine aldol from two lysine aldehydes, and the trivalent merodesmosine from reaction of the aldol with the ε-amino group of a contiguous lysine to form a Schiff base. The aldol and merodesmosine rapidly decrease with time, reacting further with another lysine aldehyde to form the tetravalent isomers desmosine and iso-desmosine cross-links (Partridge et al., 1963) as shown in Figure 16.2. Although generally depicted as only involving two peptide chains, it has been suggested that three tropoelastin chains may be involved (Brown-Augsberger et al., 1995). The ability to form these cross-links depends on the presence of an intact ε-amino group of the lysine side chain, and the third peptide chain may provide the intact lysine residue. It has been suggested that if a tyrosine or phenylalanine is adjacent to a particular lysine, then it is not oxidatively deaminated and is therefore available as the amino group donor.

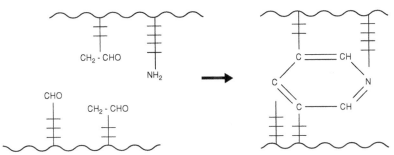

Figure 16.2: Close proximity of three aldehydes and one lysine in the α-helical alanine-rich domains allows reactions of the aldehydes, finally resulting in desmosine and iso-desmosine.

Recombinantly expressed elastin polypeptides (e.g., a tandem repeat of PGVGVA with an occasional PGVGLA for lysine cross-linking) readily coacervate to form fibrils, and following treatment with pyrroloquinoline quinone (PQQ) in the presence of copper, the fibrils cross-link through the characteristic elastin cross-links, lysinonorleucine, desmosine, and iso-desmosine (see Chapter 3 in this volume). Other components are present but have not yet been identified. PQQ fails to produce cross-links when reacted with non-coacervated elastin. This confirms that the alignment of the lysine residues only occurs during coacervation, and using these recombinant peptides, it should be possible to resolve the question whether adjacent aromatic residues prevent oxidation of lysine residues. These cross-linked fibrils derived from polypeptides possess elastic properties similar to elastin.

Collagen

The packing and cross-linking of the collagen molecule in the fibres ensure that collagen fibres are virtually inextensible. However, collagen fibres (like steel wire) are elastic, and this property is important during the movement of animals. The relative elastic properties of collagen are illustrated by Gosline et al. in Chapter 2 in this volume. The mechanism of extension (about 5%) is not fully understood, and may be due to elongation of the fibres as seen by the increase in the axial repeat in the fibre or due to an interaction between the fibre and the intrafibrillar matrix of proteoglycans (see Chapter 9). The stiff cross-linked collagen fibres have sufficient elasticity to be able to 'store energy', which on release, ensures more efficient movement (e.g., humans and kangaroos reduce the work their muscles have to do during running by storing energy which is released on rebound) (see Chapter 1).

In contrast to elastin, which is virtually metabolically inert once it is laid down in the tissue, collagen has a significant turnover rate, albeit much slower than other proteins. The rate of turnover can vary from 2 days in peridontal ligaments through bone (2–3 years) to tendon and cartilage (100 years). The mechanical properties of collagen fibres can increase for many years after maturation and the cessation of growth. This increase has been shown to be due to the formation of an increasing proportion of multivalent, intermolecular cross-links (Bailey, 2001).

The cross-linking of collagen also involves lysyl oxidase; but, in contrast to elastin where the majority of the lysines are oxidatively deaminated, in collagen the enzyme only oxidises the single lysine residue in the non-triple helical N- and C-terminal regions of each chain. These regions possess different amino acid sequences and consequently the specificity of lysyl oxidase is not clear. The enzyme acts on the fibrous aggregate, but not on the tropocollagen monomer, indicating that the enzyme attaches to the highly conserved Hyl.Gly.His.Arg.

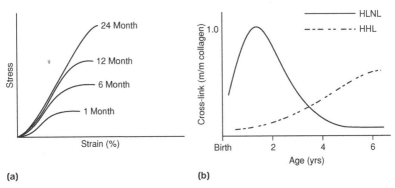

Figure 16.3: (a) Continued increasing mechanical strength of rat tail tendon collagen fibres after maturation of the animal. (b) Corresponding cross-link profile of bovine skin collagen showing the increase in mature cross-link histidino-hydroxylysinonorleucine (HHL) derived from hydroxylysinonorleucine (HLNL) after maturation of the animal.

sequence in the triple helix and oxidises the lysine in the non-helical region of an adjacent molecule which is now in close proximity because of the end-overlap alignment of the molecules in the fibre. The histidine may be involved in the stabilisation of this complex through proton donation, and the Lys/Hyl residue of this sequence acts as the receptor for the lysine aldehyde reacting to form the Schiff base/aldimine cross-link. Identification of this first intermolecular cross-link in collagen was achieved after stabilisation by borohydride reduction and acid hydrolysis (Bailey and Peach, 1968). When hydroxylysine is present in the non-helical domain, the aldimine formed undergoes a spontaneous Amadori rearrangement to a keto-imine cross-link. The aldimines and keto-imines are the predominant cross-links in immature tissues where turnover of the collagen is high. As turnover decreases, the divalent cross-links react with the lysine alde-hyde of other collagen molecules to form trivalent cross-links (Bailey, 2001). The cross-link profile of collagen therefore changes with maturation of the tissue. The proportion of newly synthesised collagen decreases as turnover falls and an increasing proportion of the divalent cross-links are converted to trivalent cross-links. This increase in the latter is consistent with the continued increase in the strength of collagen fibres with age (Bailey, 2001) as shown in Figure 16.3(a,b).

We have proposed that the divalent cross-links polymerise the molecules longitudinally head to tail and subsequently react transversely with an adja-cent molecule in register, that is, forming an interfibrillar trivalent cross-link (Bailey et al., 1980) as depicted in Figure 16.4. Such a mechanism would ac-count for the increase in mechanical strength of the collagen fibre with age. In tissues where the non-helical regions are not hydroxylated (e.g., skin, rat tail tendon), the initial aldimine reacts with histidine to form the trivalent

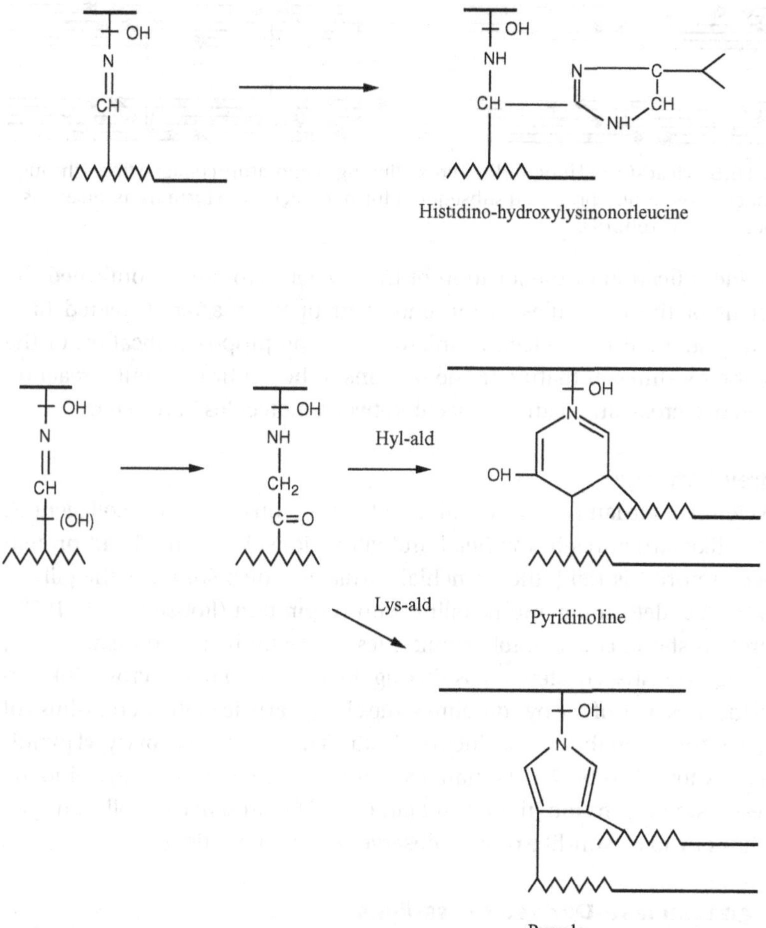

Histidino-hydroxylysinonorleucine

Hyl-ald

Pyridinoline

Lys-ald

Pyrrole

Figure 16.4: Chemical nature of the divalent collagen cross-links in immature tissue and the trivalent cross-links in mature tissue. The extent of hydroxylation of the telopeptide lysine determines the reaction mechanism. To a lessor extent, the reduced hydroxylation of the triple helical lysine can produce a non-hydroxylated version of these cross-links.

histidino-hydroxylysinonorleucine with increasing age. Surprisingly, this cross-link does not form in mouse and rat skin. In tissues such as bone and cartilage, the non-helical terminal regions are highly hydroxylated, and complete hydroxylation leads to formation of the pyridinolines, hydroxylysyl-pyridinoline, and lysyl-pyridinoline. Partial hydroxylation results in formation of the pyrroles, hydroxylysyl-pyrrole, and lysyl-pyrrole (Figure 16.5). These cross-links are predominant in the fibrous collagens (e.g., bone), where tensile strength is their important functional role (Knott and Bailey, 1998).

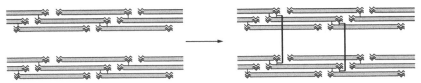

Figure 16.5: Head-to-tail longitudinal cross-linking of immature collagen fibres through divalent cross-links and proposed subsequent further reaction to form transverse cross-links between microfibrils.

The identification of the location of the divalent cross-link confirmed the alignment of the molecules in the end-overlapped, quarter-staggered fibre previously inferred from electron microscopy. The proposed location of the trivalent cross-links in mature tissue remains to be confirmed, but a reaction between two cross-links can only occur between molecules in register.

Lamprey Cartilage

The skeleton of the lamprey, a primitive vertebrate, consists of non-collagenous cartilage-like structures in the head and gill regions. While the head protein (termed lamprin) is rigid, the branchial cartilage, which supports the gills, is highly elastic, deforming and recoiling with respiration (Robson et al., 1997). The protein shows considerable similarities to elastin being non-extractable, indicating extensive covalent cross-linking. Investigation of the cross-links revealed the presence of the pyridinolines, the characteristic mature cross-links of collagen rather than the desmosines of elastin. The lysyl to hydroxylysyl pyridinoline was found to be 7:1 (Fernandes and Eyre, 1997). Neither pyridinoline nor desmosine was found in the head lamprin. The presence of collagen-type cross-links in an elastin-like protein deserves further investigation.

Transglutaminase-Derived Cross-links

Fibrillin

As a component of elastic tissue, fibrillin possesses elastic properties and Kielty (see Chapter 5) proposed that this is due to the folding of the molecule. Fibrillin is cross-linked by both disulphide and transglutaminase-induced glutamyl–lysine bonds, but the relative importance of each in restraining excessive stretching is unknown.

The transglutaminase cross-linking sites have been identified in fibrillin, two of which are at the site of molecular folding proposed by Kielty et al. (see Chapter 5). However, their importance in restraining the elastic unfolding of the molecule remains to be confirmed.

The γ-glutamyl-ε-lysine cross-links have been long established in fibrin and in keratin (Lorand et al., 1968; Lorand and Conran, 1984). However, the

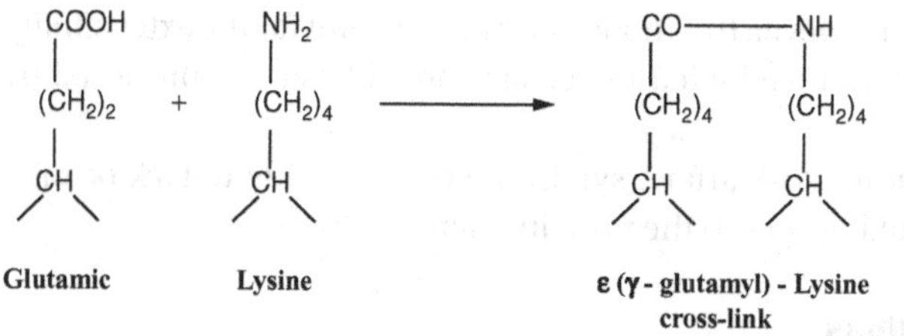

Figure 16.6: Formation of γ-glutamyl-ε-lysine cross-links mediated by transglutaminase (TGc).

transglutaminases play a diverse role in biology (Aeschlimann and Thomazy, 2000) forming cross-links in tissues from marine egg envelopes after fertilisation to the development of animal organs and in bone mineralisation. Seven different genes have been characterised to date, and the enzyme has been shown to undergo a number of posttranslational modifications (Esposito et al., 1996). One member of the transglutaminase family, TGc, is involved in the extracellular space and is best known for cross-linking fibrillin, keratin, and key structures (such as type IV collagen forming the dermal-epidermal basement membranes). Fibrillin monofibrils are formed in many tissues (see Chapter 5); for example, they occur perpendicular to the dermal basement membrane and are also thought to act as a scaffold for elastin deposition during elastic tissue formation. Transglutaminases catalyse the Ca^{2+}-dependent amine γ-glutamyl transferase reaction which leads to the formation of γ-glutamyl-ε-lysine isopeptide cross-links within or between peptide chains (Lorand and Conran, 1984). Various primary amines may react, although reaction with the ε-amino group of lysine is the most common (Figure 16.6). The reaction involves an acyl-enzyme intermediate and is driven by the release of ammonia and its subsequent protonation which occurs readily under physiological conditions. The central core domain of the enzyme is the active site which contains a Ca^{2+} binding site. The core of the enzyme is related to the cysteine proteases. The reaction centre is formed by H-bonding of the active site cysteine to histidine and aspartic residues to form a catalytic core similar to that found in the cysteine proteases.

The isopeptide cross-link is destroyed by acid and alkaline hydrolysis and consequently isolation of the cross-link involves extensive enzymic digestion of the protein. However, its presence may be detected indirectly by amino acid analysis before and after cyanoethylation or fluorodinitrobenzaldehyde (FDNB) treatment of the free ε-amino lysine groups of the unbound lysines.

Based on determination of the location of the cross-linking sites, Qian and Glanville (1997) defined the alignment of the fibrillin molecules in the elastic microfibrils as parallel and one-third overlapped, with the cross-links occurring in the overlap regions. However, this model does not account for the extensibility of fibrillin, and an alternative folded molecular model has been proposed by Kielty (see Chapter 5).

The mutations leading to Marfan's syndrome could be due to lack or mis-alignment of cross-linking sites in the fibrillin microfibrils.

Disulphide Cross-links

Gluten

The central domain of the high molecular weight (HMW) subunits of wheat gluten comprises up to three repeats, which are thought to be responsible for the elastic properties through the formation of β-turns (see Chapter 14). These independent flexible regions are linked by disulphide bonds possibly within α-helical regions (in the N- and C-terminal domains) analogous to the cross-link domain in elastin, with the disulphide bonds clearly acting as a restraining force to the elastic properties of gluten. Cleavage of the disulphide bonds reduces the elasticity of gluten, thus demonstrating their importance as a restoring force for the elastic properties of gluten. The number and distribution of the interchain disulphide bonds play a role in the relative properties of glutenin elastomers (Figure 16.7).

The network of gluten proteins are brought together during dough mixing. Although the precise changes during mixing are not completely understood, an increase in dough stiffness occurs and this probably includes some exchange of disulphide bonds since the effects of oxygen and nitrogen of the air have different effects on the sulphydryl and disulphide content of the dough. Additionally, it is probable that the disulphide bonds are of different stability, depending on their local environment of contiguous amino acids.

Work has focused on the (HMW) subunits of wheat gluten which are believed to be responsible for determining dough elasticity. Cysteine residues occur only rarely in the repetitive sequences, with single cysteine residues being present in the repetitive domains of only some subunits. In contrast, single cysteine residues are present in the short C-terminal domains of all subunits and 3 or 5 cysteine residues in the N-terminal domains. As in the case of elastin, the N- and C-terminal domains are globular with α-helices. The formation of the important functional disulphide cross-links may be facilitated by the alignment of β-spiral structures formed by the repetitive sequences initially stabilised by hydrogen bonding (see later under Hydrogen Bonding).

The mapping of the disulphide bonds has been of value in developing a model for the structure of the HMW subunit polymer. However, knowledge is not yet sufficiently complete to permit a correlation between bond distribution and biomechanical properties.

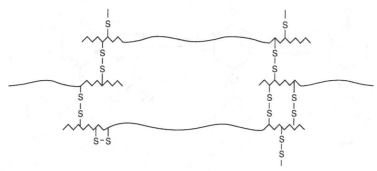

Figure 16.7: Schematic representation of interchain disulphide bonds forming a network of gluten molecules.

Catechol Oxidase–Quinones

Byssus Threads

Byssus threads, which attach mussels to rocks, possess a gradient of mechanical properties from stiff collagen to silk fibroin and elastin-like domains (see Chapter 10) They have been shown to undergo an increase in tensile strength of four to five times during aeration by seawater, but not in the presence of nitrogen, suggesting an oxidation-induced cross-link. Lysyl oxidase does not appear to be present in this system, since the lysyl aldehyde cross-links could not be detected (van Ness et al., 1988).

However, the high levels of catechol oxidase activity detected in 1 M NaCl extracts of both the byssus threads and the foot suggested that quinone cross-linking was occurring. The mechanism involves the oxidation of phenols to reactive quinones which readily react by coupling and addition reactions (Figure 16.7). Byssus threads possess diphenols in the form of peptide-bound 3,4-dihydroxyphenylalanine (DOPA) in the protein precurors and in the terminal sequences of the preCols (Qin et al., 1997). Enzyme-catalysed oxidation of these DOPA residues leads to quinones, which form Michael-type adducts with lysine (Wang et al., 1996). Unfortunately, no specific quinonelysine (Figure 16.8) cross-links have been characterised to date, probably due to the complexity of the chemistry involved. Indeed, C^{13} and N^{15} nuclear magnetic resonance (NMR) studies of newly secreted byssus threads did not reveal DOPA-lysine cross-links (Klug et al., 1996), but their presence in native threads needs to be confirmed chemically. On the other hand, McDowell et al. (1999) used rotational echo double resonance (REDOR) NMR to identify an alternative pathway involving the formation of 5,5-di-DOPA. The extent of cross-linking appears to correlate with the stress produced by the flow rate of the seawater, increasing from one di-DOPA per 1,800 amino acid residues [Figure 16.8(b)].

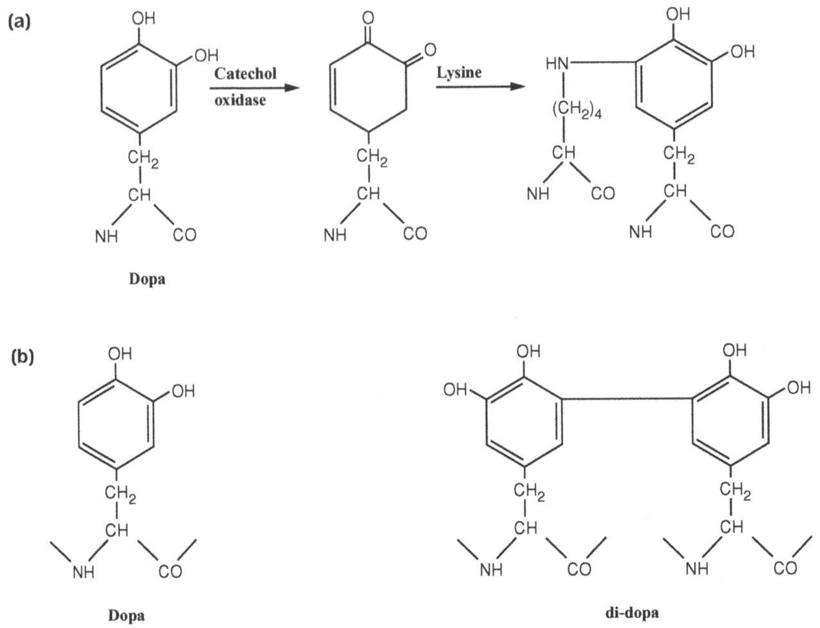

Figure 16.8: (a) Oxidation of phenols to quinones and subsequent reaction with peptide-bound lysine to form a cross-link. (b) Formation of di-DOPA.

Although cross-linking of collagen involving DOPA can be achieved in vitro by the addition of DOPA to pepsinised type I collagen (Gade et al., 1991), and also by the enzyme catalysed conversion of tyrosine to DOPA and DOPA-quinone, this does not mean that such cross-links are present in collagenous tissues in vivo.

The maturation of byssus threads by cross-linking appears to be complex and may initially be stabilised by polyfunctional metal bonds (see later), and later reinforced by quinone oxidation products. Clearly, further detailed studies of this complex chemistry are required. As an interesting aside, the production of excess byssus threads has been observed to 'tie down' a predator welk and prevent it from moving around the mussel bed and attacking the mussels, which it achieves by boring a hole in their shells before injecting enzymes to digest the tissue.

Byssus threads normally act as shock absorbers containing high strength and elastic domains, but the threads have to be securely fastened to the rocks. This is achieved by a highly efficient water-proof 'glue' secreted from the foot as a foam. The glue proteins are highly basic; the most frequently represented sequence is the decapeptide Ala, Lys, Pro, Ser, Tyr, Hyp, Hyp, Thr, DOPA-lys. The adhesiveness may be related to the DOPA residue.

NON-COVALENT CROSS-LINKS

Co-ordinate Metal–Ion Complexes

Byssus Threads

The presence of metal complexes as stabilising cross-links is suggested by the high levels of iron, copper, and zinc in the byssus threads. Peptidyl DOPA exhibits a high affinity for metals, particularly iron, and these complexes may also be involved as cross-links. The bis and tri-histidyl zinc complexes are also likely candidates, since histidine-rich regions occur at both terminii (Figure 16.9). A similar complex has been described at the active site of zinc metalloproteases (Lipscombe and Strater, 1996).

Byssus threads may initially be cross-linked by polyfunctional metal–ion bonds, but later reinforced by oxidation products involving catechol oxidases and amines. These metal–ion cross-links have the advantage of being formed immediately on secretion and of being totally reversible.

The sequential formation of different cross-links provides an excellent example of their different functions. The egg shell membrane of the sea urchin must be refractory to chemical, enzymic, and mechanical disruption to allow the embryo to develop in a protected environment. Initially, covalent cross-linking through γ-glutamyl-ε-lysine bonds occurs. Protoliaisin then attaches ovoperoxidase to the envelope which produces hydrogen peroxide and forms di-tyrosine cross-links between the various proteins of the envelope, effectively hardening it within 10 minutes of fertilisation. The process is finely regulated,

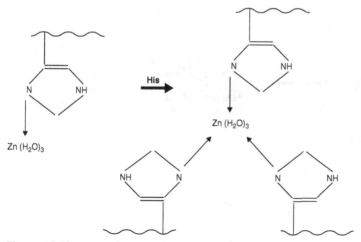

Figure 16.9: Formation of histidine-zinc co-ordination complexes as stabilising cross-links in byssus threads. (Modified from Waite et al., 1998.)

the ovoperoxidase only becoming active after fertilisation and subsequent exposure to seawater.

Hydrophobic Bond Cross-links

Silks

Both the dragline and the flagelliform spider silks contain crystalline and elastic domains that determine their extraordinary mechanical properties (see Chapter 7). The crystalline regions have been shown by X-ray diffraction and NMR to form linked antiparallel β-sheets (Simmons et al., 1996; Parkhe et al., 1997). These regions, which may constitute about 30% of the thread, are rich in alanine side chains that bind the protein together and provide tensile strength through the successive alanines interlocking with adjacent chains through hydrophobic interactions (Hayashi et al., 1999) (Figure 16.10).

Simmons et al. (1996) proposed the presence of two different alanine-rich regions, one highly orientated crystallite and the other a poorly orientated crystallite. Whether the highly orientated alanine regions interact to form the 'crystalline' domains or whether the poorly orientated crystalline domains bind two highly crystalline regions is not known On swelling in water, the stiffness of the fibre decreases 1000-fold and the extensibility increases to give a rubberlike elasticity. However, the 'crystalline' regions are retained because of the hydrophobic bonding between the alanine residues, supporting the presence of hydrophobic rather than hydrogen bonding. The increase in elasticity is due

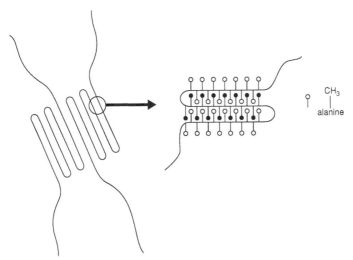

Figure 16.10: Schematic representation of stacked antiparallel β-sheets from alanine-rich domains of silks. [Modified from Simmons et al. (1996) and Hayashi et al. (1999).]

to the glycine-rich regions (70% of the fibre) now being free and independent, largely due to cleavage of some hydrogen bonding in these amorphous domains. Large residues, such as glutamic acid or lysine, could not be accommodated in these alanine β-sheets, and cross-links involving these residues would not be expected. However, it cannot be ruled out that a linking region could be present within a different sequence, which could readily accommodate a covalent cross-link. Indeed, 5% of the protein is glycosylated, and this sequence could contain other more reactive groups.

Hydrogen Bond Cross-links

Gluten
Gluten, as described previously, is cross-linked by disulphide bonds. However, NMR studies have indicated that hydrogen bonding between the subunits may also be important in stabilising the structure of gluten (Belton et al., 1995). The chains form β-spiral structures rich in proline and regularly spaced glutamine residues which can cross-link through hydrogen bonds between the glutamine residues and the amide groups of the peptide backbone of an adjacent chain to form an interchain β-sheet-like structure. This alignment of the HMW subunits may facilitate the formation of the disulphide cross-links at the N- and C-terminii of the subunits (Humphris et al., 2000). The relative importance of the disulphide bonds and the hydrogen bonds in acting as restraining cross-links remains to be determined, although current data indicate the predominance of the disulphide cross-links as the restraining force.

Silk
Hydrogen bonding also occurs in the noncrystalline elastic domains of both dragline and flagelliform silks. The proline-glycine rich elastic domains form β-turn spirals similar to elastin, and it has been proposed by Hayashi et al. (1999) that the spirals are stabilised by hydrogen bonding. The hydrogen bonds occur between the glycine hydrogen and the carbonyl group of the protein backbone. However, these hydrogen bonds, although providing some stability to the elastic amorphous regions, in the context of this chapter, they are not present as restraining bonds to prevent over-extension of the protein chains on stretching.

CONCLUDING REMARKS
Nature has clearly evolved a multiplicity of types of intermolecular cross-links to act as restraining bonds under the particular environmental conditions in which the protein elastomer exists. These chemical investigations are by no means complete and other cross-linking mechanisms remain to be identified and, indeed, some protein elastomers have not yet been investigated. An understanding

of the various types of cross-links could be of considerable value in the controlled cross-linking of recombinant polypeptides that mimic the properties of a particular native elastomer.

REFERENCES

Aeschliman, D., and Thomazy, V. (2000) Protein crosslinks in assembly and remodelling of extracellular matrices: The role of transglutaminases. *Connect. Tissue Res.* **41,** 1–27.

Andersen, S. O. (1964) The crosslinks in resilin identified as dityrosine. *Biochim. Biophys. Acta.* **93,** 213–15.

Ardell, D. H., and Andersen, S. O. (2001) Tentative identification of a resilin gene in *Drosophila melanogaster. Insect. Biochem. Mol. Biol.* **31,** 965–70.

Bailey, A. J., and Peach, C. M. (1968) Isolation and structural identification of a labile intermolecular cross-link in collagen. *Biochem. Biophys. Res. Communs.* **33,** 812–19.

Bailey, A. J., Light, N. D., and Atkins, E. D. T. (1980) Chemical cross-linking restrictions in models for the molecular organization of the collagen fibres. *Nature* **288,** 408–10.

Bailey, A. J. (2001) Molecular mechanisms of ageing in connective tissues. *Mech Ageing Dev.* **122,** 735–55.

Belton, P. S., Colquhoun, J. J., Field, J. M., Grant, A., Shewry, P. R., Tatham, A. S., and Wellner, N. (1995) FT-IR and NMR studies on the hydration of a high molecular weight subunit of gluten. *J. Biol. Macromol.* **17,** 74–80.

Brown-Augsburger, P., Tisdale, C., Brokelmann, T., Sloan, C., and Mecham, R. (1995) Identification of an elastin cross-linking domain that joins three peptide chains. *J. Biol. Chem.* **270,** 17778–83.

Esposito, C., Pucci, R., Amonesano, A., Marino, G., Cozzolino, A., and Park, R. (1996) Transglutaminase from rat coagulating gland secretion: Post-translational modifications and activation by phosphatidic acid. *J. Biol. Chem.* **271,** 27416–23.

Fernandes, R. J., and Eyre, D. R. (1999) The elastin-like protein matrix of Lamprey branchial cartilage is cross-linked by lysyl-pyridinoline. *Biochem. Biophys. Res. Communs.* **261,** 635–40.

Foerder, C. A., and Shapiro, B. M. (1977) Release of ovoperoxidase from sea-urchin eggs hardens the fertilisation membrane with dityrosine crosslinks. *Proc. Natl. Acad. Sci. U.S.A.* **74,** 4214–18.

Fujimoto, D., Horiuchi, K., and Hirama, M. (1981) Iso-tyrosine, a new cross-linking amino acid isolated from Ascaris cuticle collagen. *Biochem. Biophys. Res. Commun.* **99,** 637–700.

Gade, J. N., Fellman, J. H., and Bentley, J. P. (1991) The stabilisation of fibrillar collagen matrices with 3,4-dihydroxylphenylalanine. *J. Biomed. Mater. Res.* **25,** 799–811.

Hayashi, C. Y., Shipley, N. H., and Lewis, R. V. (1999) Hypotheses that correlate the sequence, structure and mechanical properties of spider silk protein. *Int. J. Biol. Macromol.* **24,** 271–75.

Humphris, A. D. L., McMaster, T. J., Miles, M. J., Gilbert, S. M., Shewry, P. R., and Tatham, A. S. (2000) Atomic force microscopy study of interactions of HMW subunits of wheat glutenin. *Cereal Chem.* **77,** 107–10.

Klug, C. A., Burzio, L. A., Waite, J. H., and Schaefer, J. (1996) In situ analysis of peptidyl DOPA in mussel byssus using rotational-echo double resonance NMR. *Arch. Biochem. Biophys.* **333,** 221–24.

Knott, L., and Bailey, A. J. (1998) Collagen cross-links in mineralising tissues: a review of their chemistry, function and clinical relevance. *Bone* **22**, 181–87.

Lipscombe, W. N., and Strater, N. (1996) Recent advances in zinc enzymology *Chem. Rev.* **96**, 2375–2433.

Lorand, L., Downey, J., Gotoh, T., Jacobsen, A., and Tokura, I. (1968) The transpeptide system which cross-links fibrin by γ-glutamyl-ε-lysine bonds. *Biochem. Biophys. Res. Commun.* **31**, 222–30.

Lorand, L., and Conran, S. M. (1984) Transglutaminases. *Mol. Cell Biochem.* **58**, 9–35.

MacDowell, L. M., Burzio, L. A., Waite, J. H., and Schaefer, J. (1999) REDOR detection of cross-links formed in mussel byssus under high flow stress. *J. Biol. Chem.* **274**, 20293–95.

Maki, J. M., and Kivirikko, K. I. (2001) Cloning and characterisation of a fourth lysyl oxidase isoenzyme. *Biochem. J.* **355**, 381–87.

Miles, C. A., Sionkowska, A., Hulin, S., Sims, T. J., Avery, N. C., and Bailey, A. J. (2000) Identification of an intermediate state in the helix-coil degradation of collagen by UV light. *J. Biol. Chem.* **275**, 33014–20.

Partridge, S. M., Elsden, D. F., and Thomas, J. (1963) Constitution of cross-linkages in elastin. *Nature* **197**, 1297–98.

Parkhe, A., Seeley, S., Gardner, K., Thompson, L., and Lewis, R. (1997) Structure states of spider silk proteins in the fibres are interspersed among the amorphous glycine-rich regions. *J. Mol. Recog.* **10**, 1–6.

Qian, R.-Q., and Glanville, R. W. (1997) Alignment of fibrillin molecules in elastic micro-fibrils as defined by transglutaminase-derived crosslinks. *Biochemistry* **36**, 15841–47.

Qin, X. X., Coyne, K. J., and Waite, J. H. (1997) Tough tendons: Mussel byssus has collagen with silk-like domains. *J. Biol. Chem.* **272**, 32623–27.

Robson, P., Wright, G. M., Youson, J. H., and Keeley, F. W. (1997) A family of non-collagen based cartilages in the skeleton of the sea lamprey, *Petromyzon marinus. Comp. Biochem. Physiol.* **118B**, 71–78.

Simmons, A. S. H., Mechal, C. A., and Jelinski, L. W. (1996) Molecular orientation and two-compound nature of the crystalline fraction of spider dragline silk. *Science* **271**, 84–87.

van Ness, K., Koob, T. J., and Eyre, D. R. (1988) Collagen cross-links distribution of hydrox-ypyridinium cross-links among invertebrate phyla and tissues. *Comp. Biochem. Physiol.* **91B**, 531–34.

Waite, J. H., Qin, X.-X., and Coyne, K. J. (1998) The peculiar collagens of mussel byssus. *Matrix Biol.* **17**, 93–106.

Wang, S. X., Mure, M., Medzihradsky, K. F., Burligame, A. L., Brown, D. E., Dooley, D. M., Smith, A. J., Kagan, H. K., and Klinman, J. P. (1996) A cross-linked cofactor in lysyl oxidase: Redox function for amino acid side-chains. *Science* **273**, 1078–84.

Waykole, P., and Heidemann, E. (1976) Dityrosine in collagen. *Connect. Tissue Res.* **4**, 219–22.

SEVENTEEN

Comparative Structures and Properties of Elastic Proteins

Arthur S. Tatham and Peter R. Shewry

INTRODUCTION

Elastic proteins possess rubber-like elasticity, in that they are capable of undergoing high deformation without rupture, storing the energy involved in deformation, and then returning to their original state when the stress is removed. The latter phase is passive (i.e., does not require an energy input), and the most efficient mechanisms return all (or nearly all) of the energy used in deformation. This latter requirement is not a prerequisite for elastomeric materials, as their biological requirements for energy storage/dissipation may be different.

The ability of proteins to exhibit rubber-like elasticity relates to their structure. Rubber-like materials must satisfy certain criteria: the individual components must be flexible and conformationally free, so that they can respond quickly to the applied stress, and they must be cross-linked to form a network, to distribute the stress throughout the system. These cross-links may be covalent or non-covalent, and examples of both types are found. Thus, the elastic properties of proteins are influenced by the nature of the elastomeric domains, their size, and the degree of cross-linking.

SEQUENCES OF ELASTOMERIC PROTEINS

Elastomeric proteins are widely distributed in the animal kingdom; however, only a few have been characterised in detail. This is due in part to their chemical/physical characteristics (non-globular nature, insolubility, cross-linking etc.) which make detailed characterisation difficult. More recently, gene sequences have become available, which have allowed sequence comparisons to be made and structure–function relationships to be studied. Figure 17.1 shows the schematic structures of representative elastomeric proteins for which sequences are available. Most have distinct domain structures, with at least one domain consisting of elastomeric repeat motifs and other non-elastic domains where cross-links can be formed. Exceptions to this are resilin and abductin, where cross-links occur within the elastic repeat motifs.

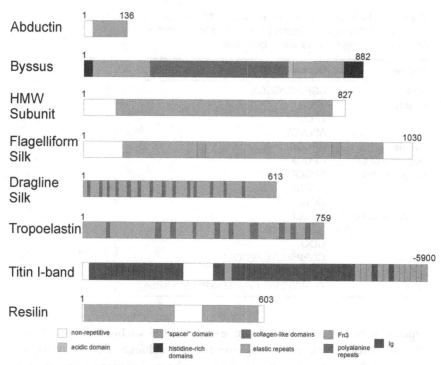

Figure 17.1: Domain structures of the elastomeric proteins: abductin (Cao et al., 1997), byssus (Coyne et al., 1997), HMW subunit (Shewry et al., 1992), flagelliform silk (Hayashi and Lewis, 1998), dragline silk (Guerrette et al., 1996), tropoelastin (Indik et al., 1987), titin I-band (Labeit and Kolmerer, 1995), and resilin (Ardell and Andersen, 2001). I$_g$, immunoglobulin. Redrawn with permission from 'Elastomeric Proteins: Biological Roles, Structures and Mechanisms', by A. S. Tatham and P. R. Shewry, 2000, *Trends Biol. Sci.*, **299**, 567–71. © 2000. (See color section.)

Abductin is present in the inner hinge ligament of bivalve molluscs, acting as an elastic pivot that antagonises the action of the adductor muscle. It also acts as an energy store that opens the shell when the adductor muscle relaxes. In scallops, this action has developed into a swimming mechanism, allowing them to swim a few metres at a time by opening and closing their shells about four times per second (Bowie et al., 1993). Cao et al. (1997) reported five abductin cDNA sequences from *Argopecten* (the bay scallop) (Figure 17.1), all of which encoded proteins of 136 amino acid residues. These proteins consist of two domains. An alanine-rich N-terminal domain (residues 1–20) contains two conserved tyrosine residues that could be involved in cross-linking. However, this domain has also been proposed to form a signal sequence for secretion and may be cleaved during processing. The second domain consists of eleven glycine/methionine-rich decapeptide repeats (Table 17.1), the methionine residues having been

Table 17.1. Sequences of the repeat motifs in the elastomeric domains of figure 17.1. Single letter amino acid codes, X represents any amino acid

Protein	Repeat Motif
Abductin	GGFGGMGGGX
Elastin	VPGG
	VPGVG
	APGVGV
Byssus	GPGGG
Flagelliform silk	GPGGX
Dragline silks	GPGQQ
	GPGGY
	GGYGPGS
HMW subunits	PGQGQQ
	GYYPTSPQQ
	GQQ
Resilin	GGRPSDSYGAPGGGN
	GYSGGRPGGQDLG
Titin	PPAKVPEVPKKPVPEEKVPVPVPKKPEA

reported to be converted to methionine sulphoxide (Kikuchi and Tamiya, 1981). This domain also contains two conserved lysine residues and a conserved tyrosine at the C-terminus, and these residues and have been proposed as possible cross-linking sites.

Byssal threads ('beards') attach mussels to hard surfaces in water. They display a gradient of mechanical properties, along the length of the fibre, from stiff to elastic, that provide sufficient flexibility to prevent brittle fracture in tidal areas (Waite et al., 1998). Two forms of byssus are present in the thread of *Mytilus edulis*, Col-D, and Col-P (Figure 17.1), which are predominantly present in the distal (i.e., close to the surface) and proximal (close to the shell) parts of the thread, respectively. Col-P consists of a central collagen-like domain (~430 residues), flanked N- and C-terminally by elastic domains (~102 and 160 residues, respectively) containing a pentapeptide repeat motif and histidine-rich domains (~106 and 62 residues, respectively) (Table 17.1) (Qin and Waite, 1995; Demming, 1999). Col-D has a similar arrangement of domains, but the elastic domains are absent and replaced by alanine-rich domains, which are non-elastic (Waite et al., 1998). The mechanism of cross-linking is unknown, but may involve metal complexation in histidine-rich domains and/or be via tyrosine residues (Waite et al. 2002).

The high molecular weight (HMW) subunits of wheat gluten are seed storage proteins, their sole function being to act as a store of carbon, nitrogen, and sulphur for the developing seedling. Their elastic properties have probably arisen as a consequence of structures that facilitate the efficient packaging and storage of the proteins in protein bodies in the endosperm. The HMW subunits are characterised by a large central repetitive domain (~630–830 residues) consisting of hexa-, nona-, and, in certain subunits, tri-peptide repeats (Table 17.1). This domain is flanked by non-repetitive N- and C-terminal domains (81–104 and 42 residues, respectively), which contain cysteine residues that may form intermolecular bonds. The elasticity of wheat dough is influenced by both the amount and type of HMW subunits present (Shewry et al., 1992).

Resilin occurs in insect cuticles, conferring long-range elasticity to the tissue and functioning as an energy store and damper of vibrations in the flight systems of many insects. Ardell and Andersen (2001) have tentatively identified a resilin gene in *Drosophila melanogaster* by searching the *Drosophila* genome for sequences related to tryptic peptides from locust (*Scistocerca gregaria*) resilin. They identified an open reading frame encoding a 603 residue protein (Figure 17.1), with an N-terminal domain comprising 18 pentadecapeptide repeats, a central non-repetitive domain, and a C-terminal domain comprising 11 tridecapeptide repeats (Table 17.1). Cross-linking occurs between tyrosine residues, which are present throughout the protein.

Elastin is widely distributed in vertebrate tissues, acting both statically, as in the dermis, to resist long-term forces and dynamically, and in such tissue as arteries, where it provides an efficient energy storage system. Elastin is one of the most extensively studied elastomeric proteins. It is secreted as a soluble precursor, tropoelastin, which consists of alternating repetitive hydrophobic domains of variable length (the elastic repeats) and alanine-rich, lysine-containing domains that form cross-links (Indik et al., 1987) (Figure 17.1). The elastic domains consist of tetra-, penta-, and hexapeptide motifs (Table 17.1).

Spiders produce several different silk proteins. The dragline silks have high tensile strength and will extend by some 30%, and form the dropping lines and frameworks of webs. The flagelliform silks form the capture spirals of webs and exhibit much greater extensibility (>200%), with similar properties to lightly cross-linked rubbers (Gosline et al., 1986). The silks are processed to produce high-performance fibres with a range of mechanical properties, allowing the web to absorb the impact of the insect and capture it. The two types of silk differ in their domain structures (Figure. 17.1); the dragline silks contain polyalanine domains (which form non-covalent cross-links between proteins) separated by elastic domains comprised of pentapeptide repeats (Table 17.1) (Guerrette et al., 1996). The flagelliform silks (Figure 17.1) contain longer elastic domains, based on pentapeptide repeats GPGGX (Table 17.1), and contain 'spacer' domains of unknown function (Hayashi and Lewis, 1998, 2000). The cross-linking mechanism in the flagelliform silks is unknown, but may involve interactions between the spacer domains (Hayashi and Lewis, 1998).

Titin (or connectin) is a sarcomeric protein that is responsible for the elasticity of striated muscle myofibrils and is also involved in muscle assembly. It occurs in a range of isoforms, which have different properties in different tissues. The elastic properties of titin are responsible for the force that is generated when passive muscle is stretched. The human cardiac form of titin comprises ~26,900 residues (Labeit and Kolmerer, 1995) divided into the Z-disc (~2,000 residues), I-band (~5,900 residues), A-band (~17,200 residues), and M-line (~1,100 residues) regions. However, only a fraction of the molecule within the I-band is functionally extensible (Figure 17.1). The elastic domain contains repeating motifs of 26–28 amino acid residues – rich in proline, glutamate, valine, and lysine residues – termed the PEVK domain (Greaser, 2001) (Table 17.1).

Spectrin is a major component of the cytoskeleton of membranes in animal cells. It contributes to molecular flexibility and long-range elasticity of the membrane skeleton (Thomas and Discher, 2002). Spectrin is composed of two subunits, α and β, that are structurally distinct and encoded by separate genes. The dimers are assembled into tetramers by tight association between antijuxtaposed C-termini of each subunit with the N-termini of its opposite (α-β, β-α) subunits. The erythrocyte β-form contains two calponin-homology domains at the N-terminal actin-binding region, 16 complete spectrin repeats, a partial spectrin repeat, and a serine-rich domain at the C-terminus. The α-form consists of a partial spectrin repeat at the N-terminus, 20 complete spectrin repeats, a Src-homology 3 domain in the ninth repeat, and a calmodulin-like domain at the C-terminus. The spectrin repeat varies from 99 to 110 residues. Although sequence identity between repeats is not great, alignment studies indicate conserved residues, and it is thought that the repeats have similar three-dimensional structures.

The fibrillins are found throughout connective tissue as components of microfibrils and impart long-range elasticity to the tissue, with the elasticity of microfibrils being similar to that of elastin. It is thought that they play a role in elastic fibre formation, providing a scaffold for the deposition of tropoelastin. Fibrillin is a 350 kDa extracellular acidic protein with a high cysteine content and an extended thread-like shape. Sequences of fibrillin isoforms (1 and 2) from different species indicate that most of the sequence is composed of 47 epidermal growth factor-like repeats, 43 of which have a consensus sequence for calcium binding; other domains include seven 8-cysteine motifs known as TB modules. The difference between fibrillin-1 and -2 is that fibrillin-1 contains a proline-rich domain close to the N-terminus, which is replaced by a glycine-rich repeat in fibrillin-2 (Kielty et al., 2002).

STRUCTURAL FEATURES OF ELASTOMERIC PROTEINS

The presence of regularly repeated sequences implies the formation of a regular structure. Although the direct determination of the structures of elastic proteins has proved problematic, the limited information that is available indicates that the repetitive sequences do form regular structures and that these may be important in the elastic mechanisms.

Studies of elastin indicate that the backbone is highly mobile (Ellis and Packer, 1976) and that the alanine-rich domains that contain lysine residues for cross-linking are predominantly α-helical (Foster et al., 1976). Urry and coworkers have carried out detailed studies of α-elastin and synthetic polypeptides corresponding to the major tetra-(VPGG), penta-(VPGVG), and hexa-(APGVGV) peptide consensus repeat motifs, demonstrating that β-turns were the predominant structural feature present. Urry and coworkers have proposed that the β-turns present in the synthetic polypentapeptide (Urry, et al., 1984) and polyhexapeptide (Rapaka et al., 1978) are assembled in multiple repeats and form a β-spiral, a loose water-containing helical structure (Urry, 1984). Only the polypentapeptide was found to be elastic with an elastic modulus similar to that of elastin. The proposed β-spiral formed by the polypentapeptide contained no significant hydrogen bonding between repeats, allowing flexible movement; whereas in the polyhexapeptide, intramolecular hydrogen bonding between repeats and the interlocking of adjacent β-spirals by hydrophobic interactions resulted in structural rigidity. The polytetrapeptide motif appeared to form a cross β-structure and exhibited little elasticity (Urry et al., 1986). The β-spiral structure formed by the polpentapeptide was proposed to contribute to the elastic mechanism of elastin. However, this has been questioned by other workers who have proposed that the proline-glycine dipeptides in the elastin sequences are sites that can bend to form tight turns and make the backbone flexible, behaving like a random coil (Gosline, 1987).

The spider silks contain crystalline and non-crystalline regions (Figure 17.1). The alanine-rich domains have been shown by X-ray diffraction and nuclear magnetic resonance (NMR) spectroscopy studies to form linked antiparallel β-sheet structures (Hijirida et al., 1996; Simmons et al., 1996), which are thought to correspond to the crystalline regions that cross-link the proteins and provide silks with their tensile strength. The structures of the glycine-rich (non-crystalline) domains are poorly understood, but β-turns are predicted to form within the domain. (Table 17.1). In the flagelliform silks, the GPGGX motif has been proposed to form β-spiral type structures composed of repetitive β-turns, which act as springs to form an elastic network (Hayashi et al., 1999; Hayashi and Lewis, 2000). In dragline silks, the GGPGQ and YPGQQ motifs are also predicted to form β-turns (Table 17.1) with Pro.Gly as the central residues of the turns.

NMR experiments have indicated the presence of 3_{10} helical structures in the glycine-rich domains, oriented in-line with the threads (Kümmerlen et al., 1996; van Beek et al., 1999). The greater elasticity of the flagelliform silks could be related to their longer repetitive regions and the lower levels of cross-linking compared with the dragline silks (Figure 17.1) and to their more hydrophobic nature.

The repetitive domains of the HMW subunits of glutenin contain three repeat motifs that are thought to be responsible for the elastic properties of wheat doughs and glutens. Structural prediction studies suggested the presence of β-turns (Table 17.1) (Tatham et al., 1990) which are predicted to form within and between the repeat motifs. Similarly, spectroscopic studies of linear and cyclic peptides corresponding to the repeat motifs also indicated the presence of β-turns (Tatham et al., 1990; van Dijk et al., 1997). Characterisation of a recombinant repetitive peptide using circular dichroism (CD) and Fourier transform-infra red (FT-IR) indicated the presence of β-turns in equilibrium with poly-L-proline II-like structure in solution with less β-turns and more intermolecular β-sheet structure in the hydrated solid (the environment in doughs and gluten) (Gilbert et al., 2000). Hydrodynamic studies indicated a rod-like structure in solution (Field et al., 1987) and scanning tunnelling microscope images showed that the central repetitive domains adopt a helical structure with a diameter of ~1.9 nm and a pitch of ~1.5 nm in the hydrated solid state (Miles et al. 1991). Thus, it has been proposed that the central repetitive domains form a β-spiral type structure. Further atomic force microscope images showed rod-like molecules that aggregated side-by-side and end-to-end to form branched networks (Humphris et al., 2000).

The repeat motif of abductin has not been studied in detail, but prediction indicates that the domain adopts a random coil conformation (Cao et al., 1997).

Resilin contains two repeat motifs of 15 and 13 residues, termed A and B repeats, respectively. Both contain conserved PG motifs associated with β-turns and an additional turn over Pro.Ser was predicted in the A motif (Ardell and Andersen, 2001). Ardell and Andersen (2001) suggested that both repeats may form β-spirals, but with larger and more irregular loops than in other elastic proteins. Spectroscopic evidence is required to confirm the presence of β-turns and/or other structures.

Titin contains a number of repeated motifs, including two different types of 100 residue motifs with similarities to immunoglobulin and fibronectin III domains (Figure 17.1). The unique feature of titin is the PEVK domain, which varies in length between 163 and 2,174 residues, for cardiac and soleus muscles, respectively (Labeit and Kolmerer, 1995). Antibody labelling of muscle at different extensions showed that the PEVK domain lengthened on stretching (Trombitas et al., 1998) implying that the domain has an elastic function in muscle. It was initially proposed that titin elasticity arose from unfolding of the immunoglobulin-like domains, but it is now thought to arise from the PEVK domain (Linke et al., 1998, 1999). The PEVK domain is more easily stretched than the immunoglobulin-like domains and is predicted to have an unstable secondary structure, as a consequence of high-charge density and the high content of proline residues, preventing the formation of a stable tertiary structure (Labeit and Kolmerer, 1995). Greaser (2001) identified repeating motifs of 26–28 amino acid residues (Table 17.1), which contained regions rich in glutamic acid. Secondary structure prediction indicated little propensity to form α-helical or β-sheet structures in the PEVK repeats, but some propensity to form α-helix in the glutamic acid regions and the presence of poly-L-proline II structure in the PEVK motifs. Structure prediction would suggest the presence of β-turns (Table 17.1), but their presence needs to be confirmed by CD or FT-IR spectroscopy.

Structure prediction indicates that the spectrin repeat comprises three α-helices separated by two loop regions rich in proline and serine. Solution NMR studies indicate that the basic unit of spectrin folds into a left-handed antiparallel triple helical coiled coil (Pascual et al., 1997). The homologous α- and β-chains form antiparallel heterodimers, that self-associate to form a tetramer, and to a certain extent higher order oligomers. This exposes actin binding sites at its termini and enables cross-linking of actin filaments.

Fibrillin-containing microfibrils are complex polymers comprising many different proteins. The fibrillins associate to form extensible beaded microfibrils, and these structures have been the focus of studies of elastic behaviour (Kielty et al., 2002).

ELASTIC MECHANISM AND FUNCTION

The sequences and structures of the different elastomeric proteins would imply that there may be differences in their elastic mechanisms.

Although elastin is the most highly characterised elastomeric protein, a number of different mechanisms have been proposed to account for its elasticity. One proposed mechanism is based on classical rubber theory. Elastin is often considered to be a network of random chains of high entropy; stressing orders the chains and decreases the entropy of the system by limiting the conformational freedom of the chains. This decrease in entropy provides the restoring force to return to the initial state (Hoeve and Flory, 1974). Urry and coworkers have proposed an alternative model, in which elasticity arises from a regular structure, a β-spiral comprising repetitive β-turns (Urry, 1988). In the polypentapeptide repeat, the β-turns are repeated regularly and act as spacers between the turns of the spiral, suspending chain segments in a kinetically free state (conformationally free). On stretching, this conformational freedom is reduced, resulting in a decrease in entropy; with this loss of entropy providing the restoring force

(Urry et al., 1986). Thus, in both the random chain and β-spiral models, the restoring force is entropic. Gosline (1978) has proposed that hydrophobic interactions account for some of the restoring force. Stretching elastin exposes hydrophobic side chains to an aqueous environment, decreasing the entropy of the surrounding water molecules, the restoring force arising from the reestablishment of the hydrophobic interactions. Thus, whereas it is agreed that the driving force for elastic recoil is entropic in origin, the precise mechanism (and structures involved) is not known.

The elastic domains of the byssal proteins have repeat motifs similar to those of elastin (Table 17.1), and an entropic mechanism can also be invoked to account for the elasticity of the protein PreCol-P. However, the byssal threads do not themselves show instant recoil. This is due to energy dissipation and forms a mechanism that stops the mussel from being dashed against the rocks after the wave has passed (Bell and Gosline, 1996).

There have been limited studies on abductin, but the amino acid sequence indicates a hydrophobic character. Jiminez et al. (2000) have synthesised a pentapeptide (FGGMG) mimic of abductin and proposed a hydroelasticity model, in which elastic recoil results from a hydrophobic mechanism and peptide self-assembly produces an aggregate with rubber-like properties, even in the absence of covalent crosslinks.

Resilin behaves as an entropic elastomer (Weis-Fogh, 1960), and was initially proposed to consist of randomly coiled protein chains linked by stable covalent cross-links, the elastic force being accounted for by a decrease in conformational entropy when the material was strained (Weis-Fogh, 1961). Unlike elastin, the repetitive domains of resilin are not hydrophobic, but dominated by hydrophilic residues, suggesting that the domains will tend not to interact by hydrophobic interactions. Ardell and Andersen (2001) suggested that the presence of the β-turns results in a β-spiral, as proposed for other proteins, and that the structure can act as a readily deformed spring. At low extension, disruption between the loops (which are stabilised by hydrophobic interactions, hydrogen bonds, and/or electrostatic attractions) occurs, while at higher extensions elastic recoil is mainly due to a decrease in conformational entropy.

The elastic repeat motifs of flagelliform spider silks, which form the capture part of spider webs, show sequence similarity with elastin (Table 17.1), are highly extensible and are covered in droplets of glue that capture the insect and hold it. The high extensibility gives the insect no solid structure to push against, and an entropic mechanism is consistent with such a function (Vollrath and Edmonds, 1989). The sequences of dragline silks, which form the dropline and radial threads of the web, contain glutamine residues (Table 17.1). The mechanism of elastic recoil is not entirely entropic, but this accounts for 85% of the total force (Gosline et al., 1984). Part of the energy of an insect impacting is absorbed and dissipated as heat; the energy is, therefore, unavailable to break the silk threads, and the insect is not catapulted out of the web. In view of their content of glutamine residues, which have a capacity to hydrogen bond, non-covalent cross-linking between elastic repeats (via hydrogen bonding) may act to modulate the elastic mechanism. Both flagelliform and dragline silks have been proposed to contain β-turns and to form β-spirals which account for their elasticity.

HMW subunits of glutenin that have been cross-linked by γ-radiation show a similar modulus of elasticity to the cross-linked polypentapeptide of elastin. However, the mechanism of elasticity appears complex (Tatham et al., 2001). The high content of glutamine residues in the repetitive domains suggests that elastic recoil, in part, may be associated with extensive hydrogen bonding within and between subunits. Thus, entropic and energetic mechanisms contribute to the observed elasticity of the HMW subunits. The elastic properties have no known in vivo role, so there has not been selective pressure to attain mechanisms that are energy efficient. The proteins contain repetitive β-turns which may form β-spirals, as proposed for other elastic proteins, but the spirals may interact via hydrogen bonding, perhaps making their behaviour more similar to dragline silks and resilin than to elastin.

Titin elasticity has been proposed to arise from a number of different mechanisms. Recent studies of the PEVK domain indicate a flexible coiled structure, comprising polyproline-II helical structures interspersed with unordered regions (Gutierrez-Cruz et al., 2001; Li et al., 2001; Ma et al., 2001). Repeating motifs of 26–28 residues, termed PPAK, have been identified in the PEVK domain, where they occur in groups of 2–12 motifs separated by regions rich in glutamic acid. It has been proposed that the difference in charge between the PPAK motifs and the glutamic acid-rich regions suggests ionic interactions between the positively charged PPAK regions, and negatively charged glutamic acid-rich regions are involved in the elastic mechanism (Greaser, 2001). Both the immunoglobulin and PEVK domains are thought to contribute to passive tension; at low tension, the immunoglobulin domains extend without unfolding, whereas at a higher stretch, the PEVK domain extends.

The elastic properties of spectrin are probably based on the conformational flexibility of the repeated region (Thomas and Discher, 2002). Bloch and Pumplin (1992) suggested that residues connecting the repeats act as a hinge region and are responsible for elasticity. However, others – including Hansen et al. (1997) – have proposed an entropic spring model, spectrin behaving as a flexible chain, or a model based on a helical coiled-coil, which extends by increasing the helical pitch.

The elasticity of fibrillin-containing microfibrils has been discussed by Kielty et al. (2002). Untensioned microfibrils show repeating interbead organisation, and stretching results in a proportion of the microfibrils appearing extended. This suggested that interbead unfolding occurred at low extensions (<100 nm) and that bead unravelling occurred beyond this, with the extended arrays appearing to be irreversibly stretched. Thus, the fibrils elongate initially as folded domains or loops are pulled out, and there is an entropic drive to return to the folded structure.

CONCLUSIONS

The advantage of an entropic mechanism of elasticity is that it depends on the decrease in numbers of accessible states on extension and can be considered ideal for biological systems. Entropic elasticity provides for a durable elastomer, with elastic fibres that can last the lifetime of an individual. Several protein systems (i.e., elastin and titin) have evolved elastic proteins, which have similar entropic mechanisms, where energy is required to stress the proteins and relaxation is non-energy–consuming. In other systems, however, it would be detrimental to restore all the energy to the system (i.e., byssal threads, dragline silk), as the function of the material is, in part, to dissipate the energy. Spectrin and fibrillin do not contain the proline-rich domains; elasticity would appear to arise from the unfolding of domains with defined secondary structure and/or from the interactions of those domains. Thus, similarities and differences in the sequences, structures, and mechanisms of elasticity show that precise biological requirements for elastic behaviour can be satisfied with different mechanisms.

REFERENCES

Ardell, D. H., and Andersen, S. (2001) Tentative identification of a resilin gene in *Drosophila melanogaster*. *Insect Biochem. Mol. Biol.* **31**, 965–70.

Bell, E. C., and Gosline, J. M. (1996) Mechanical design of mussel byssus: Material yield enhances attachment strength. *J. Exp. Biol.* **199**, 1005–17.

Bloch, R. J., and Pumplin, D. W. (1992) A model of spectrin as a concertina in the erythrocyte membrane skeleton. *Trends Cell Biol.* **7**, 186–89.

Bowie, M. A., Layes, J. D., and Demont, M. E. (1993) Damping in the hinge of the scallop *Placopecten magellanicus*. *J. Exp. Biol.* **175**, 311–15.

Cao, Q., Wang, Y., and Bayley, H. (1997) Sequence of abductin the molluscan 'rubber' protein. *Curr. Biol.* **7**, R677–78.

Coyne, K. J. (1997) Extensible collagen in mussel byssus: a natural block copolymer. *Science* **277**, 1830–32.

Demming, T. J. (1999) Mussel byssus and biomolecular materials. *Curr. Opin. Chem. Biol.* **3**, 100–5.

Ellis, G. E., and Packer, K. J. (1976) Nuclear spin-relaxation studies of hydrated elastin. *Biopolymers* **15**, 813–32.

Field, J. M., Tatham, A. S., and Shewry, P. R. (1987) The structure of a high M_r subunit of durum wheat (*Triticum aestivum*) gluten. *Biochem. J.* **247**, 215–21.

Foster, J. A., Bruenger, E., Rubin, L., Imberman, M., Kagan, H., Mecham, R., and Franzblau, C. (1976) Circular dichroism studies of of an elastin crosslinked peptide. *Biopolymers* **15**, 833–41.

Gilbert, S. M., Wellner, N. K., Belton, P. S., Greenfield, J. A., Siligardi, G., Shewry, P. R., and Tatham, A. S. (2000) Expression and characterisation of a highly repetitive peptide derived from a wheat seed storage protein. *Biochim. Biophys. Acta* **1479**, 135–46.

Gosline, J. M. (1978) Hydrophobic interaction and a model for the elasticity of elastin. *Biopolymers* **17**, 677–95.

Gosline, J. M. (1987) Structure and mechanical properties of rubberlike proteins in animals. *Rubber Chem. Technol.* **60**, 417–38.

Gosline, J. M., Demont, M. E., and Denny, M. W. (1986) The structure and properties of spider silk. *Endeavour* **10**, 37–43.

Gosline, J. M., Denny, M. W., and Demont, M. E. (1984) Spider silk as rubber. *Nature* **309**, 551–52.

Greaser, M. (2001) Identification of new repeating motifs in titin. *Proteins Struct. Funct. Genet.* **43**, 145–49.

Guerrette, P. A., Ginzinger, D., Weber, B., and Gosline, J. (1996) Silk properties determined by gland-specific expression of a spider fibroin gene family. *Science* **272**, 112–15.

Gutierrez-Cruz, G., van Heerden, A. H., and Wang, K. (2001) Modular motif, structural folds and affinity profiles of the PEVK segment of human skeletal muscle titin. *J. Biol. Chem.* **276**, 7442–49.

Hansen, J. C., Skalak, R., Chien, S., and Hoger, A. (1997) Influence of network topography on the elasticity of red blood cell membrane skeleton. *Biophys. J.* **72**, 2369–81.

Hayashi, C. Y., and Lewis, R. V. (1998) Evidence from flagelliform silk cDNA for the structural basis of elasticity and modular nature of spider silks. *J. Mol. Biol.* **275**, 773–84.

Hayashi, C. Y., Shipley, N. H., and Lewis, R. V. (1999) Hypotheses that correlate the sequence, structure, and mechanical properties of spider silk proteins. *Int. J. Biol. Macromol.* **24**, 271–75.

Hayashi, C. Y., and Lewis, R. V. (2000) Molecular architecture and evolution of a modular spider silk protein gene. *Science* **287**, 1477–79.

Hijiridia, D. H., Do, K. G., Michal, C., Wong, S., Zax, D., and Jelinski, L. W. (1996) C13 NMR of *Nephila clavipes* major ampullate silk gland. *Biophys. J.* **71**, 3442–47.

Hoeve, C. A. J., and Flory, P. J. (1974) The elastic properties of elastin. *Biopolymers* **13**, 677–86.

Humphris, A. D. L., McMaster, T. J., Miles, M. J., Gilbert, S. M., Shewry, P. R., and Tatham, A. S. (2000) Atomic force microscopy (AFM) study of interactions of HMW subunits of wheat glutenin. *Cereal Chem.* **77**, 107–11.

Indik, Z., Yeh, H., Ornstein-Goldstein, N., Sheppard, P., Anderson, N., Rosenbloom, J. C., Peltonen, L., and Rosenbloom, J. (1987) Alternative splicing of human elastin mRNA indicated by sequence analysis of cloned genomic and complementary DNA. *Proc. Natl. Acad. Sci. U.S.A.* **84**, 5680–84.

Jimenez, F., Banco, M., Goddard, W. A., and Sandberg, L. (2000) Scallop abductin translates hydrophobicity to elasticity. Available from: http://www.wag.caltech.edu/anmeeting/2000/presentations/mario2/

Kielty, C. M., Baldock, C., Lee, D., Rock, M. J., Ashworth, J. L., and Shuttleworth, C. A. (2002) Fibrillin: From microfibril assembly to biomechanical function. In: Elastomeric

Proteins: Structures, Biochemical Properties, and Biological Roles. P. R. Shewry, A. J. Bailey and A. S. Tatham (eds.) Cambridge University Press, pp. 125–57.

Kikuchi, Y., and Tamiya, N. (1981) Methionine sulfoxide in the resilin of surf clams. *J. Biochem.* **89**, 1975–76.

Kümmerlen, J., van Beek, J., Vollrath, F., and Meier, B. (1996) Local structure in spider dragline silk investigated by two-dimensional spin-diffusion nuclear magnetic resonance. *Macromolecules* **29**, 2920–28.

Labeit, S., and Kolmerer, B. (1995) Titins; giant proteins in charge of muscle ultrastructure and elasticity. *Science* **270**, 293–96.

Li, H., Oberhauser, A. F., Redick, S. D., Carrion-Vazquez, M., Erickson, H. P., and Fernandez, J. M. (2001) Multiple conformations of PEVK proteins detected by single-molecule techniques. *Proc. Natl. Acad. Sci. U.S.A.* **98**, 10682–86.

Linke, W. A., Ivemeyer, M, Mundel, P., Stockmeier, M. R., and Kolmerer, B. (1998) Nature of PEVK-titin elasticity in skeletal muscle. *Proc. Natl. Acad. Sci. U.S.A.* **95**, 8052–57.

Linke, W. A., Rudy, D. E., Centner, T., Gautel, M., Witt, C., Labeit, S., and Gregorio, C. C. (1999) I-Band titin in cardiac muscle is a three element molecular spring and is critical for maintaining thin filament structure. *J. Cell Biol.* **146**, 631–44.

Ma, K., Kan, L.-S., and Wang, K. (2001) Polyproline II helix is a key structural motif of the elastic PEVK segment of titin. *Biochemistry* **40**, 3427–38.

Miles, M. J., Carr, H. J., McMaster, T. J., I'Anson, K. J., Belton, P. S., Morris, V. J., Field, J. M., Shewry, P. R., and Tatham, A. S. (1991) Scanning tunneling microscopy of a wheat storage protein reveals details of an unusual supersecondary structure. *Proc. Natl. Acad. Sci. U.S.A.* **88**, 68–71.

Pascual, J., Pfuhl, M., Walther, D., Saraste, M., and Nilges, M. (1997) Solution structure of the spectrin repeat: A left-handed antiparallel triple-helical coiled-coil. *J. Mol. Biol.* **273**, 740–51.

Qin, X., and Waite, J.H. (1995) Exotic collagen gradients in the byssus of the mussel *Mytilus edulis. J. Exp. Biol.* **198**, 633–44.

Rapaka, R. S., Okamoto, K., and Urry, D. W. (1978) Non-elastomeric polypeptide models of elastin. *Int. J. Peptide Prot. Res.* **11**, 109–27.

Shewry, P. R., Halford, N. G., and Tatham, A.S. (1992) The high molecular weight subunits of wheat glutenin. *J. Cereal Sci.* **15**, 105–20.

Simmons, A., Michal, C., and Jelinski, L. (1996). Molecular orientation and two component nature of the crystalline fraction of spider dragline silk. *Science* **271**, 2920–28.

Tatham, A. S., Drake, A. F., and Shewry, P. R (1990) Conformational studies of peptides corresponding to the repetitive regions of the high molecular weight (HMW) glutenin subunits of wheat. *J. Cereal Sci.* **11**, 189–200.

Tatham, A. S., Hayes, L., Shewry, P. R., and Urry, D. W. (2001) Wheat seed proteins exhibit a complex mechanism of protein elasticity. *Biochim. Biophys. Acta* **1548**, 187–193.

Tatham, A. S., and Shewry, P. R. (2000) Elastomeric proteins: Biological roles, structures and mechanisms. *Trends Biol. Sci.* **299**, 567–71.

Thomas, G. H., and Discher, D. E. (2002) Conformational compliance of spectrins in membrane deformation, morphogenesis, and signalling. In: Elastomeric Proteins: Structures, Biochemical Properties, and Biological Roles. P. R. Shewry, A. J. Bailey and A. S. Tatham (eds.) Cambridge University Press, pp. 276–306.

Trombitas, K., Greaser, M., Labeit, S., Jin, J. P., Kellermeyer, M., Helmes, M., and Granzier, H.

(1998) Titin extensibility in situ: entropic elasticity of permanently folded and permanently unfolded molecular segments. *J. Cell Biol.* **140**, 853–59.

Urry, D. W. (1988) Entropic elastic processes in protein mechanisms. 1. Elastic structures due to an inverse temperature transition. *J. Prot. Chem.* **7**, 1–34.

Urry, D. W., Harris, R. D., Long, M. M., and Prasad, K. U. (1986) Polytetrapeptide of elastin: Temperature correlated elastomeric force and structure development. *Int. J. Peptide Prot. Res.* **28**, 649–60.

Urry, D. W., Henze, R., Harris, R. D., and Prasad, K. U. (1984) Polypentapeptide of elastin: Temperature dependent correlation of elastomeric force and dielectric permittivity. *Biochem. Biophys. Res. Commun.* **125**, 1082–88.

van Beek, J. D., Kümmerlen, D., Vollrath, F., and Meier, B. H. (1999) Supercontracted spider dragline silk: A solid state NMR study of the local structure. *Int. J. Biol. Macromol.* **24**, 173–78.

van Dijk, A., DeBoef, E., Bekkers, A., van Wijk, L. L., van Swieten, E., Hamer, R. J., and Robillard, G. T. (1997) Structure characterisation of the central repetitive domain of high molecular weight gluten proteins. II. Characterization in solution and the dry state. *Protein Sci.* **6**, 649–56.

Vollrath, F., and Edmonds, D. T. (1989) Modulation of the mechanical properties of spider silk by coating with water. *Nature* **340**, 305–7.

Waite, J. H., Qin, X. X., and Coyne, K. J. (1998) The peculiar collagens of mussel byssus. *Matrix Biol.* **17**, 93–106.

Waite, J. H., Vaccaro, E., Sun, C., and Lucas, J. M. (2002) Collagens with Elastin- and Silk-like Domains. In: Elastomeric Proteins: Structures, Biochemical Properties, and Biological Roles. P. R. Shewry, A. J. Bailey and A. S. Tatham (eds.) Cambridge University Press, pp. 240–275.

Weis-Fogh, T. (1960) A rubber-like protein in insect cuticle. *J. Exp. Biol.* **37**, 889–907.

Weis-Fogh, T. (1961) Molecular interpretation of the elasticity of resilin, a rubber-like protein. *J. Mol. Biol.* **3**, 648–67.

EIGHTEEN

Mechanical Applications of Elastomeric Proteins – A Biomimetic Approach

J. F. V. Vincent

INTRODUCTION

The range of application of any science is restricted as much by commerce as by the inherent possibilities of the science and the resulting technology. This is especially true of areas where the immediate profitability is high: typically medicine, sport, and defence. In these areas, the profit motive disappears beneath hopes of survival, on which there is no rational price. Biologically, this is understandable; intellectually, it can be restricting. Equally restricting is the perception of commerce that an improvement of less than 10% is not worth chasing.

Therefore, when trying to see what applications might be suggested by a biomimetic approach to the mechanical properties of proteins, applications in medicine, sport, and defence are the first to be tried. Medicine is seen to be (relatively!) intellectually easy, since the typical working environment will be similar to that in which the protein evolved – temperatures ranging between $-20°$ and $+40°C$ and an aqueous environment. A few proteins (silk and keratin are the obvious examples; collagen as tendon and processed animal skin should be included) are mechanically interesting in the absence of liquid water, and could therefore be considered as paradigms for applications in sport and defence. But elastomeric proteins – such as elastin, titin, and resilin – rely on water not only as a plasticiser, keeping the protein below its glass transition at ambient conditions, but also to allow expression of the hydrophobicity which lies at the heart of the mechanism of long-range ('rubbery') elasticity.

To increase their range of use, we need to develop ways in which the environmental resistance of proteins (or their analogues) can be increased. In the first instance, the competition is from thermoplastic materials, which melt at about $120°C$. This is well above the temperature which most hydrated proteins can withstand, at least at normal pressures, although of course elastin is routinely prepared by autoclaving tissue at $110°C$, at which temperature collagen, other proteins, and polysaccharides melt out of the network. The insolubility of the elastin arises directly from its extensive cross-linking, which also contributes to the thermal stability. The hottest metazoa known, worms living around hydrothermal vents (260 bar, 350°C), can withstand temperatures up to 105°C. The stability of collagen at this temperature has been ascribed to an increase in the proline content in the helical domain (as deduced from the cDNA sequence) (Sicot et al., 2000). The mechanism for the stability is unclear, though of course proline is not only the amino acid in which the side chain curves back to the main backbone, and thus confers main chain stability. It would be possible to add artificial amino acids to a polypeptide, thus conferring non-biological properties which could include increased stability at high temperature (Rutjes et al., 2000). Another important mechanism by which proteins can be protected from denaturation at high temperatures is surface adsorption. For instance, α-chymotrypsin and cutinase, adsorbed onto Teflon, showed no structural changes up to 100°C (the highest temperature reached in the experiments) (Zoungrana et al., 1997).

All these figures are for the hydrated protein. When water is removed, the dry protein has a much higher temperature resistance, better than most plastics, and can withstand temperatures up to 300°C or so. This is the limit for catgut, horn, and hair – all durable and useful materials – which probably contain about 12% of water, representing the tightly bound fraction.

The interest is increased if the protein is allowed to be part of a composite material/structure, but these materials have not been fully explored and explained at the biological level (we still do not understand how to make bone, mother-of-pearl, or insect cuticle de novo) nor exploited at the technological level.

PROTEINS IN CERAMICS

Proteins influence ceramics at a number of different levels. Mother-of-pearl, or nacre, leads studies in biomineralisation, probably because its geometry of relatively large platelets (polygons about 8 μm across and 0.5 μm thick) arranged in sheets is so beguilingly simple. In one model, these platelets are connected together by small holes in the matrix which separate the layers of platelets (Addadi and Weiner, 1997). This neatly solves the problem of how to start each individual platelet growing as a crystal (it does not have to, after all) and how to ensure crystallographic, and probably hence mechanical, perfection from one layer to the next (it is just one big crystal). The matrix between the sheets is very tenuous (allowing for less than 5% of the total volume) and is made of protein. Simple calculation shows it to be about 25 nm thick. In the abalone (*Haliotis rufescens*) shell, three protein fractions have been identified and assigned roles in mineralisation: an insoluble silk-like protein forms a basic structure, then soluble polyanionic proteins control crystal growth. A further fraction has been implicated in nucleation of the calcium carbonate crystals. The a- and b-axes of the silk-like protein are aligned with the a- and b-crystallographic directions of the aragonite, suggesting that there is some form of geometric matching between the functional groups of the protein and lattice spacing of the ions in the crystal. The silk-like fraction has been characterised by X-ray diffraction (Weiner and Traub, 1984) and been found to contain blocks composed of alanine or glycine alone, which are longer than those found in spider silk (Sudo et al., 1997). It may well also contain chitin. When mother-of-pearl fractures, the matrix is spun into fibres across the gaps between the platelets. It seems reasonable that these fibres should be silk-like and therefore capable of taking high loads. They probably contribute at least half of the toughness of nacre (Jackson et al., 1988). The silk-like nature of the fibres suggests that the matrix material is a liquid crystal, which would also assist the initiation and epitaxy of the aragonite, giving as it would a highly ordered surface. This idea has not yet been investigated. These fibres have been stretched in an atomic force microscope, when they elongate in a stepwise manner as folded domains or loops are pulled open. The steps occur with forces of a few hundred pN, smaller than the force (more than a nanoNewton) which would be required to break the main chain peptide bonds (Smith et al., 1999). This behaviour is similar both to some dragline silks and to titin (Kellermayer et al., 1998).

In the initial stages of the mineralisation of abalone shells, a primer layer of orientated calcite crystals is nucleated and grows on a protein sheet. The deposition of this primer is followed by an abrupt transition to c-axis–oriented crystals of aragonite. The formation of each of the two crystal types is accompanied by the synthesis of specific polyanionic proteins. These proteins alone are sufficient to control the crystal phase, suggesting that interactions between these proteins and the inorganic ions during crystal nucleation control the phase of the deposited mineral and that differential expression of the proteins allows the organism to induce phase changes. Thus, soluble organic components can control hierarchical biomineral growth offering the prospect of similar phase control in materials (Belcher et al., 1996). In some instances (more will doubtless emerge), protein has been found to form a molecular composite with the ceramic phase. Sea urchin skeletons are made of single crystals of calcite which do not have well-developed cleavage planes and so are more resistant to fracture. This is due in part to acidic glycoproteins intercalated within the lattices of single crystals. In synthetic calcite, the protein only slightly decreases the coherence of the molecular arrangement, but significantly increases the angular spread of perfect domains of the crystals. In biogenic calcite, the coherence length is one-third to one-fourth as much as that in synthetic calcite, and the angular spread is 20–50 times as wide. One idea is that the macromolecules are concentrated at mosaic boundaries oblique to the cleavage planes and thus change the fracture properties (Berman et al., 1990). The sea urchin tooth, a composite with a matrix of high-magnesium calcite and reinforcing calcite fibers, and organic material at the fibre–matrix interface, is much harder and tougher than most other calcitic materials. Comparison of the fracture morphology of intact teeth and teeth from which the organic component has been removed shows that the organic material resists fibre pull-out and contributes significantly to the toughness of the teeth (Wang et al., 1997; Wang, 1998).

The biomimetic application of these results would require expansion of the temperature-stability range of tough ceramics from the normal biological range (mainly below about 35°C) to the higher working temperatures, in which engineering ceramics are more routinely used. It would be instructive to examine the temperature response of echinoderm calcite to determine the breakdown temperature of the protein (a) when confined within the crystal structure and (b) when extracted and purified. Meanwhile, it is possible to deposit calcite onto elastin as a heat-stable matrix (Manoli and Dalas, 1999), although no attempt has been made to assess the mechanical properties of the resulting composite.

Possibly more encouraging is the development of methods for the design and production of a heat-stable adhesive protein (Kobatake et al., 2000) based on combining a stable peptide unit with a biologically active peptide unit. A peptide sequence, Arg-Gly-Asp, which acts as a cell adhesive in vitro, was incorporated into the elastin-based rigid polyhexapeptide, Ala-Pro-Gly-Val-Gly-Val. The resulting protein retained more than 90% of cell adhesive activity after autoclaving at 120°C for 20 minutes. The conclusion to be drawn is that it would be possible to engineer a heat-stable protein which could act as a substrate for the deposition of calcite or aragonite. If one could understand how the protein controls and interacts with the ceramic, proteins might be developed which could interact with a range of ceramics, including man-made ones, thus toughening a material which is intrinsically very brittle. Such a protein (or related polymer) would be stabilised by its intimate association with the relatively inert ceramic and thus be able to withstand higher temperatures (Zoungrana et al., 1997). In this way, one could increase the work of fracture of engineering ceramics from the few tens of J m^{-2} to the 1.5 kJ m^{-2}, which is the minimum for nacre and cow leg bone.

In another experimental approach, vaterite, calcite, and aragonite crystals were grown on a substrate made from *Loligo* chitin onto which silk fibroin has been deposited, followed by protein extracted from mollusc shell. To test further the individual effects of the most frequently occurring amino acids in the shell protein, some artificial polypeptides were also tested in the system, showing that poly(Asp-Leu) specifically induces the formation of aragonite. The silk proteins orientated themselves on the chitin surface, presumably binding onto β-spaced hydroxyls on the chitin (see later), and generally stimulated crystallisation. Thus, the specificity of the polymorph produced is dependent upon the amino acid sequence, the conformation of specific protein(s) in the mollusc shell, and the microenvironment in which crystal nucleation and growth takes place (Levi et al., 1998).

FIBROUS COMPOSITES

In a composite material, it is useful to compare the basic morphology and see whether it is the chemistry or structure which is giving rise to some of the good mechanical properties observed. It is possible to compare insect cuticle (a 'simple' fibrous composite containing nanofibrils of α-chitin which are 3 nm in diameter and 100–200 nm long, composed of 19 molecular chains arranged antiparallel, embedded in a protein matrix). The crystalline structure of α-chitin has a number of surprises. Along the 010 face (the c-axis), the hydrogen chains are laterally spaced by 0.475 nm, the same as the lateral distance between the adjacent protein chains of an antiparallel β-sheet, such as silk. Along the length of each chitin chain, the space between adjacent residues is 1.032 nm. Twice this repeat (2.064 nm) is almost exactly the same as three times the repeat of a protein β-sheet (0.69 nm) (Fraenkel and Rudall, 1940). Thus, the lattice which appears on the 010 face of the chitin nanofibril can be related to the spacings in a chain-folded sheet of protein β-sheet (Atkins, 1985) and the chitin is bonded to half the amino acid side chains. Consider then the surface area available for bonding between matrix and fibre. In a man-made composite, the minimum practical size for a fibre (carbon) is a diameter of 5 μm. Below this size, the fibres pose a health hazard. Although only half of the surface area of the chitin nanofibril is available for matrix bonding (Atkins, 1985), the surface area of chitin per unit volume of cuticle is 10^6 greater than the carbon fibre material. Although the interfacial shear strength of the protein–chitin interface is unknown, guessing that the shear strength of a single bond is of the order of 30 pN, and calculating that the area around one bond is about 10^{-18} square metres, then the shear strength will be about 30 MPa, or about half that measured for carbon fibres. However, with the huge surface area difference, it is fairly obvious that the total fibre–matrix interaction per unit volume in insect cuticle is still 5×10^5 greater than carbon fibre composites. How should a protein, or its analogue, be designed to make use of this huge specific surface area? In general, the proteins of stiffer cuticles tend to be more hydrophobic (Andersen et al., 1995) so one might expect them to be globular. However, the effect of denaturants on ANS-fluorescence suggests that the proteins of insect cuticle do not have a folded and packed hydrophobic core like that of native globular proteins, implying that they do not denature in the same way as globular proteins. Instead, they are more like unordered, randomly coiled, thermally agitated polymer chains, whose hydrodynamic volumes depend upon the composition of the medium. It seems, therefore, that most of the peptide chains of insect cuticular proteins form unordered, random structure, when the proteins are in solution and when they are present in the intact cuticle (Andersen, 2001). It is probable that only the chain regions involved in binding the proteins to chitin will have a well-defined spatial organisation. This is suggested by the observation that many insect cuticular proteins include a 35–36 amino acid motif (Rebers and Riddiford, 1988) which probably binds chitin. Its sequence does not resemble the cysteine-containing, chitin-binding domain found in chitinases and some peritrophic membrane proteins. However, an extended form of this motif from proteins of hard insect cuticle is necessary and sufficient for chitin binding, shown both by direct binding studies and by disrupting the motif to eliminate chitin binding. (Rebers and Willis, 2001). Once again, there is evidence for the presence of elastin-like structures in insect cuticle proteins [e.g., resilin (Ardell and Andersen, 2001)], suggesting the possibility of a thermostable protein matrix which can interact with the chitin with the addition of a silk-like binding domain.

There are several issues here which could be transferred into technology. For example, chitin nanofibres can be separated relatively easily from crab cuticle and are safely handled. They can then be used in aqueous suspension, since the matrix protein is also water-miscible and interacts in a specific hydrophobic manner with the fibrous phase. We could then use nanofibres with their vastly increased surface area for interfacial reactions, increasing toughness in proportion. Since the nanofibres are stiff, it is highly likely that the material will show some self-assembly characteristics. For example, in biological tissues, the three-dimensional arrangements of chitin-protein or of collagen fibrillar networks appear to be very similar to those described in certain liquid crystals (Neville, 1993). In crab cuticle and compact bone, mobile fringes are seen in polarised light, periodic cleavage is seen in scanning electron microscopy, and arced patterns are seen in the transmission electron microscope (Bouligand, 1987). In vitro, highly concentrated solutions of purified collagen form ordered assemblies characterised by polarising microscopy as cholesteric phases (Giraud-Guille, 1994). The ease with which proteins can move between these states to make stiff structures has been shown in the egg capsule, a complex laminated structure, of the dogfish *Scyliorhinus canicula*. Each lamina is secreted from a spinneret in one of six discrete zones of tubular glands within the oviducal gland and is composed of a number of layers of uniaxially orientated collagen fibres, each layer laid at an angle of 45° or 90° to its neighbour. The collagen passes through a series of ordered phases, many of which have a well-defined liquid crystalline structure. When first synthesised, the collagen appears isotropic, but becomes assembled into a smectic A or lamellar phase granules in the Golgi cisternae. As these granules mature, collagen passes through a cholesteric mesophase before adopting a columnar hexagonal arrangement. These phases seem to involve progressive dehydration and so are probably more concerned with storage of the protein rather than any modification. On secretion, the contents of the granules revert rapidly to the smectic A-lamellar and micellar phases, suggesting that they are rehydrating. As it passes along the nidamental gland tubules, the collagen is first converted into a second distinct micellar phase before assembling into the final fibrils that constitute the egg capsule (Knight et al., 1993). Orientation of the fibrils seems to be as much due to flow as intrinsic self-assembly; this seems to be a common thread occurring in the production of silk: the intrinsic order of the liquid crystalline structures is aided by more macro, flow, conditions (Knight et al., 1996a, 1996b, 2000). The capsule wall, together with its surface coatings and content of jelly-like material, is formed as a continuous concentric coextrusion. The spinnerets are arranged circumferentially along the length of the nidamental gland, each contributing a layer to the egg capsule. Thus, the posterior spinnerets contribute the outermost layers while the anterior ones deliver the innermost layers. Thus, starting anteriorly, the first zone (A) secretes a highly hydrated jelley and some collagen; the second (B) is a single row of spinnerets secreting the inner wall of the capsule; the third (C) are thought to secrete the makings of a sealant varnish; the fourth (D) secrete the bulk of the capsule wall (collagen); the fifth (E) secretes tendrils which anchor the capsule to strands of weed; and the sixth (F) contributes an outer varnish and some lubricant to help the capsule on its way down the oviduct (Knight et al., 1996a, 1996b). Eventually, the entire assembly is phenolically tanned and turns brown. This design of extruder for multilayered laminated materials with defined orientations of fibres or molecules could have some interesting lessons for us, and a series of patents is in preparation at the time of writing.

Very similar systems, which are much less complex, but rely on similar physical chemistry, occur amongst the insects, especially the Orthoptera, such as *Periplaneta* and *Mantis*, which produce a capsule to protect their eggs from predators and desiccation. These capsules are made of liquid crystalline protein which is subsequently stabilised and waterproofed with phenolic tanning (Pryor, 1940; Brunet, 1951).

The phenolic residues make the proteins insoluble. The general chemistry is to introduce the phenol as single molecules into the material to be stabilised, and to polymerise it and link it to the surrounding material as a single event. The phenol is rendered more reactive by having two OH groups placed next to each other on the ring (ortho-positioning) and then converting them into much more reactive $=O$ groups with a metal-containing phenoloxidase. The latest clues to the way phenols work to drive water out come from food science and the bitter drying taste of strong tea and cheap red wines which have been 'over-oaked' and have too much polymeric phenol absorbed from oak chips or the wooden walls of the storage barrels. These polyphenols are largely responsible for the 'mouthfeel' since they react with basic proline-rich proteins in the saliva.

Astringency arises from precipitation of polyphenol/peptide complexes, which is an important protective mechanism in animals that consume polyphenols. It has been shown the polyphenolic rings stack onto planar hydrophobic surfaces stabilised by cooperative binding. The peptides become increasingly coated with polyphenols and eventually precipitate (Charlton et al., 2002). As the water is lost, the remaining water-binding sites are brought together and interact, rendering the material not only more highly cross-linked, but insoluble, since the water cannot penetrate back into the network which is held together by these secondary bonds acting in co-operation with each other. In insect cuticle, for example, the proteins tend to form large amounts of β-sheet conformation. This chemically driven dehydration can happen under water, which explains why the mussel byssus thread, secreted as a watery collagenous fibre, can become strong and stiff. It also means phenolic tanning can be used for another function, as waterproof varnish, which it appears to fulfil on the outside of the dogfish egg case (Knight et al., 1996a, 1996b). In passing, it can be noted that there is little or no evidence for covalent phenolic cross-linking, even if it occurs, being important in stabilising these materials. This is because the addition of formic acid, which breaks secondary, but not primary, bonds in these systems, causes the material to hydrate and swell once the water-binding sites are re-exposed. The swollen cuticle has long-range elasticity, showing that the protein chains are kinetically free (Andersen, 1981). The implication is that the primary links, if any of these are contributed by phenolic residues, are not the primary cause of the stiffness of the protein; this must be due to the secondary bonds formed on dehydration. Therefore, any cross-linking provided by the phenolic residues is trivial. The main function of the phenols is dehydration. Until this lesson is learned, we shall not progress with our understanding of water control in protein and protein-like matrices.

Similar principles apply to biological adhesives based on phenolics. The mussel byssus adhesive system is probably a special case in that the phenol is built into the structural protein rather than being added later. Perhaps this is to prevent it from being washed away into the surrounding sea. The protein commonly consists of 80 or more repeats of a peptide 7–15 residues long which contains two residues of 3,4-dihydroxyphenylalanine (DOPA) and a lysine. The DOPA may have two roles: it competes with water molecules for strong adsorption to surfaces (illustrated by the chemical sorption of catechol, the simplest ortho-dihydroxyphenol, to the surface of metal) and, when the OH groups have been converted to =O, can react with lysine to (theoretically) cross-link the protein. These cross-links have not been identified. The reacted DOPA also increases the hydrophobicity of the protein; thus, the DOPA sticks and solidifies (Holl et al., 1993). These systems should also be tested by exposure to formic acid. Do they swell in the same way as insect cuticle? If so, the same arguments apply regarding the function of the phenols. Comparing the amino acid sequences of these proteins across a number of related species of mussel reveals surprisingly little similarity. This may be because the taxonomy of these molluscs is not very well understood, but also because the role of environmental pressures on evolutionary selection of adhesives is not well known, and also because adhesive proteins are not well understood anyway. There seems no reason why mussel adhesive should not work in the same way as Charlton's model. Barnacle adhesives rely on much greater hydrophobicity of the protein and sulphur cross-links (Naldrett, 1993).

Another protein adhesive has been identified in the centipede *Henia* (Hopkin et al., 1990), where it constitutes the main defensive material, comprising more than 10% of the animal's body weight. When the centipede is attacked, it exposes its undersurface to the attacker and exudes the glue from segmental glands. The glue sticks to the mandibles, palps, antennae, etc., of the attacker, stopping it from moving properly. The attacker then retreats to clean itself and the centipede escapes. The glue appears to be a two-part protein system with a fibrous (silk-like) component which can bridge large gaps, and a globular component which could be the adhesive. It sets without the need for phenolic tanning, and can adhere to wet and dirty surfaces. Its performance equals that of commercial cyanoacrylates, when it is confined between two plates held closely together. Clearly, it is a versatile adhesive. The concept of mixing spinnable polymers with a contact adhesive does not appear to have been explored commercially.

MECHANICAL PROPERTIES
Most proteins are still inscrutable mechanically. We do not know enough to be able to derive useful materials using the information available about amino acid sequence or the implications of that sequence for the generation of all but the simplest secondary and tertiary structures. The notable exception is elastin and other proteins which appear to use the same molecular mechanisms, such as resilin. Toughness and stiffness, two important characteristics, are not self-evident from molecular studies. Classical engineering requires that, for most applications, materials should be stiff, strong, and tough. This facilitates the formation of structures which can carry loads and require little or no maintenance and leads to a traditional use of stone or wood for compressive structures and of metal or wood for structures which withstand bending or tension. To some extent, the same can be said of biology: a compressive structure, such as mollusc shell or bone, is typified by calcium salts and is fairly stiff, though much tougher than the structural ceramics used by man. A biological tensile structure is typically fibrous (silk, mussel byssus thread) and, again, is quite stiff and strong, but is mostly much tougher than man-made materials. Indeed, the specific properties of biological materials (stiffness, strength, etc., as a function of density) are very competitive with those of man-made materials.

This emphasis on toughness needs to be elaborated. In materials engineering, there are two main ways of quantifying resistance to fracture. The most commonly used is the critical stress intensity factor, K_C, which quantifies the resistance to growth of a crack, but for most biological materials this is not used much since it assumes non-biological phenomena, such as isotropy. More useful is the work to fracture, which is the work done when propagating a crack. It is related to the area cracked and so is expressed in $J\ m^{-2}$. More usefully, it can be measured directly from the graphical output of a suitably arranged experiment, needing the minimum of calculation (Atkins and Mai, 1985). It is thus a relatively easy parameter to derive. Toughness ensures that when the material fractures, it will not only require a relatively large amount of energy, but also that the fracture progresses slowly and allows time for the structure to be unloaded to a safe level before fracture is complete and the structure is destroyed. Another set of questions arises. Do we know the range of mechanical properties available from any single protein? Consider collagen. The accepted value for the Young modulus of collagen fibres is only about 1.5 GPa. This value is typical of amorphous polymers, but one would expect a higher value from a relatively crystalline material in which the tensile stress is borne directly by the main-chain covalent bonds. For example, silk has a modulus nearer 100 GPa. One type of collagen which is obviously very stiff and strong is found in the wall of the nematocyst, the hydraulic capsule which powers the sting of coelenterates (sea anemones, corals, and jelly fish). In the freshwater polyp, *Hydra vulgaris*, the capsule contains a 2 M salt solution and so reaches a turgor pressure of 150 atmospheres (15 MPa) before it shoots out the dart in the first phase of stinging. If the nematocyst is assumed to be a sphere with walls 200 nm thick, then the stress in the collagen will be 190–375 MPa (Holstein et al., 1994). If one adds a reasonable safety factor into this design, the fracture stress will need to be something like 500 MPa. If, as seems reasonable for a fibre which is under continuous tension, this collagen stretches over the straight part of the 'normal' collagen stress-strain curve, then the fracture stress will be reached at a strain of about 2%, putting the stiffness at about 25 GPa. Although far higher than the stiffness of collagens so far measured, this figure seems much more in line with the stiffness to be expected of a covalently bonded fibre, and is still probably too low by a factor of 5 or so. Interestingly, using a rheological approach, the stiffness of the helical part of type I collagen in solution was found to be five times larger than that of a coiled-coil, α-helical protein (Amis et al., 1985). If that is 6–10 GPa (the highest estimate from measurements of mammalian keratins (Kitchener and Vincent, 1987), then the value for the collagen triple helix is 30–50 GPa. Both these estimates are about ten times the 'normal' measured stiffness of tendon collagen in the steeply rising part of the curve, suggesting that this shortfall is a result of the relatively low-energy bonding between the various components in the hierarchical structure of tendon.

However, many proteins are highly hydrated (plasticised) and are therefore very soft and fail only at high strains. Can these materials be of use in any context outside of medicine? Is there any opportunity for the invention of a new technology in engineering, or are there niches in our current technology in which hydrated stretchy materials offer new possibilities? One area might be robotics (and associated areas of prosthetics), where there is increasing interest in the production of robots capable of a variety of facial expressions (Hara and Endo, 2000). Since these are designed by engineers rather than biologists (or, for that matter, people such as skin graft surgeons who know something of the fibre orientations of skin), the distribution of directional compliance in facial skin has never been examined or modelled. Skin analogues based on the components found in real skin would make the expressions more realistic and more easily attained, since the localisation of deformation due to Poisson ratio effects would be more controlled. Also, the distribution of compliance through the thickness of the skin would make for better touch sensitivity of prosthetic fingers (also useful in robots required to handle delicate objects) and of protective clothing, such as gloves.

USE OF PROTEINS IN MACROSTRUCTURES

The cuticular protein resilin has long been touted as a 'wonder' elastomer, and its occurrence in the structures of insects and other arthropods can suggest ways in which such highly elastic materials can be used. For example, the legs of many animals and a few walking machines (robots) contain elastic elements that contribute to their efficient and adaptable use in locomotion. It has long been known that resilin amplifies the muscular power of insects for jumping (Bennet-Clark and Lucey, 1967; Bennet-Clark, 1975). Resilin is also found in the legs of several insect species (cockroaches, crickets, and grasshoppers) associated with specific ligaments of limb joints and at some muscle insertions. Resilin ligaments function as muscle antagonists, simplifying the construction and control of movement. For instance, the tibio-tarsal joint of the cockroach leg is an eccentric ball and socket. In walking, the leg is probably raised by stored elastic energy in a band of resilin in the joint and the levator muscle helps in placing it down (Neff et al., 2000). Resilin is also found at the base of sensory hairs, where its ability to retain its shape for long periods ensures that the hairs always return to the same position after moving, obviating the need for a 'null position' signal from the nervous system (Neville, 1975).

It is impossible to review biomimetic applications of proteins and their structures without mentioning silk. Silks are covered in detail elsewhere in this volume, but most of the applications are forecast to be in the area of personal armour or crowd control – for instance, trapping or tangling fibres or gels for catching things or people, or super-silly string to spray on something so it can move away and still be caught and reeled in. This reflects both the origin of silk as a trap material and the military sources of most of the funding. As a caution, remember that Nylon, probably the first biomimetic material, found its most important use providing bearing surfaces for use with aqueous lubricants. Biomimetics would then suggest that we should look at all the other uses of silk in nature. There are many types of silk. Some are used to make a protective casing for a cocoon and form the basis of the silk industry (Brunet and Coles, 1974); others are used for adhesion both to anchor the web of the spider and to stick together the components of the case of the underwater caddis larva (Rowlands and Hansell, 1987). Perhaps the ultimate value of silk will be as an adhesive rather than as an absorber of impacts.

REFERENCES

Addadi, L., and Weiner, S. (1997) A pavement of pearl. *Nature* **389,** 912–15.

Amis, E. J., Carriere, C. J., Ferry, J. D., and Veis, A. (1985) Effect of pH on collagen flexibility determined from dilute solution viscoelastic measurements. *Int. J. Biol. Macromol.* **7,** 130–34.

Andersen, S. O. (1981) The stabilization of locust cuticle. *J. Insect Physiol.* **27,** 393–96.

Andersen, S. O. (2001) Matrix proteins from insect pliable cuticles: Are they flexible and easily deformed? *Insect Biochem. Mol. Biol.* **31,** 445–52.

Andersen, S. O., Hojrup, P., and Roepstorff, P. (1995) Insect cuticular proteins. *Insect Biochem. Mol. Biol.* **25,** 153–76.

Ardell, D. H., and Andersen, S. O. (2001) Tentative identification of a resilin gene in *Drosophila melanogaster. Insect Biochem. Mol. Biol.* **31,** 965–70.

Atkins, A. G., and Mai, Y.-W. (1985) *Elastic and Plastic Fracture: Metals, Polymers, Ceramics, Composites, Biological Materials.* Chichester, UK: Ellis Horewood.

Atkins, E. D. T. (1985) Conformations in polysaccharides and complex carbohydrates. *Proc. Int. Symp. Biomol. Struct. Interact., Suppl. J. Biosci.* **8,** 375–87.

Belcher, A. M., Wu, X. H., Christensen, R. J., Hansma, P. K., Stucky, G. D., and Morse, D. E. (1996) Control of crystal phase switching and orientation by soluble mollusc-shell proteins. *Nature* **381,** 56–58.

Bennet-Clark, H. C. (1975) The energetics of the jump of the locust, *Schistocerca gregaria. J. Exp. Biol.* **63,** 53–83.

Bennet-Clark, H. C., and Lucey, E. C. A. (1967) The jump of the flea. *J. Exp. Biol.* **47,** 59–76.

Berman, A., Addadi, L., Kvick, A., Leiserowitz, L., Nelson, M., and Weiner, S. (1990) Intercalation of sea urchin proteins in calcite: Study of a crystalline composite material. *Science* **250,** 664–67.

Bouligand, Y. (1987) Cuticular morphogenesis of crustacea. Meeting of Carcinology, Concarneau, France.

Brunet, P. C. J. (1951) The formation of the ootheca by *Periplaneta americana.* I. The microanatomy and histology of the posterior part of the abdomen. *Q. J. Microsc. Sci.* **92,** 113–27.

Brunet, P. C. J., and Coles, B. C. (1974) Tanned silks. *Proc. R. Soc. Lond. B* **187,** 133–70.

Charlton, A. J., Baxter, N. J., Khan, M. L., Moir, A. J. G., Haslam, E., Davies, A. P., and Williamson, M. P. (2002) Polyphenol/peptide binding and precipitation. *J. Agric. Food Chem.* 1593–601.

Fraenkel, G., and Rudall, K. M. (1940) A study of the physical and chemical properties of insect cuticle. *Proc. R. Soc. Lond. B* **129**, 1–35.

Giraud-Guille, M. M. (1994) Liquid crystalline order of biopolymers in cuticles and bones. *Microsc. Res. Technol.* **27**, 420–28.

Hara, F., and Endo, K. (2000) Dynamic control of lip-configuration of a mouth robot for Japanese vowels. *Robot. Auton. Sys.* **31**, 161–69.

Holl, S. M., Hansen, D., Waite, J. H., and Schaefer, J. (1993) Solid-state NMR analysis of cross-linking in mussel protein glue. *Arch. Biochem. Biophys.* **302**, 255–58.

Holstein, T. W., Benoit, M., Herder, G. V., Wanner, G., David, C. N., and Gaub, H. E. (1994) Fibrous mini-collagens in *Hydra* nematocysts. *Science* **265**, 402–4.

Hopkin, S. P., Gaywood, M. J., Vincent, J. F. V., and Mayes-Harris, E. L. V. (1990) Defensive secretion of proteinaceous glues by *Henia* (=*Chaetechelyne*) *vesuviana* (Chilopoda, Geophilomorpha). Proceedings of the seventh International Congress on Myriapodology. E. J. Brill, Lelden, pp. 175–81.

Jackson, A. P., Vincent, J. F. V., and Turner, R. M. (1988) The mechanical design of nacre. *Proc. R. Soc. Lond. B* **234**, 415–40.

Kellermayer, M. S. Z., Smith, S. B., Bustamante, C., and Granzier, H. L. (1998) Complete unfolding of the titin molecule under external force. *J. Struct. Biol.* **122**, 197–205.

Kitchener, A. C., and Vincent, J. F. V. (1987) Composite theory and the effect of water on the stiffness of horn keratin. *J. Mater. Sci.* **22**, 1385–89.

Knight, D. P., Feng, D., Stewart, M., and King, E. (1993) Changes in macromolecular organization in collagen assemblies during secretion in the nidamental gland and formation of the egg capsule wall in the dogfish *Scyliorhinus canicida*. *Phil. Trans. R. Soc. Lond. B* **341**, 419–436.

Knight, D. P., Feng, D., and Stewart, M. (1996a) Structure and function of the Selachian egg case. *Biol. Rev.* **71**, 81–111.

Knight, D. P., Hu, X. W., Gathercole, L. J., Rusaouen, M., Ho, M.-W., and Newton, R. (1996b) Molecular orientations in an extruded collagenous composite, the marginal rib of the egg capsule of the dogfish, *Scyliorhinus canicula*; a novel lyotropic liquid crystalline arrangement and its origin in the spinnerets. *Phil. Trans. R. Soc. Lond. B* **351**, 1205–22.

Knight, D. P., Knight, M. M., and Vollrath, F. (2000) Beta transition and stress-induced phase separation in the spinning of spider dragline silk. *Int. J. Biol. Macromol.* **27**, 205–10.

Kobatake, E., Onoda, K., Yanagida, Y., and Aizawa, M. (2000) Design and gene engineering synthesis of an extremely thermostable protein with biological activity. *Biomacromolecules* **1**, 382–86.

Levi, Y., Albeck, S., Brack, A., Weiner, S., and Addadi, L. (1998) Control over aragonite crystal nucleation and growth: An in vitro study of biomineralization. *Chemistry—A Eur. J.* **4**, 389–96.

Manoli, F., and Dalas, E. (1999) Calcium carbonate overgrowth on elastin substrate. *J. Crystal Growth* **204**, 369–75.

Naldrett, M. J. (1993) The importance of sulphur cross-links and hydrophobic interactions in the polymerisation of barnacle cement. *J. Mar. Biol. Assn. UK* **73**, 689–702.

Neff, D., Quimby, L., Frazier, S. F., Wang, R., and Zill, S. (2000) Elasticity, mobility and adaptable locomotion: Structure and function of resilin in insect legs. *Soci. Neurosci. Abstr.* **26.**

Neville, A. C. (1975) *The Biology of Arthropod Cuticle.* Berlin, Germany: Springer.

Neville, A. C. (1993) *Biology of Fibrous Composites.* Cambridge, UK: The University Press.

Pryor, M. G. M. (1940) On the hardening of the ootheca of *Blatta orientalis. Proc. R. Soc. Lond. B* **128,** 378–98.

Rebers, J. E., and Riddiford, L. M. (1988) Structure and expression of a *Manduca sexta* larval cuticle gene homologous to *Drosophila* cuticle genes. *J. Mol. Biol.* **203,** 411–23.

Rebers, J. E., and Willis, J. H. (2001) A conserved domain in arthropod cuticular proteins binds chitin. *Insect Biochem. Mol. Biol.* **31,** 1083–93.

Rowlands, M. L. J., and Hansell, M. H. (1987) Case design, construction and ontogeny of building in *Glyphotaelius pellucidus* caddisfly larvae. *J. Zool. Lond.* **211,** 329–35.

Rutjes, F. P. J. T., Wolf, L. B., and Schoemaker, H. E. (2000) Applications of aliphatic unsaturated non-proteinogenic alpha-H-alpha-amino acids. *J. Chem. Soc., Perkin Trans. 1.* (24) 4197–212.

Sicot, F.-X., Mesnage, M., Masselot, M., Exposito, J.-Y., Garrone, R., Deutsch, J., and Gaill, F. (2000) Molecular adaptation to an extreme environment: Origin of the thermal stability of the pompeii worm collagen. *J. Mol. Biol.* **302,** 811–20.

Smith, B. L., Schaffer, T. E., Viani, M., Thompson, J. B., Frederick, N. A., Kindt, J., Belcher, A., Stucky, G. D., Morse, D. E., and Hansma, P. K. (1999) Molecular mechanistic origin of the toughness of natural adhesives, fibres and composites. *Nature* **399,** 761–63.

Sudo, S., Fujikawa, T., Nagakura, T., Ohkubo, T., Sakaguchi, K., Tanaka, M., Nakashima, K., and Takahashi, T. (1997) Structures of mollusc shell framework proteins. *Nature* **387,** 563–64.

Wang, R. (1998) Fracture toughness and interfacial design of a biological fiber-matrix ceramic composite in sea urchin teeth. *J. Am. Ceram. Soc.* **81,** 1037–40.

Wang, R. Z., Addadi, L., and Weiner, S. (1997) Design strategies of sea urchin teeth: Structure, composition and micromechanical relations to function. *Phil. Trans. R. Soc. Lond. B* **352,** 469–80.

Weiner, S., and Traub, W. (1984) Macromolecules in mollusc shells and their functions in biomineralization. *Phil. Trans. R. Soc. Lond. B* **304,** 425–34.

Zoungrana, T., Findenegg, G. H., and Norde, W. (1997) Structure, stability, and activity of adsorbed enzymes. *J. Colloid Interface Sci.* **190,** 437–48.

NINETEEN

Biomimetics of Elastomeric Proteins in Medicine

Allen J. Bailey

INTRODUCTION

A characteristic feature of elastomeric proteins is the presence of repeating motifs in the amino acid sequence which provide the freely mobile peptide chains that are the basis of their elasticity (see Chapter 17). These repetitive domains are highly ordered and mainly hydrophobic in nature and show an ability to self-assemble. The elastic domains are normally interspersed with short structured or 'crystalline' regions, which act as restraining domains. In some cases, the highly ordered 'crystalline' domains may also undergo conformational changes during self-assembly to allow subsequent precise intermolecular interactions. When isolated from the protein, these short amino acid sequences have been shown to have the ability to mimic many of the properties of the original protein, such as elastin or silk. More recently, amino acid sequences of both the 'elastic' and 'crystalline' domains have been probed by modern molecular biology techniques and synthetic polypeptides prepared. These peptides allow researchers to study the control mechanisms of their assembly and the basis of their elastomeric properties, knowledge of which can lead to the evolution of designer peptides for specialised applications. Further modification can be achieved by alteration of the amino acid sequence, or alteration of the amino acid side chains, generally the basic groups of lysine and arginine, and the carboxyl groups. Although several examples have been reported to date, only a few have found application; but this is an exciting field which is expanding rapidly with the advent of recombinant peptides.

ELASTIN

The platelet adhesion and aggregation activity of elastin is extremely low, compared with the other components of the extracellular matrix in the vascular system (Barnes and MacIntyre, 1979; Legrand and Fauvel, 1992), and this property is an important aspect of the aetiology of atherosclerosis. Indeed, platelet aggregation at a damaged surface of the arterial wall is one of the earliest stages of atherosclerosis. Furthermore, this unique property allows the use of composites of solubilised elastin, with collagen and fibrin for use as a scaffold for tissue repair, for example, of the major blood vessels (Lefebvre et al., 1989) and tympanic membranes (Bonzon et al., 1995). It is highly probable that the availability of recombinant elastin-type polypeptides tailored to exhibit specific properties will enhance and expand this particular application in medicine.

Recombinant expression has been used to produce specific elastin peptides corresponding to the amorphous domains (e.g., PGVGVA, which is repeated seven times in human elastin), or containing at least one cross-linking domain with the 'crystalline' polyalanine sequence and pairs of lysine separated by two or three alanine residues (e.g., AAAKAAKAA). The cross-linking peptides must not only be capable of self-assembly, but also possess the further property of specifically aligning the monomers to achieve 'crystalline' cross-linking domains. Indeed, these peptides were found to be capable of coacervation behaviour analogous to the intact elastin protein (Bellingham et al., 2001), and once coacervated could be enzymatically cross-linked to give elastic properties similar to elastin, demonstrating that precise alignment of the lysine residues in the peptides had been achieved. This technique allows the fabrication of elastic protein membranes. Varying the length of the hydrophobic region and altering the cross-linking by varying the lysine residues could significantly modify the properties of the membranes (see Chapter 3).

The polypeptides have been made into membranes which could be used to present a less thrombogenic surface to the flowing blood. For example, the use of Dacron and expanded polytetra fluorethylene (PTFE) grafts poses difficulties due to thrombosis and internal hyperplasia, hence coating with elastin peptides could greatly reduce these undesirable problems as shown in Figure 19.1.

Preliminary studies on the potential of polypeptide sequences to reproduce other unique properties of elastin are also proving successful. For instance, short peptide sequences had increased efficiency, compared with the whole solubilised molecule in preventing smooth muscle cell migration whilst allowing proliferation of endothelial cells.

A series of peptides based on the elastic hydrophobic regions of elastin – from poly-tetrapeptides (VPGG)n to the poly-hexapeptide (PGVGVA)n, where n may be up to 256 – have been prepared and coacervated and subsequently stabilised by cross-linking through γ-irradiation by Urry and colleagues (1988) as potential vascular prosthetic materials. The hexapeptide possessed a high tensile strength and has a known binding site to the elastin-binding protein found on the membrane of smooth muscle cells and fibroblasts (Urry et al., 1988).

Although elastin does not calcify in vivo, which would obviously destroy its optimal elasticity, some of the synthesised polypeptides have been reported to

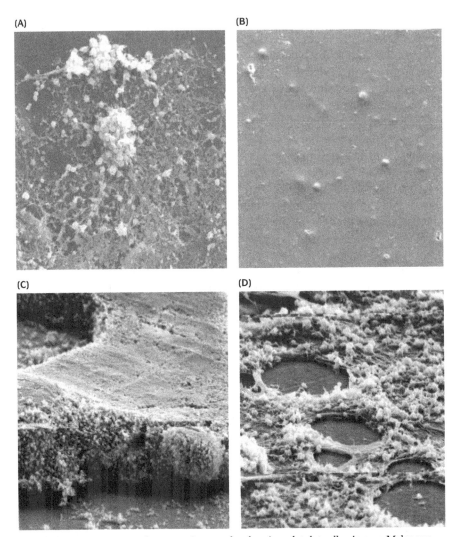

Figure 19.1: Scanning electron micrographs showing platelet adhesion on Mylar surfaces after exposure to platelet-rich human plasma (1 hr, 37°C). (A) Non-coated surface. (Mag ×2K) (B) Elastin recombinant polypeptide (EP-20-24-24)-coated surface. Peptide was absorbed in 0.5 mg/ml EP-20-24-24 from phosphate-buffered saline (2 hr at 37°C). (Mag ×1.0K) Scanning electron micrographs showing platelet adhesion on Tefzel surfaces after incubation in platelet-rich human plasma (1 hr at 37°C). (C) Non-coated surface. (Mag ×1.0K) (D) Elastin recombinant polypeptide (EP-20-24-24)-coated surface. Polypeptide adsorbed in 0.5 mg/ml EP-2 from phosphate-buffered saline (2 hr at 37°C). (Mag ×2.0K) (Electron micrographs kindly provided by Vivien Chen, Department of Cardiovascular Research, Hospital for Sick Children, Toronto, Ontario, Canada.)

bind calcium very strongly (Hollinger et al., 1988; Urry et al., 1988), which is a surprising and clearly an undesirable property. However, the mechanism is unclear, and the peptides may not actually calcify under physiological conditions. On the other hand, several proteins are involved in the control of biomineral nucleation and growth, most of which contain acidic domains rich in aspartic acid, glutamic acid, and phosphorylated serine. One of these phosphoproteins, stratherin, is found in saliva and functions biologically to inhibit nucleation and growth of calcium phosphate mineral. The N-terminus contains an aspartic acid, two phosphorylated serines, and two glutamic acids, and these have been shown to be important in the recognition of hydroxyapatite surfaces (Long et al., 2001). Calcification of elastin polypeptides can therefore be prevented. Alternatively, the calcifying polypeptides could find a role in other biological applications.

Elastin sequences may be based on the hydrophobic 'elastic' domains or the 'crystalline' α-helical rich domains which present specific, but very different properties. It may also be necessary to stabilise these polypeptides against rapid enzymic degradation when studying their efficacy as *in vivo* implants, and this is generally achieved by cross-linking of the peptide chains. The hydrophobic domains studied by Urry and colleagues do not possess reactive side chains suitable for chemical cross-linking and have been stabilised by γ-radiation. But such cross-links tend to be random and occur as carbon–carbon bonds through the peptide backbone. However, peptide sequences from the more crystalline regions of elastin or from proteins such as collagen or silks contain some hydrophilic groups (e.g., lysine or carboxyl groups) and can therefore be cross-linked using specific cross-linking reagents. This can readily be achieved with di-aldehydes, such as, glutaraldehyde, or to provide longer cross-links, with hexamethylenediisocyanate, both reagents reacting specifically with the ε-amino group of lysine residues present in different chains. Cross-linking involving the carboxyl groups can be formed by reaction with carbodiimide. The many different cross-linking agents provide a wide range of thermal stability and enzyme resistance essential for in vivo applications (Paul and Bailey, 2002).

(a) Covalent cross-linking by glutaraldehyde through reaction with the ε-NH$_2$ lysine groups to form a complex network of cross-links.

$$R-N{=}C{=}N-R' \longrightarrow Protein-CO-NH-CO-Protein$$

(b) Covalent cross-linking by carbodiimide through carboxyl groups to form a single short cross-link.

It may also be possible to utilise the unusual ability of elastin peptides to coacervate from a random soluble form to a more structured insoluble form on raising the temperature. Polymers of the repeating sequence VPGVG, which occurs 50 times in tropoelastin, coacervate on raising the temperature changing from an extended chain to a β-spiral with three VPGVG units per turn (Urry et al., 1998). This conversion has been shown to occur with a single pentamer unit, suggesting that interactions between units make only a small contribution to the energetics of the transition (Reiersen et al., 1996). Therefore, changes in the terminal regions could be used to regulate the transition. For example, it may therefore be possible to 'trigger' a change of such a peptide introduced into an enzyme from random chain to structured conformation in response to temperature, thereby modifying the activity and thermal stability of the enzyme. For example, the addition of hydrophobic residues lowers the transition temperature by 20°C, whereas the addition of glycine residues increases the transition temperature by 6°C.

SILKS

The extremely high tensile strength of silks has tended to push the applications of silk mimetics into mechanical structures, such as strength and energy-absorbing fabrics, rather than medical applications. However, some interesting modifications are being made that could eventually lead to such applications.

Sequences based on spider silk fibroin, GAGAGS for the 'crystalline' β-sheet regions and GAGAGY for the 'amorphous' region, have been prepared and used by Kaplan and colleagues (Valluzzi et al., 1999; Winkler et al., 2000). They have synthesised variants of these sequences to give two peptides with different 'triggers' which control polymer assembly and therefore the properties of the polymer. One trigger was chemically controlled with a methionine oxidation-reduction system providing an on–off switch, with the oxidised state being on and the reduced state off. The second was enzymically controlled and driven by a phosphorylation-dephosphorylation system for an on–off switch, with the phosphorylated state being on and dephosphorylated state off. In addition to controlling polymer assembly, these modifications also control solubility, the oxidised form of the methionine-containing silk peptide possessing higher solubility in water. X-ray diffraction demonstrated that the oxidised methionine system had a randon coil structure while, in contrast, the reduced peptide assembled to form β-pleated sheet 'crystallites'. The extent of the crystallinity was also shown to be dependent on the level of chemical modification, high levels of phosphorylation tending to increase the proportion of the random coil state. The phosphate trigger clearly blocks β-sheet conformation. In addition, the location of the 'triggers' can be varied in order to influence the assembly of the 'crystallites'.

Drying on hydrophilic surfaces, or with methanol, can also be used to modify the conformation of the silk-like peptides. Employing GAGAGS as a crystalline conformation, and GAGAGY as an amorphous state, rearrangements between β-strand conformation and different β-sheet precursor conformations can readily be achieved on drying (Wilson et al., 2000).

Polypeptide-based silk sequences have also been copolymerised with synthetic polymers to modify their properties. For example, the polypeptides comprising the repeating sequence GAGAGA form about 90% β-pleated sheet structure, and this can be regulated by block copolymerisation with polyethylene glycol (PEG) to provide tailored mechanical properties. The inclusion of PEG not only lowers the tendency of the peptides to form β-sheets, but also most interestingly sequesters isolated β-sheets giving a two-phase system. Consequently, the copolymer contains 20–50 nm peptide domains dispersed in a continuous PEG phase, the presence of these so-called nanostructures being confirmed by transmission electron microscopy and atomic force microscopy. The excellent mechanical properties provided by the silk-based peptide can therefore be modulated in a controlled fashion by PEG, modifying the extent of the β-sheet content from 90% to 40% (Rathmore and Sogah, 2001).

COLLAGEN

Collagen is widely used in the pharmaceutical and medical industries. For example, it has been used pharmaceutically in skin ointments, cosmetics, drug capsules, and medically as a plasma extender, haemostat sponge, contact lens membrane, bone substitute, vessel prosthesis, and in tissue reconstruction. Obviously, there are problems employing the whole molecule such as cross-reaction through species differences and inflammatory responses. However, synthetic collagen structures could provide an alternative to natural collagens. In addition to its unique tensile properties, collagen also exhibits haemostatic properties, low antigenicity, low inflammatory properties, and promotes cell growth and attachment. These wide-ranging features make collagen a desirable target for the development of novel biomimetics that reproduce or even enhance these properties.

The deca-polypeptides of Gly-Pro-Hyp- and Gly-Pro-Pro- both form triple helices, and most studies have concentrated on the folding and unfolding of the triple helix and the basis of their stability. For example, the triple helix may be stabilised by hydrogen-bonded water bridges from the hydroxyl group of hydroxyproline to the peptide backbone (Ramachandran et al., 1973; Bella et al., 1995), or alternatively, by an inductive effect of the hydroxyl group to stabilise the *trans* form of the prolyl peptide bond and hence the triple helix (Holmgren et al., 1998). The introduction of charged groups – such as glutamic acid, lysine, or arginine in place of some of the Pro-Hyp residues – results in ion-pairing within the triple helix, clearly demonstrating that, although these imino acids are the major factor in conferring stability in the triple helix, the charged groups also make a significant contribution (Kramer et al., 2001).

An intact triple helix is crucial to a thermally stable and degradation resistant biomaterial. The integrity of the helix can be determined by X-ray diffraction of the fibre, circular dichroism, or nuclear magnetic resonance of the molecule in solution. We have found that differential scanning calorimetry is ideal for studying both the fibrous and molecular forms using the same technique (Miles et al., 1995).

A template has been used by several workers to ensure folding of the chains into a stable triple helix. Branched triple helical peptides have been produced by Fields et al. (1993, 1995) for studies on the thermal stability of the helix.

BRANCHED TRIPLE HELICAL PEPTIDES

$(Gly.Pro.Hyp)_n$-$(Gly.X.Y.)_n$-Ahx ⌐   (Fields et al., 1993, 1995)

$(Gly.Pro.Hyp)_n$ Ahx-Lys-Lys-Tyr-Gly

$(Gly.Pro.Hyp.)_n$-$(Gly.X.Y.)_n$-Ahx ⌐   Ahx = 6-amino-hexanoic acid

In an extension of this work, Goodman and colleagues (Feng et al., 1996) formed very short triple helices by linking the three chains to the template Kemp triacid or KTA (*cis*-1,3,5,-trimethyl cyclohexane-1,3,5,-tricarboxylic acid) and found that the tripolypeptide $(Gly-Pro-Hyp)_3$ is the shortest chain capable of forming a triple helix in water at room temperature

KTA

More recently, Slatter et al. (2002) have achieved a separation of unlinked hetero triple helices of both decapeptides of $(Pro-Pro-Gly)_{10}$ and $(Pro-Hyp-Gly)_{10}$, the heterotrimers $(PPG)_2POG$ and $(POG)_2$ PPG mimicing type I collagen and permitting a more precise study of the role of hydroxyproline in stabilising the triple helix and its binding capacity for other proteins.

The major benefit of studying these branched and non-branched peptides is the significant enhancement of the triple helical stability to increase our understanding of the mechanisms of stabilisation and denaturation of the collagen triple helix. However, the incorporation of other specific recombinant sequences into these structures will allow investigations in the rapidly expanding field of collagen cellular interactions relating to medical research.

Biomaterials made directly from collagen often stimulate a mild inflammatory response in tissue, resulting in rapid enzymic degradation of the implant (Bailey, 2000). An increase in residence time in the tissue can be achieved by cross-linking the collagen with glutaraldehyde or hexamethylene diisocyanate. However, biomimetic peptides are preferable, since they are more stable to natural proteases and unlikely to stimulate an inflammatory reaction.

Although glutaraldehyde is widely used to stabilise the collagen peptides, unfortunately it can induce local cytotoxicity, hence other cross-linking compounds are now being evaluated by Herbage and colleagues, for example, hydrazine or diphenylphosphorylazide, the latter apparently possessing excellent cytocompatability (Petite et al., 1995). Treatment of pericardium with acylazides or carbodiimide results in similar thermal stability and enzyme resistant properties to glutaraldehyde, but there appear to be no cytotoxic problems.

The introduction of hydrophobic groups can also have a stabilising effect on the triple helix, and the bulky side chain of norleucine was introduced into the KTA-templated tripeptide to give $(Gly-Pro-Nleu)_9$ or $(Gly-Nleu-Pro)_9$, where Nleu is norleucine, $(CH_3)_2$ $C-CH_2-NH(COOH)$ (Malacini et al., 1997). The latter peptides form triple helices that denature around 39°C, a denaturation temperature similar to that of native tropocollagen, whilst there is no melting transition with the former. The incorporation of Nleu into these peptides expands the ability to vary the properties of the collagen mimetics, and surprisingly the Nleu incorporated peptides were found to bind calcium. These peptides also show cell-binding activity, which depend on the extent of triple helical conformation (Johnson et al., 2000). Furthermore, the peptide stimulated the attachment and growth of corneal epithelial cells and fibroblasts, together with the migration of the epithelial cells after immobilisation on a surface. The peptide KDGEA inhibited the attachment, indicating that the collagen biomimetic involves the $\alpha2\beta1$ integrin receptor. Surprisingly, the peptides containing Nleu in position 2 of the collagen peptide $(Gly-Nleu-Pro)_{10}$ did not possess cell-binding activity.

Other studies on triple helical peptides have investigated the role of collagen in platelet aggregation (Fields and Prockop, 1996), since the triple helix conformation is crucial to this particular aggregation. The role of specific sequences in cell–collagen interactions involving adhesion or migration in development and growth, as well as repair in adult tissues, is becoming a major field of research. The use of basement membrane mimetics does not appear to have been studied to date, but could be an important approach, particularly in remodelling situations, such as wound healing. The incorporation of specific sequences will enhance these studies to allow investigation of mechanisms of cell activation, cell signalling, and metallo-proteinase binding.

Collagen exists in the native state in the presence of proteoglycans which are believed to modify the properties of the fibres, possibly by regulating fibre size and their interaction with other matrix components and with cells, particularly in developing tissue and in healing dermal wounds. Attempts have been made to recombine collagen and proteoglycans, such as dermatan and chondroitin and polysaccharides such as hyaluron in order to produce novel biomaterials. The use of collagen implants as guided tissue in wound healing have generally failed due to cytotoxic effects. Reduction of such effects and more precise studies on the interactions of collagen and glycoproteins could be achieved by recombinant peptides.

Another opportunity for the involvement of collagen biomimetics is in the therapy of the late complications of diabetes mellitus, or at providing a better understanding of the biochemical mechanisms of non-enzymic glycation. In contrast to most other proteins, elastin and collagen possess low turnover rates, hence they are susceptible to ageing in terms of the non-enzymic and adventitious reaction with glucose in vivo, resulting in stiffening of the fibres, for example of the aortic collagen and elastin fibres (Paul and Bailey, 1996). The higher levels of glucose in subjects with diabetes mellitus result in many such deleterious complications and has been termed 'accelerated ageing'. The mechanisms involved and the therapeutic interventions are being intensively investigated. Stiffening of the fibres is due to intermolecular glycation cross-links which are formed through reaction of glucose with the ε-amino group of lysine and arginine to form complex compounds, only one of which has been identified and termed pentosidine (Sells and Monnier, 1989). However, in view of its low concentration, it is unlikely to account for the increase in stiffness of the fibres and other glycation cross-links remain to be identified. Glycation cross-linking can also occur in the amorphous basement membranes resulting in increased permeability, reduced flexibility, and reduced interaction with cells. Perhaps just as importantly, glycation can also result in modification of cell–collagen interactions (Paul and Bailey, 1999) and consequently could affect collagen-cellular regulation of development and turnover in adults. This effect is obviously most important in the basement membrane where the cells are in direct and continuous contact with the glycated collagen in the membrane and thereby affect development and turnover in adults. The glycation-induced inhibition is partly due to the glycation of arginine in the sequence arginine-glycine-aspartate (RGD) tripeptide integrin-binding motif, although clearly other sequences containing arginine have been shown to interact with the integrins acting as cellular receptors.

Biomimetric peptides of collagen would be ideal in any attempts to study the inhibition of the glycation reaction, and the mechanism of reaction with cells through the integrins could also be studied by glycating the short collagen peptides and following the conformational and chemical changes during their interactions.

CONCLUDING REMARKS

These biomimetic studies are moving towards increased thermal stability, resistance to biodegradation and ever wider variation in properties, and clearly open up new opportunities for the design of elastomeric protein biomimetics. The incorporation of non-standard amino acids, such as Norleucine into collagen, can lead to unexpected properties. This approach could be valuable with recombinant elastin and silk sequences in the development of novel biomaterials. Eventually, this continuing exploration using mimetics should result in discriminatory therapeutics and provide a better understanding of the biological effects.

REFERENCES

Bailey, A. J. (2000) The fate of collagen implants in tissue defects. *Wound Repair Regen.* **8,** 5–12.

Barnes, M. J., and MacIntyre, D. (1979) Platelet activity of isolated constituents of the blood vessel wall. *Haemostasis* **8,** 158–70.

Bella, J., Brodsky, B., and Berman, H. M. (1995) Hydration structure of a collagen peptide. *Structure* **3,** 893–906.

Bellingham, C. M., Woodhouse, K. A., Robson, P., Rothstein, S. J., and Keeley, F. W. (2001) Self-aggregation characteristics of recombinantly expressed human elastin polypeptide. *Biochim. Biophys. Acta* **1550,** 6–19.

Bonzon, N., Carrat, X., Deminiere, C., Daculsi, G., Lefebvre, F., and Rabaud, M. (1995) New artificial connective tissue matrix made of fibrin monomers, elastin peptides and type I + II collagens: A structural study, biocompatability and use as a tympanic membrane in rabbits. *Biomaterials* **16,** 881–85.

Feng, Y., Melacini, G., Taulane, J. P., and Goodman, M. (1996) Collagen based structures containing peptoid residue N-isobutylglycine (Nleu); synthesis and biophysical studies of Gly-Pro-Nleu sequence by circular dichroism. *Biopolymers* **39,** 659–72.

Fields, G. B., and Prockop, D. J. (1996) Perspectives on the synthesis and application of triple collagen-model peptides. *Biopolymers* **40,** 345–57.

Fields, C. S., Lovdahl, C. M., Miles, A. J., Matthais-Hagen, V. L., and Fields, G. B. (1993) Solid phase synthesis and stability of triple helical peptides incorporating native collagen sequences. *Biopolymers* **33,** 1695–1707.

Fields, C. G., Grab, B., Lauer, J. L., and Fields, G. B. (1995) Purification and analysis of synthetic, triple-helical mini-collagens by RP-HPLC. *Anal. Biochem.* **231,** 57–64.

Hollinger, J. O., Schmitz, J. P., Yaskovich, R., Long, M. M., Prasad, K. U., and Urry, D. W. (1988) A synthetic polypeptide of elastin for initiating calcification. *Calcified Tissue Int.* **42,** 231–36.

Holmgren, S. K., Taylor, K. M., Bretscher, L. E., and Raines, R. T. (1998) Code for collagen stability deciphered. *Nature* (Lond.) **392,** 666–67.

Johnson, G., Jenkins, M., McLean, K. M., Griesser, H. J., Kwak, P., Goodman. M., and Steele, J. G. (2000) Peptide containing collagen mimetics with cell binding activity. *J. Biomed. Mater. Res.* **51,** 612–24.

Kramer, R. Z., Bella, J., Brodsky, B., and Berman, H. M. (2001) The crystal and molecular structure of a collagen-like peptide with a biologically relevant sequence. *J. Mol. Biol.* **311,** 131–37.

Lefebvre, F., Droullet, F., Savin de Larclasuse, A. M., Aprahamian, M., Midy, D., Bordenave, L., and Rabaud, M. (1989) Repair of experimental arteriotomy in rabbit aorta using a new resorbable elastin-fibrin biomaterial. *J. Biomed. Mater. Res.* **23,** 1423–32.

Legrand, Y., and Fauval, L. (1992) Molecular mechanisms of the interaction of subendothelial microfibrils with blood platelets. *Nouv. Rev. Fr. Hemat.* **34,** 17–25.

Long, J. R., Shaw, W. J., Stayton, P. S., and Drobny, G. P. (2001) Structure and dynamics of hydrated statherin on hydroxyapatite as determined by solid state NMR. *Biochemistry* **40,** 15451–55.

Malacini, G., Feng, Y., and Goodman, M. (1997) Collagen based structures containing the peptoid residue N-iso-butylglycine (Nleu): Conformational analysis of Gly-Nleu-Pro sequences by 1H-NMR and molecular modelling. *Biochemistry* **36,** 8725–32.

Miles, C. A., Burjanadze, T. V., and Bailey, A. J. (1995) The kinetics of the thermal denaturation of collagen in unrestrained rat tail tendon determined by differential scanning calorimetry. *J. Mol. Biol.* **245,** 437–46.

Paul, R. G., and Bailey, A. J. (1996) Glycation of collagen: The basis of its central role in the late complications of ageing and diabetes. *Int. J. Biochem. & Cell Biol.* **28,** 1297–1310.

Paul, R. G., and Bailey, A. J. (1999) The effect of advanced glycation end-product formation upon cell-matrix interactions. *Int. J. Biochem. Cell Biol.* **31,** 653–60.

Paul, R. G., and Bailey, A. J. (2002) Cross-link stabilization of collagen as a biomimetic. Submitted.

Petite, H., Duval, L. L., Frei, V., Abdulmalak, N., Sigotluizard, M. F., and Herbage, D. (1995) Cytocompatibility of calf pericardium treated by glutaraldehyde and by the acyl azide methods in an organotypic culture model. *Biomat.* **16,** 1003–8.

Ramachandran, G. N., Bansal, M., and Bhatnagar, R. S. (1973) A hypothesis on the role of hydroxyproline in stabilising the collagen structure. *Biochim. Biophys. Acta* **323,** 166–71.

Rathmore, O., and Sogah, D. Y. (2001) Nanostructure formation through β-sheet self assembly in silk-based materials. *Macromolecules* **34,** 1477–86.

Reiersen, H., Clarke, A. R., and Rees, A. R. (1996) Short elastin-like peptides exhibit the same temperature-induced structural transitions as elastin polymers: Implications for protein engineering. *J. Mol. Biol.* **283,** 255–64.

Sells, D. R., and Monnier, V. M. (1989) Structural elucidation of a senescent cross-link from human extra-cellular matrix. Implication of pentoses in the ageing process. *J. Biol. Chem.* **264,** 21597–21602.

Slatter, D. A., Miles, C. A., and Bailey, A. J. (2002) Asymmetry in the triple helix collagen-like heterotrimers confirms that external bonds stabilize collagen structure. *J. Mol. Biol.* (in press).

Sims, T. J., Rasmussen, L. M., Oxlund, H., and Bailey, A. J. (1996) The role of glycation cross-links in diabetic vascular stiffening. *Diabetologia* **39,** 946–51.

Slatter, D. A., Bolton, C. H., and Bailey, A. J. (2000) The role of lipid-derived malondialde-hyde in diabetes mellitus. *Diabetologia* **43,** 550–57.

Urry, D. W., Prasad, K., Long, M., and Harris, R. (1988) Elastomeric polypeptides as potential vascular prosthetic materials. *Polym. Mater. Sci. Eng.* **59,** 648–89.

Valluzzi, R. S., Szela, P., Avtges, D. A., Kirschner, D. L., and Kaplan, D. L. (1999) Methionine redox triggered crystallisation of biosynthetic silk fibroin. *J. Phys. Chem.* **103,** 11382–92.

Wilson, D., Valluzzi, R., and Kaplan, D. L. (2000) Conformational transitions in silk-like peptides. *Biophys. J.* **78,** 2690–2701.

Winkler, S. D., Wilson, D. L., and Kaplan, D. L. (2000) Controlling β-sheet assembly in genetically engineered silk by enzymatic phosphorylation/dephosphorylation. *Biochemisty* **39,** 12739–46.

Index